Advances in Neuroscience, Neuropsychiatry, and Neurology

Cândida Lopes Alves
Federal University of Maranhão, Brazil

Katie Moraes Almondes
Federal University of Rio Grande do Norte, Brazil

Gilberto Sousa Alves
Federal University of Maranhão, Brazil

A volume in the Advances in Medical Technologies and Clinical Practice (AMTCP) Book Series

IGI Global
PUBLISHER of TIMELY KNOWLEDGE

Published in the United States of America by
IGI Global
Medical Information Science Reference (an imprint of IGI Global)
701 E. Chocolate Avenue
Hershey PA, USA 17033
Tel: 717-533-8845
Fax: 717-533-8661
E-mail: cust@igi-global.com
Web site: http://www.igi-global.com

Library of Congress Cataloging-in-Publication Data

Names: Alves, Cândida, 1983- editor. | Almondes, Katie Moraes de, 1974- editor. | Alves, Gilberto Sousa, 1976- editor.
Title: Advances in neuroscience, neuropsychiatry, and neurology / edited by Cândida Alves, Katie Moraes de Almondes, Gilberto Sousa Alves.
Description: Hershey, PA : Medical Information Science Reference, [2024] | Includes bibliographical references and index. | Summary: "This book aims to provide greater technical and scientific knowledge about advances in Neuroscience, Neuropsychiatry and Neurology, through original research contributions and scientific advances and relevant clinical results"-- Provided by publisher.
Identifiers: LCCN 2023051306 (print) | LCCN 2023051307 (ebook) | ISBN 9798369308516 (hardcover) | ISBN 9798369308523 (ebook)
Subjects: MESH: Nervous System Diseases--diagnosis | Nervous System Diseases--therapy | Neurologic Manifestations | Neuropsychiatry | Diagnosis, Differential | Research
Classification: LCC RC348 (print) | LCC RC348 (ebook) | NLM WL 141 | DDC 616.8/0475--dc23/eng/20231207
LC record available at https://lccn.loc.gov/2023051306
LC ebook record available at https://lccn.loc.gov/2023051307

This book is published in the IGI Global book series Advances in Medical Technologies and Clinical Practice (AMTCP) (ISSN: 2327-9354; eISSN: 2327-9370)

British Cataloguing in Publication Data
A Cataloguing in Publication record for this book is available from the British Library.
All work contributed to this book is new, previously-unpublished material.
The views expressed in this book are those of the authors, but not necessarily of the publisher.
For electronic access to this publication, please contact: eresources@igi-global.com.

Advances in Medical Technologies and Clinical Practice (AMTCP) Book Series

ISSN:2327-9354
EISSN:2327-9370

Editor-in-Chief: Srikanta Patnaik, SOA University, India, Priti Das, S.C.B. Medical College, India

MISSION

Medical technological innovation continues to provide avenues of research for faster and safer diagnosis and treatments for patients. Practitioners must stay up to date with these latest advancements to provide the best care for nursing and clinical practices.

The **Advances in Medical Technologies and Clinical Practice (AMTCP) Book Series** brings together the most recent research on the latest technology used in areas of nursing informatics, clinical technology, biomedicine, diagnostic technologies, and more. Researchers, students, and practitioners in this field will benefit from this fundamental coverage on the use of technology in clinical practices.

COVERAGE

- Biometrics
- Telemedicine
- Nutrition
- Clinical Studies
- Nursing Informatics
- Neural Engineering
- Medical imaging
- Biomechanics
- Clinical Data Mining
- Clinical High-Performance Computing

IGI Global is currently accepting manuscripts for publication within this series. To submit a proposal for a volume in this series, please contact our Acquisition Editors at Acquisitions@igi-global.com or visit: http://www.igi-global.com/publish/.

The Advances in Medical Technologies and Clinical Practice (AMTCP) Book Series (ISSN 2327-9354) is published by IGI Global, 701 E. Chocolate Avenue, Hershey, PA 17033-1240, USA, www.igi-global.com. This series is composed of titles available for purchase individually; each title is edited to be contextually exclusive from any other title within the series. For pricing and ordering information please visit http://www.igi-global.com/book-series/advances-medical-technologies-clinical-practice/73682. Postmaster: Send all address changes to above address.

Titles in this Series

For a list of additional titles in this series, please visit:
http://www.igi-global.com/book-series/advances-medical-technologies-clinical-practice/73682

Approaches to Human-Centered AI in Healthcare
Veena Grover (Noida Institute of Engineering and Technology, India) Balamurugan Balusamy (Shiv Nadar University, India) Nallakaruppan M.K. (Vellore Institute of Technology, India) Vijay Anand (Vellore Institute of Technology, India) and Mariofanna Milanova (University of Arkansas at Little Rock, USA)
Medical Information Science Reference • © 2024 • 328pp • H/C (ISBN: 9798369322383) • US $380.00

Future of AI in Medical Imaging
Avinash Kumar Sharma (Sharda University, India) Nitin Chanderwal (University of Cincinnati, USA) Shobhit Tyagi (Sharda University, India) and Prashant Upadhyay (Sharda University, India)
Medical Information Science Reference • © 2024 • 312pp • H/C (ISBN: 9798369323595) • US $390.00

Multisector Insights in Healthcare, Social Sciences, Society, and Technology
Darrell Norman Burrell (Marymount University, USA)
Engineering Science Reference • © 2024 • 392pp • H/C (ISBN: 9798369332269) • US $545.00

Change Dynamics in Healthcare, Technological Innovations, and Complex Scenarios
Darrell Norman Burrell (Marymount University, USA)
Medical Information Science Reference • © 2024 • 331pp • H/C (ISBN: 9798369335550) • US $495.00

For an entire list of titles in this series, please visit:
http://www.igi-global.com/book-series/advances-medical-technologies-clinical-practice/73682

IGI Global
PUBLISHER of TIMELY KNOWLEDGE

701 East Chocolate Avenue, Hershey, PA 17033, USA
Tel: 717-533-8845 x100 • Fax: 717-533-8661
E-Mail: cust@igi-global.com • www.igi-global.com

Table of Contents

Detailed Table of Contents

Electrocardiogram (ECG) analyses can only be performed by health professionals whose demand for care is often greater than the availability. In this context, this work consists of the development of an application capable of processing long-lasting ECG signals to assist health professionals in making decisions. The application has an interactive interface that allows view the entire ECG signal in a single image generated by all overlapping cardiac cycles. The proposed application still has email communication between users with the objective of facilitating patient follow-up. The application was tested on three different ECG signals, one artificial and two real. The first signal was an artificial signal generated in software Matlab. The second ECG signal has normal sinus rhythm, available in the MIT-BIH normal sinus rhythm database. The third ECG sign diagnosed with arrhythmia can be found in the MIT-BIH arrhythmia database. The results obtained by the proposed method can be used to support decision-making in clinical practice.

The first episode of psychosis is the interval between the onset of the first positive psychotic symptom above the threshold for clinical psychosis for at least 1 week until the first 5 years of starting treatment. The at-risk mental state is subdivided into three ultra high-risk populations for psychosis: brief intermittent psychotic symptoms; attenuated positive symptoms syndrome; genetic risk and deterioration syndrome. The incidence of psychotic disorders varies between 15-34/100,000 person-years at risk, most of the specific diagnoses are schizophrenia. The duration of untreated psychosis is the most studied variable and closely related to the assessment of the impact of early treatment on the patient's prognosis. Psychosis can be divided into primary (affective and non-affective) or secondary causes. Intervening early in the course of psychotic illness is important as centers specialized in FEP aim reduce DUP, achieve remission of the psychotic condition, reduce recurrence, and reduce the number of hospital admissions.

Gilberto Sousa Alves, Federal University of Rio de Janeiro, Brazil
Romulo Kunrath Pinto Silva, Nina Rodrigues Hospital, Brazil
Marielia Barbosa Freitas Leal, Federal University of Ceará, Brazil
Bianca de Melo Ferro, Federal University of Maranhao, Brazil
Leandro de Oliveira Trovão, Nina Rodrigues Hospital, Brazil

The experience of the psychosis can open a critical window into primary psychiatric conditions, particularly schizophrenia or bipolar disorder. More recently, the role of environment in early adolescence or late childhood, including psychological abuse or physical violence, has been associated with a greater likelihood of developing psychosis years later. Evidence has also shown, even in symptomatic remission, the occurrence of cognitive symptoms, particularly in complex attention, executive functioning, episodic memory, and verbal and procedural IQ. The current chapter discusses the epidemiology of first episode psychosis, neurobiological characteristics, and main risk factors, including early exposure to the trauma situation and early interventions.

Felippe Mendonca, D'Or Institute for Research and Education, Brazil
Paulo Mattos, D'Or Institute for Research and Education, Brazil
Felipe Kenji Sudo, D'Or Institute for Research and Education, Brazil

Although it has now been accepted that most individuals with attention-deficit/hyperactivity disorder (ADHD) remain symptomatic in adulthood, the idea that the condition may persist in those aged 60+ has only recently garnered consideration from clinicians and researchers. Consequently, little is known about how the condition manifests in late life, and diagnosis is challenging in those in this age group. This review provides a comprehensive examination of the current state of knowledge on ADHD in old age, shedding light on its epidemiology, diagnostic challenges, clinical manifestations, comorbidities, and treatment options.

This chapter offers a comprehensive review of Parkinson's disease (PD), spanning from epidemiology to advanced treatment options. It delves into the neuropathology and the evolution of etiological theories, elucidating the genetic mechanisms and clinical subtypes of PD. The text outlines criteria for clinical diagnosis, including essential markers and exclusion factors, while introducing innovative diagnostic frameworks. The chapter is also an integrated guide discussing future research avenues for optimizing patient treatment and outcomes.

All 29 patients underwent structural neuroimaging examination. With the advent of analysis cerebrospinal fluid biomarkers, the authors were able to perform this in 38% of cases, with 45% having Alzheimer's pathology, 36% tau protein pathology, 9% having normal biomarkers, and 9% having increased β-amyloid protein alone. With technological developments, 55% of cases underwent brain single photon perfusion scintigraphy (SPECT) studies (81%) and brain PET/SCAN (19%). The clinical results were 43% compatible with the logopenic variant, 36% with the semantic variant, and 21% with the non-fluent variant. Around 3% of cases occurred due to a mutation in exon 5 of the VCP gene. Around 10% of cases underwent post-mortem studies, with Alzheimer's disease (logopenic variant) confirmed in 33%, Pick's disease (non-fluent variant) in 33%, and cortico-basal degeneration (variant not fluent).

Social cognition is an umbrella term for the brain process underlying social behavior. Neurodegenerative dementia might affect the social cognition domains in different ways and degrees. Therefore, each type of dementia has a distinct profile of social cognitive impairments. This chapter presents a discussion about social cognition in healthy aging and in the most prevalent dementia subtypes, Alzheimer's disease (AD) and frontotemporal dementia (FDT). The available data about mentalizing, emotional contagion, and perception of social cues in AD and FTD variants are analyzed based on the correlation between behavioral and neuroimaging findings.

Preface

In the ever-evolving landscape of Neuroscience, Neuropsychiatry, and Neurology, the convergence of scientific breakthroughs and technological innovations has propelled these fields into new frontiers. This edited reference book, *Advances in Neuroscience, Neuropsychiatry, and Neurology*, curated by Cândida Lopes Alves, Katie Moraes Almondes, and Gilberto Sousa Alves, serves as a testament to the transformative impact of cutting-edge research and data processing facilitated by artificial intelligence.

The dynamic interplay between technology and the realms of Neuroscience, Neuropsychiatry, and Neurology has ushered in a new era of unified problem-solving approaches. This book encapsulates the profound advancements made by researchers and professionals in these fields, marking a departure from century-long endeavors towards feasible and beneficial solutions for society. The collective wisdom presented within these pages embraces a comprehensive "brain health" perspective, spanning from diagnostic methodologies to the treatment of diverse conditions and disorders.

The integration of technology and scientific expertise has yielded numerous examples and applications, unraveling the intricacies of brain functions through in-depth learning about cortical and subcortical areas, correlated functions, network alterations, and psychopathologies. This, in turn, has paved the way for early detection, differential diagnosis, and increasingly specialized treatment modalities.

Advances in Neurosciences, Neuropsychiatry, and Neurology embarks on a journey of theoretical contemplations and the presentation of relevant clinical outcomes. This work aspires to contribute to the enhancement of therapeutic strategies, fostering their seamless integration into clinical practice. Through original research contributions, scientific advances, and clinical insights, this book aims to enrich the technical and scientific knowledge surrounding the vast domains of Neuroscience, Neuropsychiatry, and Neurology.

Designed for researchers, professionals, and students in the fields of Neuroscience, Medicine, Psychology, and other health-related disciplines, this book encompasses a diverse array of contributions, including case reports, literature reviews, empirical studies, theoretical discussions, and systematic reviews. It is our sincere hope that the contents of this book not only serve as a source of knowledge but also inspire further exploration, innovation, and collaboration in the pursuit of advancements in the understanding and treatment of neurological and neuropsychiatric conditions.

As editors, we extend our gratitude to the contributors for their valuable insights and commitment to advancing the frontiers of knowledge in these critical fields. May this compilation foster a deeper appreciation for the ongoing advancements in Neuroscience, Neuropsychiatry, and Neurology, ultimately contributing to the improvement of patient care and the betterment of global brain health.

IN SUMMARY

As the editors of *Advances in Neuroscience, Neuropsychiatry, and Neurology*, we find great satisfaction in presenting this comprehensive compilation of cutting-edge research and insights at the intersection of science, technology, and clinical practice. The journey through the pages of this edited reference book has been a testament to the remarkable progress achieved in the fields of Neuroscience, Neuropsychiatry, and Neurology.

The collaborative efforts of esteemed contributors have resulted in a rich tapestry of knowledge, spanning from theoretical considerations to practical clinical applications. The amalgamation of scientific advancements and technological breakthroughs showcased in this book reflects the dynamic nature of these disciplines, offering a unified approach to solving intricate problems that have perplexed researchers and practitioners for over a century.

The integration of artificial intelligence and data processing technologies has not only deepened our understanding of neural functions but has also paved the way for more precise diagnostic methodologies and targeted therapeutic interventions. The exploration of cortical and subcortical areas, network alterations, and psychopathologies has illuminated new avenues for early detection and specialized treatments.

This book aspires to be more than a compilation of research findings; it is a beacon for researchers, professionals, and students in Neuroscience, Medicine, Psychology, and related fields. The diverse contributions, including case reports, literature reviews, empirical studies, theoretical discussions, and systematic reviews, collectively form a repository of knowledge that can inspire further inquiry and innovation.

As we conclude this preface, we express our gratitude to the contributors for their dedication and commitment to advancing the frontiers of knowledge in these critical disciplines. We hope that the insights shared within these pages stimulate ongoing dialogue, collaboration, and exploration in the ever-evolving landscape of Neuroscience, Neuropsychiatry, and Neurology.

May this book serve as a source of inspiration for future endeavors, fostering a continuous cycle of learning, discovery, and improvement in patient care. We invite readers to engage with the contents, reflect on the advancements presented, and join us in the collective pursuit of enhancing our understanding of the intricacies of the brain and improving the lives of those affected by neurological and neuropsychiatric conditions.

Cândida Lopes Alves
Federal University of Maranhão, Brazil

Gilberto Sousa Alves
Federal University of Maranhão, Brazil

Chapter 1
Artificial Intelligence in Neuropsychology:
Advances and Challenges

Katie Moraes de Almondes
Federal University of Rio Grande do Norte, Brazil

Gilberto Sousa Alves
https://orcid.org/0000-0002-0463-6183
Federal University of Maranhão, Brazil

Candida Lopes Alves
https://orcid.org/0000-0002-6161-2938
Federal University of Maranhão, Brazil

ABSTRACT

This chapter reflects the challenge of discussing the possibility of a paradigm shift in neuropsychology to cognitive and behavioral science based on technology, especially with the use of artificial intelligence, considering all continents and their sociocultural differences. With this purpose, it presents the history of neuropsychology in the digital universe, the advancement of artificial intelligence/technology in neuropsychological assessment, and the creation of standardized tests through digital means. It discusses psychological intervention and the use of artificial intelligence. Lastly, there are monitoring applications for neurocognitive and technology in rehabilitation.

DOI: 10.4018/979-8-3693-0851-6.ch001

In the mid-20th century, there was an unimaginable network of global information flows. This period marked the emergence of the third industrial revolution characterized by technological and scientific advances used in industry, agriculture, commerce, and livestock. The famous digital era was born!

Undoubtedly, great progress has been made, with various resources and tools emerging to make our lives more agile and comfortable, modifying the structure and logic of information, means of communication, and the provision of services. In the meantime, technology has demonstrated that it is always at the forefront, revolutionizing systems, enabling discoveries, crossing different areas, and, consequently, contributing to improving the quality of life of societies.

This tone is no different for health, despite today's great challenges in dealing with chronic diseases, an aging population, and diseases with no or reduced therapeutic possibilities, rehabilitation, and progressive diseases such as amyotrophic lateral sclerosis, and dementia. Technology has been applied to support life expectancy with excellence, responding to the challenges posed by these limiting conditions.

The technological wave also intervened in neuropsychology, an area that aims to study the relationship between the brain, cognition, and behavior, providing conceptual and substantive advances in training, research, and the practice of assessment and rehabilitation, providing an evidence-based area.

Unquestionably, this path has not been easy, as the perspective is of a possible paradigm shift towards a cognitive and behavioral science based on technology, which presupposes more accuracy and effectiveness, faster, fewer errors in data correction, with a unique contribution to the structuring and cohesion of the information collected, and with a substantial contribution to respond to the demands of large-scale population cohorts.

We are moving slowly towards this paradigm, considering all continents and their socio-cultural differences. Nevertheless, emphasis is placed on the progressive technical-scientific development of Neuropsychology. In this sense, this chapter aims to present the advancement of artificial intelligence/technology in neuropsychological assessment and the creation of standardized tests through digital means, presenting some applications of neurocognitive monitoring and technology in rehabilitation.

TECHNOLOGICAL NEUROPSYCHOLOGICAL ASSESSMENT

The development of artificial intelligence was influenced by the exact sciences and the humanities, with the help of human imagination, but sometimes the words used by professionals and users of this technology do not understand the current new capabilities of the various algorithms and procedures included in the generic term "artificial intelligence", as well as the associated limitations that it entails (Iliescu, Greiff, Ziegler, & Fokkema, 2022).

Since the first regulations on carrying out psychological assessments, many changes have been proposed and, in the last 15 years, we have come across several debates in the world. It is important to highlight that the use of these tools in the field of psychology initially consisted of the use of software to help correction tests and not in their application. Thus, it was proposed that the patient could answer the test with the evaluator, but the correction was carried out with the help of calculations carried out in a computerized manner. In the 1980s, there was a great demand for computerization of tests, to change the application of pen and paper instruments to their computerized version (Joly et al., 2005). The use of new technologies is comprehensive in all areas of science, and in Psychology it is no different.

In the United States, the guidelines that propose the carrying out of Psychological Assessment and online testing have been in force since 2013 at the American Psychological Association (APA). However, after the COVID-19 pandemic, new guidelines were proposed so that the psychologist's work with psychological assessment would not cease due to the restrictions imposed at this time of pandemic. Among the guidelines we find proposals aimed at maintaining the quality of the data obtained, the safety of the instruments, and the ethical and technical standards used (Marasca, Yates, Schneider, Feijó, & Bandeira, 2020).

Since 2013, in the United States, psychological assessments have been carried out online. After the outbreak of COVID-19, new debates were held on the topic, due to social isolation (Marasca, Yates, Schneider, Feijó & Bandeira, 2020).

In Europe, after the Bologna Declaration, in 1999 multiple changes were made to Psychology training. Thus, there are differences in different European countries regarding the possibility of carrying out psychological assessments in both face-to-face and online formats, and in several countries, it is mandatory to carry out a supervised internship for the area, regardless of the format in which the assessment is carried out.

Online tests in Europe are used regularly, however, studies indicate the use of online testing and not the carrying out of a more comprehensive psychological assessment in this same format.

Psychological Assessment carried out using Information and Communication Technologies was regulated in Brazil in 2018, through Resolution No. 11/2018 issued by the Federal Council of Psychology (Conselho Federal de Psicologia - CFP). Worldwide, since the 1990s, new technologies have been used in this context, in the beginning, by phone (Marasca, Yates, Schneider, Feijó & Bandeira, 2020).

Despite the ease in logistics, such as greater flexibility in schedules and less travel time to the office, when we talk about carrying out an online psychological assessment, the professional must pay attention to aspects such as how the setting occurs and confirm that the patient feels comfortable, with privacy and the environment allows there to be no distractions (Marasca, Yates, Schneider, Feijó & Bandeira, 2020).

By carefully analyzing the literature published on the topic, we found that there are advantages, disadvantages, and limitations in using these online techniques. The advantages cited include the standardization of specifications for stimulus presentation and response collection, leading to more severe control of assessment conditions, which leads to greater psychometric reliability (Conklin et al., 2013); the possibility of reaching a larger sample and the possibility of large-scale studies; data recording is easier, more effective and faster (Brooks & Sherman, 2012); there is a reduction in the examiner's subjectivity; greater accuracy of response time and processing speed measurements, in addition to the possibility of automatic performance correction.

Regarding the disadvantages, we can mention poorly designed interaction interfaces, face-to-face interaction between the evaluator and the patient is lost, and there may be practical and technical limitations, in addition to the limitation that the use of some electronic devices may have with people with certain limitations. physical, such as motor or sensory difficulties (Browndyke, Albert, Malone, Schatz, Paul, Cohen, & Gouvier, 2002).

In the area of psychological assessment, numerous questionnaires, and other tests are already in use, and it is sometimes argued that the experience of the professional using these resources would be much better if the tedious work of developing and completing such a traditional tool could be avoided. In other words, the time required to apply and fill out certain forms would be much shorter. This is one of the reasons why artificial intelligence and machine learning techniques seem, at first glance, so attractive to the field of psychological assessment. However, we cannot forget the basic principles applied to evaluate the quality of the tools, such as evidence of objectivity, reliability, and validity (Bandeira, Andrade, & Peixoto, 2021).

Currently, we are faced with new demands and new challenges, namely with the use of Artificial Intelligence (AI) as an ally in data collection. All areas of science are being subjected to new adaptations and psychological assessment is no exception, namely, through the use of methods such as deep learning, and machine learning, attempts are made to discover patterns through the creation of useful representations that allow the prediction of some relevant variable (Primi, 2018).

Advances in Standardized Neuropsychological Tests Through Digital Means

Fundamental Concepts

Neuropsychologists in the digital age have increasing access to emerging technologies. Initiatives associated with behavioral and cognitive sciences involve the use, for example, of neuropsychological tests allocated on digital platforms, assessments based on virtual reality, and new automatic correction programs for reliable characterization of neurocognitive, neurobehavioral, affective, and social processes.

The process of generating, integrating, storing, and using data is a necessary element in understanding the integration between neuropsychological tests and technology. Networks, data, and algorithms come into play.

Networks are significant data structure tools for collaboratively solving complex problems, using a mathematical language to communicate something under investigation (Emmert-Streib, Dehmer, & Shi, 2016). It does not mean reducing cognition and/or behavior to a small set of variables, but highlighting the magnitude of the depth of the processes evaluated holistically and systemically, which we would not achieve manually as effectively.

The data are those considered focuses of interest in the context of Neuropsychology, such for example, the test scores, the measurement phases of the phenomenon, and the output variables corresponding to the clinical diagnosis.

The algorithm comprises the finite sequence of reasoning or operations with the data in question, which allows for solving problems and developing new knowledge about many aspects of patients and how we can improve their lives.

The interrelationship between networks, data, and algorithms helps to develop new computerized assessment methods, more robust statistical measures for these data, and sophisticated patterns of theoretical description between cognitive and behavioral processes and brain processes.

This entire discussion is encompassed by artificial intelligence, which is the general system and refers to "the scientific field within the discipline of computer science, concerned with the construction of systems or machines (computers) to perform tasks that normally require human intelligence, such as decision making" (Graham et al., 2020).

These sets of knowledge and skills mentioned above promote the technical-scientific-technological training of neuropsychologists in the new digital era, who must approach the areas of computer science, modify their training, as well as reevaluate and discuss with the entire scientific community on the topics ethical and legal issues raised by neuropsychological research and practices in the field of technology.

Some Recent Examples of Digital Neuropsychological Testing Advances

In an extensive review of the relationship between virtual reality and cognitive assessment for clinical application in neuropsychology, Rizzo et al (2017) indicate that there is construct validity and clinical group discrimination for healthy controls.

Several research studies have been developed to validate digital psychological tests through various formats, to enable cognitive and behavioral assessment through technology.

The feasibility of using virtual reality, for example, as a tool for cognitive assessment has been demonstrated in studies targeting various cognitive processes, such as attention, executive functions, memory, and spatial abilities (Kim et al, 2021; Negut et al, 2016; Pieri, Tosi, and Romano, 2023).

Gagnon Shaigetz et al (2021) developed a technological platform called bWell, which simultaneously assesses multisensory cognitive tasks and rehabilitation. The platform offers exercises that combine standardized cognitive assessment with adaptive and gamified variants for therapy, a user interface aimed at the therapist for managing tasks with systematization of complexity levels to advance and automated recording of activity data. bWell aims to target general aspects of cognition present in various disorders, not accentuating a specific cognitive process.

Goulart et al (2022) developed a digital game (MentalPlus®) applied to 60 patients, aged between 29 and 82 years, targeting cognitive dysfunction that can occur in the postoperative period after anesthesia. Standardized pencil and paper tests were used to compare with the processes measured by the game (short and long-term memory, visuospatial process, selective and alternating attention, inhibitory control, and visual perception). The digital game provided scores in agreement with standardized tests for evaluating visual perception, but conflicting data for other cognitive processes, requiring an increase in the sample and continuity of the research.

Lunardini et al (2020), to provide a monitoring tool that could be used to support the screening of cognitive decline, helping to identify decline early, developed a digitalized and supervised version of two neuropsychological tests - Trail Making Test and Bells Test – applied to 83 elderly people who were unfamiliar with the technology. The study consisted of developing a web application for a tablet version of the tests with a vocal assistant acting as a virtual supervisor for carrying out the test. A replay functionality has been added to allow inspection of user performance after testing is complete. The researchers investigated the concurrent validity of the computerized assessment in comparison with corresponding paper-and-pencil tests. The system software was successfully validated, demonstrating the good ability of digitalized tests to retain the same predictive power as corresponding paper tests.

In the same direction of validating online psychological tests and using these tests for self-assessment of cognitive decline in the elderly, Chan et al (2021) carried out a survey of data in the literature on the diagnostic performance of digital cognitive tests for mild neurocognitive disorder and dementia in the elderly. Among the 56 studies surveyed with 46 digital cognitive tests, researchers demonstrated that the majority of digital cognitive tests had diagnostic performance comparable to paper-

and-pencil tests for detecting mild neurocognitive disorder and dementia, with sensitivity and specificity above 0.80.

Another interesting piece of data comes from research validating cognitive tests with patients suffering from Unilateral Spatial Neglect by Massetti et al (2023). These patients are unable to pay attention to, respond to, and report sensory events occurring on the contralesional side of the space. Massetti et al (2023) constructed a modified digital version of three paper-and-pencil tests widely used to detect signs of Unilateral Spatial Neglect, titled Neurit.Space. They applied it to twelve patients with right brain injury and compared it with 12 healthy participants of the same ages and education, using the computerized version and the paper and pencil version. The results demonstrated good sensitivity and specificity of Neurit.Space.

It appears that most of the studies mentioned above transform the common and traditional pencil and paper tests used in the clinic and are biased by the socio-cultural context (as not all tests are available in all countries) and transport them to a virtual reality or platform digital. Few studies present the development of a digitally created test or set of tests.

Along these lines, two studies respond to this point of novelty. Yuan et al (2021) created the digital Clock Drawing Text (dCDT) based on the Clock Drawing Test (TDR), which is widely used in the clinic for cognitive assessment. This test captures the entire sequence of behaviors in creating a watch design. Then, the researchers evaluated the characteristics derived from the dCDT in a large community cohort followed longitudinally, who were not diagnosed with dementia, and tried to observe whether there was a correlation with cognitive processes. They applied the dCDT and compared it to a standard protocol of neuropsychological tests that measured a wide range of cognitive functions, to identify whether there was a diagnosis of mild neurocognitive disorder compared to healthy cognition. 2,062 participants participated with an average age of 62 years, among whom 36 were diagnosed with mild cognitive impairment. Each neuropsychological test was associated with an average of 50 dCDT features. Composite scores derived from dCDT features were significantly associated with both neuropsychological testing and mild cognitive impairment.

Singh et al (2021) developed the TMB Digital Neuropsychology Toolkit, a set of tests allocated on a platform called TestMyBrain.org (TMB), which assesses processing speed, working memory, and a broader general cognitive ability factor that included reasoning perceptual and episodic memory. The results demonstrated reliability and validity data for digital tests to assess neuropsychological functioning.

Finally, together these data indicate that neuropsychology is already characterized in many research and assistance centers as digital, providing new approaches to measure and monitor neuropsychological functioning, but with care due to some limitations and the potential of digital technology. Germine et al (2019) identifies three barriers to definitive paradigm shift: (1) variability in the perceptual, motor, and cognitive demands of the same test across classes of digital devices (e.g., personal computer, tablet, and smartphone); (2) hardware and software variability between devices in the same class that affects stimulus presentation and measurement, and (3) rapid changes over time in the ownership of hardware, software, and devices, which can lead to rapid obsolescence of certain tests and testing standards.

Training and research seem to be the way to dilute these limitations and identify solutions.

PSYCHOLOGICAL INTERVENTION AND USE OF AI

Therapy is a practice that aims to help people in the process of self-reflection, self-knowledge, and the search for solutions to better deal with emotional and psychological problems. With the advancement of technology, specifically in the field of artificial intelligence, a new approach has been explored: the use of machine learning in therapy.

Machine Learning is one of the main areas of artificial intelligence and allows computers to learn through the use of data and make decisions based on this learning. In therapy, this technology can be used to analyze large volumes of data and identify patterns and trends that can be useful in diagnosing, planning, and monitoring patients (Zhou, Zhao, & Zhang, 2022; Biermann, Arruda, & Holanda, 2022).

One of the applications of machine learning in therapy is the analysis of clinical data. By feeding an algorithm with information about patients, such as medical history, symptoms, test results, and treatment progress, it is possible to identify correlations and predictions that can help therapists in the decision-making process. This allows for a more personalized and efficient approach, adapted to the individual needs and characteristics of each patient.

Furthermore, machine learning can be used to detect warning signs and prevent emotional and psychological crises. By analyzing data from wearable devices, such as smart watches or heart rate sensors, it is possible to monitor behaviors and sleep patterns, for example. Based on this data, the algorithm can identify significant changes and alert both the patient and therapist to the need for early intervention (Elyoseph & Levkovich, 2023).

Emotional problems occur frequently in the population, especially depressive and anxiety disorders, known as emotional disorders. Mental disorders are considered a serious problem due to their high prevalence, their social, and economic consequences and the suffering caused (Nakamura, (2022).

In recent years, investment has been made in improving treatments and intervention methods for treating this type of problem. Cognitive behavioral therapy (CBT) is considered one of the best forms of therapeutic intervention, but self-help proposals have also emerged in the literature, with lower costs and the possibility of sessions held online. However, there is a need for ecological adaptation of intervention practices and Artificial Intelligence can be extremely important for its application by carrying out momentary ecological interventions with the possibility of carrying out and adjusting them in real-time. This type of intervention can be carried out during therapy, or in parallel, as a supporting process, improving treatment adherence (Nakamura, 2022).

With AI, it is possible to create interactive virtual environments that allow patients to practice skills and strategies learned during therapy sessions. These games can provide real-time feedback and adapt to individual needs, making therapy more effective and engaging.

Therapeutic chatbots are computer programs capable of chatting with patients, offering emotional support and therapeutic guidance. Based on previous interactions with the patient and data from other patients in similar situations, the chatbot can learn and adapt over time, becoming increasingly efficient and personalized (Nakamura, 2022).

In short, the use of machine learning in therapy has the potential to revolutionize the way patients are diagnosed, treated, and monitored through the analysis of large volumes of data, the detection of warning signs, and the participation of therapeutic chatbots, it is possible to offer a more efficient and personalized approach, targeting the emotional and psychological well-being of patients. However, the human element must continue to be valued and present throughout the therapeutic process (Lobo, 2018; Nakamura, 2022).

Therefore, despite all the possibilities and benefits that machine learning can bring to therapy, it is important to emphasize that technology does not replace the therapist's work. The human element continues to be fundamental to establishing empathy and creating an environment of trust, which is so important in this type of treatment.

ARTIFICIAL INTELLIGENCE IN TREATMENT AND REHABILITATION IN PSYCHIATRY

Currently, mental disorders are among the main causes of functional disability and absence from work. Diagnosis in the area of psychiatry is usually carried out using classification systems such as the DSM-5-TR and the ICD-11. Often, this review of manuals is carried out concomitantly with the use of scales and application of tests, to ensure greater accuracy in making the diagnosis (Lobo, 2018).

More and more clinical case studies are being developed to evaluate the real effectiveness of AI assistance in the treatment of psychological disorders. This is an intervention with potential impact on technological development and the health sector, directly interfering with the doctor's work, potentially saving time and energy, as well as capturing imperceptible details.

Mood disorders, for example, are generally diagnosed through clinical interviews, but in many cases, there is a loss of data and there are misdiagnosed cases. These disorders increase the risk of suicide, making it imperative to diagnose and treat these disorders quickly. Artificial intelligence (AI) has been investigated for the diagnosis of mood disorders, but the merits of the literature have not been completely confirmed, although we already have data indicating that AI increases the possibility of clinical diagnosis of mood disorders, as well as the identification of individuals at higher risk of suicide (Edavally, Miller, & Youssef, 2021).

AI and psychiatry are two distinct fields, but they may have some intersections and mutual benefits. On the one hand, AI can be used in the field of psychiatry to develop more accurate assessment and diagnostic tools. For example, algorithms can be trained to analyze large amounts of patient data and identify patterns or signs of mental disorders. This can help healthcare professionals make more accurate and personalized diagnoses.

Additionally, AI can also be used in developing chatbots or virtual assistants, which can provide emotional support and guidance to people facing mental health issues. These tools can be accessible 24 hours a day and provide a judgment-free environment for people to express their feelings (Júnior, et al., 2023; Oliveira & Junior, 2020).

Currently, AI has been increasingly used in various medical areas and advantages and disadvantages have been observed. In the area of psychiatry and mental health, the therapeutic response varies greatly from one subject to another, and there are currently, to date, no objective biological markers that can assertively predict the effectiveness of drug treatment. Furthermore, research indicates the interference of the environment, the impact of genetic factors, as well as other factors of important relevance in the therapeutic response presented by each patient individually (Kim, Sharma, & Ryan, 2015).

In medical practice, one of the broadest studied areas of AI is its use for predicting pharmacological treatment through the analysis of multiple variables and the use of algorithms has demonstrated that it is possible to predict, based on the patient's previous history, the factors associated with therapeutic success or failure and the best drugs of choice for individual cases.

Another line of study refers to the application of ML through the use of discourse analysis in patients with psychiatric diagnoses. Speech analysis and machine learning made it possible to successfully differentiate depressed patients from healthy controls and identify those who responded to treatment from those who did not, with a significant level of 85% accuracy (75% accuracy; Carrillo et al, 2019).

AI is increasingly studied and used in medical practice and, in the area of psychiatry, we have seen significant advances, both in treatment and in making diagnoses. It is possible to carry out a differential diagnosis between patients with diagnoses such as ADHD, depressive disorder, schizophrenia, obsessive-compulsive disorder, autism spectrum disorder, and isolated cases of psychosis and suicidal ideation (Oliveira & Junior, 2020).

Another application of AI is the analysis of patient data to help with clinical decision-making. Algorithms can analyze large amounts of medical data and identify patterns and trends that can be useful in evaluating a patient's progress, predicting relapse, or personalizing treatments. Furthermore, AI in combination with technologies such as sensors and wearable devices can be used to monitor patients' emotional well-being. These devices can track biomarkers related to stress, anxiety, or mood, and generate alerts or provide real-time feedback to patients and healthcare professionals (Rodrigues, 2021; Margarido, 2020).

We found that, compared to just a decade ago, the amount of data produced on the use of AI in medicine has increased significantly and it is predicted that by the end of 2025, more than 175 zettabytes of data will be produced. Therefore, in the era of big data, great computational power is needed to increase the size of analyses and help minimize the prejudice that currently exists (Videtič Paska & Kouter, 2021).

We know that Chatbots and other conversational agents have been available in medicine for several years, but it was after the launch of ChatGPT, a language model launched by OpenAI in 2022, that the general public, people who do not work in healthcare, discovered the power. This system has been powered and trained on approximately 300 billion words, can chat on a wide variety of topics, and reached over 50 million users in the first month, making it the fastest-growing consumer app ever. A study was published in JMIR Medical Education in February 2023 where the potential of ChatGPT was evaluated, intending to evaluate this application as a possible medical education tool and was found to achieve knowledge equivalent to a third-year medical student (Eysenbach, 2023).

A recent survey of 791 psychiatrists, mostly from developed countries, found that 83% of the sample believed it was unlikely that future technology could provide empathetic care and 3.8% said it could make their jobs obsolete. In another survey carried out with radiologists, 77% of participants reported favorable attitudes towards adopting AI and 89% were not afraid of losing their job. Another study carried out with pathologists showed a high acceptance of AI, with only 17.6% of respondents concerned about their job security. Similarly, 60% of neurosurgeons in one study reported using AI for outcome prediction purposes and 90% of doctors contacted from different specialties in Germany expected the future of medicine to include a mix of human intelligence and AI (Al-Medfa, Al-Ansari, Darwish, Qreeballa, & Jahrami, 2023).

The algorithms included in ML are created by humans, that is, they are prone to errors, which could mean that they could have significant implications for the health of the subjects, which is why rigorous studies carried out with clear supervision are necessary (Oliveira & Junior, 2020).

Internet-based psychosocial interventions hold considerable promise for meeting the educational and support needs of informal dementia caregivers at reduced costs. Several of them were delivered to support caregivers (Alves, Casali, Veras, Carrilho, Bruno, Rodrigues, & Dourado 2020).

The types of intervention are very varied, as is the quality of the methods used. We found proposals for electronic counseling to support personalized disease management in dementia, as well as monitoring and counseling for caregivers (Op. cit.).

AI can play a relevant role in the area of rehabilitation in psychiatry, offering innovative tools and technologies to assist in the treatment of patients. However, it is essential to balance the use of AI in psychiatric rehabilitation with human interaction. The empathy and emotional connection provided by healthcare professionals are crucial in the rehabilitation process. AI should be seen as a complementary tool to improve and support treatment, not as a substitute (Oliveira, & Junior, 2020).

In summary, AI can play a promising role in psychiatry rehabilitation, helping healthcare professionals offer more personalized and effective treatments. Combined with other technologies, AI can provide new forms of intervention and monitoring for patients seeking a better quality of life.

However, it is important to highlight that AI cannot completely replace mental health professionals. Human interaction and clinical judgment are essential for adequate treatment. AI should be seen as a complementary tool and not a replacement.

REFERENCES

Al-Medfa, M. K., Al-Ansari, A. M. S., Darwish, A. H., Qreeballa, T. A., & Jahrami, H. (2023). Physicians' attitudes and knowledge toward artificial intelligence in medicine: Benefits and drawbacks. *Heliyon*, *9*(4), e14744. doi:10.1016/j.heliyon.2023.e14744 PMID:37035387

Alves, G., Casali, M., Veras, A., Carrilho, C., Bruno, C., Rodrigues, V., & Dourado, M. (2020). A Systematic Review of Home-Setting Psychoeducation Interventions for Behavioral Changes in Dementia: Some Lessons for the COVID-19 Pandemic and Post-Pandemic Assistance. *Frontiers in Psychiatry*, *11*, 577871. doi:10.3389/fpsyt.2020.577871 PMID:33132937

Bandeira, D., Andrade, J., & Peixoto, E. (2021). O uso de testes psicológicos: Formação, avaliação e critérios de restrição. *Psicologia (Conselho Federal de Psicologia), 41*(spe1), 41. doi:10.1590/1982-3703003252970

Biermann, M., Arruda, C., & Holanda, L. (2022). Machine Learning e Processos Diagnósticos com Avaliação Neuropsicológica: Uma Revisão Integrativa. *Revista de Psicologia da IMED, 14*(1), 102–120. doi:10.18256/2175-5027.2022.v14i1.4568

Brooks, B., & Sherman, E. (2012). Computerized neuropsychological testing to rapidly evaluate cognition in pediatric patients with neurologic disorders. *Journal of Child Neurology, 27*(8), 982–991. doi:10.1177/0883073811430863 PMID:22290863

Browndyke, J., Albert, A., Malone, W., Schatz, P., Paul, R., Cohen, R., & Gouvier, K. (2002). Computer-related anxiety: Examining the impact of technology-specific affect on the performance of a computerized neuropsychological assessment measure. *Applied Neuropsychology, 9*(4), 210–218. doi:10.1207/S15324826AN0904_3 PMID:12584075

Carrillo, F., Sigman, M., Slezak, D., Ashton, P., Fitzgerald, L., Stroud, J., Nutt, D., & Carhart-Harris, R. (2018). Natural speech algorithms applied to baseline interview data can predict which patients will respond to psilocybin for treatment-resistant depression. *Journal of Affective Disorders, 230*, 84–86. doi:10.1016/j.jad.2018.01.006 PMID:29407543

Chan, J. Y. C., Yau, S. T. Y., Kwok, T. C. Y., & Tsoi, K. K. F. (2021). Diagnostic performance of digital cognitive tests for the identification of MCI and dementia: A systematic review. *Ageing Research Reviews, 72*, 101506. doi:10.1016/j.arr.2021.101506 PMID:34744026

Conklin, H., Ashford, J., Di Pinto, M., Vaughan, C., Gioia, G., Merchant, T., Ogg, R., Santana, V., & Wu, S. (2013). Computerized assessment of cognitive late effects among adolescent brain tumor survivors. *Journal of Neuro-Oncology, 113*(2), 333–340. doi:10.1007/s11060-013-1123-5 PMID:23525951

Edavally, S., Miller, D. D., & Youssef, N. A. (2021). Artificial intelligence to aid detection and diagnostic accuracy of mood disorders and predict suicide risk: A systematic review. *Annals of Clinical Psychiatry, 33*(4), 270–281. PMID:34672929

Elyoseph, Z., & Levkovich, I. (2023). Beyond human expertise: The promise and limitations of ChatGPT in suicide risk assessment. *Frontiers in Psychiatry, 14*, 1–7. doi:10.3389/fpsyt.2023.1213141 PMID:37593450

Emmert-Streib, F., Dehmer, M., & Shi, Y. (2016). Fifty years of graph matching, network alignment and network comparison. *Information Sciences, 346–347*, 180–197. doi:10.1016/j.ins.2016.01.074

Eysenbach, G. (2023). The Role of ChatGPT, Generative Language Models, and Artificial Intelligence in Medical Education: A Conversation With ChatGPT and a Call for Papers. *JMIR Medical Education, 9*, e46885. doi:10.2196/46885 PMID:36863937

Gagnon Shaigetz, V., Proulx, C., Cabral, A., Choudhury, N., Hewko, M., Kohlenberg, E., Segado, M., Smith, M. S. D., & Debergue, P. (2021). An Immersive and Interactive Platform for Cognitive Assessment and Rehabilitation (bWell): Design and Iterative Development Process. *JMIR Rehabilitation and Assistive Technologies, 8*(4), e26629. doi:10.2196/26629 PMID:34730536

Germine, L., Reinecke, K., & Chaytor, N. S. (2019). Digital neuropsychology: Challenges and opportunities at the intersection of science and software. *The Clinical Neuropsychologist, 33*(2), 271–286. doi:10.1080/13854046.2018.1535662 PMID:30614374

Goulart, A. A., Lucatelli, A., Silveira, P. S. P., Siqueira, J. O., Pereira, V. F. A., Carmona, M. J. C., Valentin, L. S. S., & Vieira, J. E. (2022). Comparison of digital games as a cognitive function assessment tool for current standardized neuropsychological tests. *Brazilian Journal of Anesthesiology, 72*(1), 13–20. doi:10.1016/j.bjane.2021.06.027 PMID:34411626

Graham, S. A., Lee, E. E., Jeste, D. V., Van Patten, R., Twamley, E. W., Nebeker, C., Yamada, Y., Kim, H. C., & Depp, C. A. (2020). Artificial intelligence approaches to predicting and detecting cognitive decline in older adults: A conceptual review. *Psychiatry Research, 284*, 112732. doi:10.1016/j.psychres.2019.112732 PMID:31978628

Iliescu, D., Greiff, S., Ziegler, M., & Fokkema, M. (2022). Artificial Intelligence, Machine Learning, and Other Demons. *European Journal of Psychological Assessment, 38*(3), 163–164. doi:10.1027/1015-5759/a000713

Joly, M., Welter, G., Martins, R., Marini, J., Montiel, J., Lopes, F., & de Carvalho, M. (2005). Sistema de avaliação para testes informatizados (SAPI): Estudo preliminar. *Psic: Revista da Vetor Editora, 6*(2), 51–60.

Júnior, J., Neto, R., Gusmão, V., Menezes, N., Silva, M., Santos, L., & Reinoso, L. (2023). O futuro da aprendizagem com a inteligência artificial aplicada à educação 4.0. *Revista Educação, Humanidades e Ciências Sociais.*

Kim, E., Han, J., Choi, H., Prié, Y., Vigier, T., Bulteau, S., & Kwon, G. H. (2021). Examining the Academic Trends in Neuropsychological Tests for Executive Functions Using Virtual Reality: Systematic Literature Review. *JMIR Serious Games*, *9*(4), e30249. doi:10.2196/30249 PMID:34822341

Lobo, L. C. (2018). Inteligência artificial, o Futuro da Medicina e a Educação Médica. *Revista Brasileira de Educação Médica*, *42*(3), 3–8. doi:10.1590/1981-52712015v42n3rb20180115editorial1

Lunardini, F., Luperto, M., Romeo, M., Basilico, N., Daniele, K., Azzolino, D., Damanti, S., Abbate, C., Mari, D., Cesari, M., Borghese, N. A., & Ferrante, S. (2020, September 21). Supervised Digital Neuropsychological Tests for Cognitive Decline in Older Adults: Usability and Clinical Validity Study. *JMIR mHealth and uHealth*, *8*(9), e17963. doi:10.2196/17963 PMID:32955442

Marasca, A., Yates, D., Schneider, A., Feijó, L., & Bandeira, D. (2020). Avaliação psicológica online: Considerações a partir da pandemia do novo coronavírus (COVID-19) para a prática e o ensino no contexto a distância. *Estudos de Psicologia*, *37*, e200085. doi:10.1590/1982-0275202037e200085

Margarido, I. T. M. M. (2020). *A Medicina do Futuro nos Dias de Hoje: Aplicações da Inteligência Artificial* [Doctoral dissertation]. Universidade de Lisboa.

Massetti, G., Albini, F., Casati, C., Toneatto, C., Terruzzi, S., Etzi, R., Tesio, L., Gallace, A., & Vallar, G. (2023). Validation of "Neurit.Space": Three Digital Tests for the Neuropsychological Evaluation of Unilateral Spatial Neglect. *Journal of Clinical Medicine*, *12*(8), 3042. doi:10.3390/jcm12083042 PMID:37109378

Nakamura, K. (2022). Information seeking criteria: Artificial intelligence, economics, psychology, and neuroscience. *Reviews in the Neurosciences*, *33*(1), 31–41. doi:10.1515/revneuro-2020-0137 PMID:33855841

Oliveira, L., & Junior, L. (2020). Aplicabilidade da inteligência artificial na psiquiatria: Uma revisão de ensaios clínicos. *Revista Debates em Psiquiatria*, *10*(1), 14–25. doi:10.25118/2236-918X-10-1-2

Pieri, L., Tosi, G., & Romano, D. (2023). Virtual reality technology in neuropsychological testing: A systematic review. *Journal of Neuropsychology*, *17*(2), 382–399. doi:10.1111/jnp.12304 PMID:36624041

Primi, R. (2018). Avaliação psicológica no século XXI: De onde viemos e para onde vamos. *Psicologia (Conselho Federal de Psicologia)*, *38*(spe), 87–97. doi:10.1590/1982-3703000209814

Rizzo, A. S., & Koenig, S. T. (2017). Is clinical virtual reality ready for primetime? *Neuropsychology, 31*(8), 877–899. doi:10.1037/neu0000405 PMID:29376669

Rodrigues, F. (2021). Inteligência Artificial E Doenças Mentais. *Recisatec - Revista Científica Saúde e Tecnologia, 1*(5).

Singh, S., Strong, R. W., Jung, L., Li, F. H., Grinspoon, L., Scheuer, L. S., Passell, E. J., Martini, P., Chaytor, N., Soble, J. R., & Germine, L. (2021). The TestMyBrain Digital Neuropsychology Toolkit: Development and Psychometric Characteristics. *Journal of Clinical and Experimental Neuropsychology, 43*(8), 786–795. doi:10.1080/13803395.2021.2002269 PMID:34907842

Videtič Paska, A., & Kouter, K. (2021). Machine learning as the new approach in understanding biomarkers of suicidal behavior. *Bosnian Journal of Basic Medical Sciences, 21*(4), 398–408. doi:10.17305/bjbms.2020.5146 PMID:33485296

Yuan, J., Libon, D. J., Karjadi, C., Ang, A. F. A., Devine, S., Auerbach, S. H., Au, R., & Lin, H. (2021). Association Between the Digital Clock Drawing Test and Neuropsychological Test Performance: Large Community-Based Prospective Cohort (Framingham Heart Study). *Journal of Medical Internet Research, 23*(6), e27407. doi:10.2196/27407 PMID:34100766

Zhou, S., Zhao, J., & Zhang, L. (2022). Application of Artificial Intelligence on Psychological Interventions and Diagnosis: An Overview. *Frontiers in Psychiatry, 13*, 811665. doi:10.3389/fpsyt.2022.811665 PMID:35370846

Chapter 2
Artificial Neural Networks for Classification of Pathologies Based on Moments of Cardiac Cycle

Jonathan Araujo Queiroz
Federal University of Maranhão, Brazil

Yonara Costa Magalhões
Federal University of Maranhão, Brazil

Gean Sousa
Federal University of Maranhão, Brazil

Allan Kardec Barros Filho
Federal University of Maranhão, Brazil

Priscila Lima Rocha
Federal University of Maranhão, Brazil

ABSTRACT

The advancement of cardiac pathology quantification hinges on the utilization of computer algorithms. To transform this vision into reality, these algorithms must distinguish among the most prevalent cardiac disorders. While some studies have leveraged the R-R interval for data extraction from ECG signals to diagnose various arrhythmias, this approach falls short in measuring changes in other ECG waves, like distortions in the P wave indicative of atrial fibrillation. This chapter introduces a new metric bi-level based on Shannon entropy to gauge the information within cardiac cycles, accounting for both the events themselves and their momentary decomposition. Experimental results reveal the method's high accuracy in classifying four distinct cardiac signal types (including one healthy signal and three pathological ones), achieving a classification rate ranging from 97.28% to 100% when employing a multilayer perceptron neural network. It holds great promise in aiding the diagnosis of cardiac pathologies.

DOI: 10.4018/979-8-3693-0851-6.ch002

INTRODUCTION

It is well known that the heartbeat dynamics are modulated by the autonomous nervous system (Victor et al., 2003; Kuhtz-Buschbeck et al., 2020). In fact, most computational methods quantify the frequency modulation in heartbeats as a method to analyze the characteristics of the electrocardiogram (ECG) signal (Parvaneh et al., 2019), in a technique known as heart rate variability (HRV) (Zarei & Asl, 2020; Ebrahimzadeh et al., 2018; Sahoo et al., 2020).

HRV has been successfully used in many studies, obtaining high accuracy rates in helping the diagnosis of cardiac anomalies. Ebrahimzadeh et. al. (2018) classified Paroxysmal Atrial Fibrillation with accuracy of 98.21%. For this, they used a combination of linear and nonlinear features extracted from the HRV, being nine linear features, including five features of the time domain and four of the frequency domain, and eight nonlinear features. Jovic & Jovic (2017) developed a methodology for classifying arrhythmias that obtained accuracy of 91.20%. To do so, they used Alphabet Entropy and HRV applied to Random Forests. Kim et al. (2016) propose to classify patients with arrhythmia using 5-second windows for HRV as input data for the Support Vector Machine (SVM) classifier and achieved accuracy of 89.13% with the MIT-BIH arrhythmia dataset. Martins et al. (2013) automatically classified five types of heartbeats using high-order statistics. The five types of ECG beats are: healthy, right branch block, left branch block, premature auricular contraction, and premature ventricular contraction, obtaining mean accuracy of 93.48%. Tsipouras et al. (2004) studied heart rate-based resources to detect arrhythmia. This method yielded sensitivity and specificity of 87.5% and 89.5%, respectively, using only features in the time domain. The combination of features in the time domain and frequency domain reached sensitivity and specificity of 90% and 93%, respectively. Despite the large number of applications, HRV-based algorithms require a series of steps and adjustment until they reach high rates of correctness in the arrhythmia classification, for example, several layers and/or activation functions in neural networks, several kernels for Support Vector Machines (SVM), numerous neighbors in k-Nearest Neighbor (k-NN) (Yao et al., 2020; Sing et al. 2018; Poddar et al. 2019; Homaeinezhad et al., 2012).

In this paper we go one level deeper in the analysis of the cardiac dynamics by quantifying the changing statistics across each trajectory in the cardiac cycle in phase space. A cardiac cycle is defined by the signal (the trajectory) between consecutive QRS complexes, which can be easily accomplished through time domain segmentation. Synchronizing the analysis per cardiac cycle is preferable over performing predefined length window-based analysis because ECG dynamics are quasi periodic, so timing information will be lost in the subsequent analysis.

In our work we are not interested in estimating heart optimal parameters using statistical moments. This is very profound and motivated our methodological approach. Statistical moments have been widely used to characterize signals such as ECG (Martis et al., 2013; Elhaj et al., 2016; Sameni et al., 2005; Afkhami et al. 2016; Qurraie & Afkhami, 2017). Normally, only the mean and/or the variance of the moment is used as features (Kim et al., 2016; Petrènas & Vaidotas, 2015; Huang et al., 2011), which is suboptimal since only first and second order information about the moment variability is captured.

Therefore, we propose to employ a bi-level moment decomposition of the ECG heartbeats. The statistical moments used in this work will be the first, second, third and fourth moments that characterize the mean, variance, skewness, and kurtosis metrics, respectively, of the model that creates the observed ECG signal. In addition, this article proposes to measure the modulation information in frequency and amplitude of cardiac cycles using a new metric based on Shannon entropy. The efficacy of the proposed method will be analyzed using four different group of ECG signals: normal sinus rhythm, atrial fibrillation, arrhythmia, and supraventricular arrhythmia. The main contributions of the proposal for a classification of cardiac pathologies based on moments of cardiac cycle are the indirect characterization of the probability distribution of cardiac signals using statistical moments and the non-dichotomous classification of four classes of cardiac signals through a Multilayer Perceptron neural network.

METHODS

Databases

In this work, the heartbeats were obtained in the following databases: MIT-BIH Normal Sinus Rhythm, MIT-BIH Atrial Fibrillation, MIT-BIH Supraventricular Arrhythmia, and MIT-BIH Arrhythmia. Each ECG signal was preprocessed and segmented into cardiac cycles of one second, characterized by 400 ms before and 600 ms after the R wave (Queiroz et al. 2017; Queiroz et al., 2019). Next, we apply the bi-level feature extraction using statistical moments in the set of cardiac cycle segments. At the first level, the mean, variance, skewness, kurtosis, and Eq. 3 metric are extracted from each cardiac cycle, thus generating six new signals for each ECG signal. At the second-level of feature extraction, the mean, variance, skewness, kurtosis, and Eq. 3 metric are again applied to each of the six signals generated in the first level.

The metrics resulting from the last level of feature extraction are used as input attributes of a Multilayer Perceptron neural network for discrimination of the four classes of cardiac signals evaluated. Finally, results for accuracy, sensitivity, and specificity of classification of healthy signals and pathological signals (atrial fibrillation, arrhythmia, supraventricular arrhythmia) are obtained.

ECG Signals Database

The ECG signals used in this study were obtained from four public databases. Each of the databases corresponds to one class in the problem of discrimination of cardiac pathologies with the proposed methodology. The description for each database used is given as follows:

- MIT–BIH Normal Sinus Rhythm (NSR) database contains 18 ECG recordings of approximately 24h duration. Subjects included in this database had no significant arrhythmias; they include 5 men, aged 26 to 45, and 13 women, aged 20 to 50 (Goldberger et al., 2000a);
- MIT-BIH Arrhythmia Database contains 48 half hour excerpts of two-channel ambulatory ECG recordings, obtained from 47 subjects (Goldberger et al., 2000b);
- MIT-BIH AF contains 319 episodes of atrial fibrillation. The individual recordings have approximately 10 hours' duration of 25 individuals (Moody & Mark, 1983);
- The MIT-BIH Supraventricular Arrhythmia Database includes 78 half-hour ECG recordings chosen to supplement the examples of supraventricular arrhythmias in the MIT-BIH Arrhythmia Database (Goldberger et al., 2000b).

ECG Signal Preprocessing

At this stage, we used the algorithm of Pan and Tompkins. This algorithm was chosen because it does not require large computational effort and has a 99.3% success rate of the QRS complexes for the MIT-BIH Arrhythmia database (Pan & Tompkins, 1985). The algorithm consists of a band-pass filter (Low Pass and High Pass Filters), derivatives, a squaring function, a moving window integration, threshold, and decision. The goal of preprocessing step is to reduce contamination of different types of noise and artifacts in the ECG signal.

Therefore, to perform this work, the following types of noise have been removed: a signal in the frequency of 60 Hz and its bandwidth below 1 Hz; baseline wander, a low-frequency (0.15 up to 0.3 Hz) noise that results from the patient inhaling and compels a baseline shifting of the ECG signals; electrode contact noise, that results from a deficiency in the contiguity between the electrode and skin, which adequately cuts off the measurement system from the subject; electrode motion artifacts, artifacts that result from variations in the electrode-skin impedance with electrode motion; muscle contractions, noise that results from the contraction of other muscles apart from the heart; electrosurgical noise, noise produced from other medical apparatus in the patient care circumstance at frequencies between 100 and 1 MHz; and instrumentation noise, noise produced by the electronic equipment utilized in the ECG measurements.

Each ECG signal record was independently labeled by two or more cardiologists and records with divergent labels were eliminated. The amplitude of the ECG signal was normalized and the sampling frequency adjusted at 256 Hz with 12-bit resolution in a range of 10 millivolts. The start and end of each ECG signal record were excluded due to the measurement error characterized by the value of 1% of the signal duration. In addition, the experiments were carried out by segmenting the ECG signal in cardiac cycles, which had a constant segment of one second where 400 ms was before and 600 ms after the R-wave (Queiroz et al. 2017; Queiroz et al. 2019).

Modeling the Cardiac Cycle by Statistical Moments

Let us define each repeating time interval extracted from the ECG signal, r(t), as an integer b=1,2,...n, for a total count of n beats. Thus, we can extract all statistics from r(t). More specifically, within the b-th beat, we can define r(t) within that beat given by r_b (t). This is carried out by taking the time which the R-wave from the b-th beat, t_b, occurs and taking N time points before and after that R-wave. We evaluated cardiac signals on two levels. We analyzed the heartbeat through two metrics at the first level. Then, we define the k-th moment at the first level for the b-th beat as,

$$m_{k,b} = \sum_{i=t_{b-N}}^{t_{b+N}} \left(r_b(i) - \mu_b\right)^k \tag{1}$$

which μ_b is the mean of $r_b(t)$. In particular, we have the following moments: a. mean- μ, m_1; b. variance- ρ^2, m_2; c. skewness - γ, m_3; and d. kurtosis- $\kappa, {m_4}/{(m_2)^2}$.

The second metric used to assess heartbeats at the first level é given by

$$c_b = -\sum_{i=t_{b-N}}^{t_{b+N}} \left(r_b(i)\right)^2 \log\left(r_b(i)\right)^2 . \tag{2}$$

For the second-level of analysis, we define two metrics, which reevaluates the signals generated in the first level of analysis. The first metric at the second-level is the following moment equation,

$$M_{l,k} = \sum_{i=1}^{n} (\omega - \mu_M)^l . \tag{3}$$

which l is the order of moment at second-level.

The second metric at the second-level is given by,

$$C = -\sum_{i=0}^{n} (\omega)^2 \log(\omega)^2 . \tag{4}$$

which μ_M is the mean of ω and $\omega = m_{k,b}$ or $\omega = c_b$.

Classifier

In this work, the proposed Multilayer Perceptron (MLP) network architecture has two hidden layers, with 100 neurons in each layer. The output layer presents four neurons, which each neuron represents one of the classes of cardiac signals under analysis. The network training stage was performed with 80% of the data set for training and 20% for testing.

The parameters used in the configuration of the Multilayer Perceptron neural network are presented as follow: 1000 epoch for convergence; Hyperbolic Tangent (Tanh) activation function; Weight optimizer with the Stochastic Gradient Descending (SGD) algorithm; Adaptive learning rate over epoch; Network regularization term, defined by $\alpha = 10^{-5}$. Let us define the MLP input parameters as

$$\mathbf{M}_l = [M_{l,1}, M_{l,2}, M_{l,3}, M_{l,4} - (M_{l,2})^2, C] \tag{5}$$

and

$$\mathbf{C} = [C(m_1), C(m_2), C(m_3), C(m_4/(m_2)^2), C(c) =$$

$$[C(m), C(v), C(s), C(k), C(c)]. \tag{6}$$

Evaluation Metrics

The results are evaluated in four categories, as follows: True positive (TP), in which a record with AF is classified with AF; True negative (TN), which in a normal record is classified as normal; False positive (FP) a normal record is classified with AF; False negative (FN), for this category, a record with AF is classified as normal.

We use sensitivity (SENS), specificity (SPEC), and accuracy (ACC) metrics to evaluate the classification stage. Sensitivity reflects how the adopted methodology is effective in correctly identifying the TP; Specificity reflects how the adopted methodology is effective in correctly identifying TN, and accuracy indicates how the method is effective in correctly performing the diagnosis.

The sensitivity, specificity, and accuracy are defined, respectively, by (7), (8), and (9):

$$\text{SENS} = \frac{VP}{VP + FN} * 100 \tag{7}$$

$$\text{SPEC} = \frac{VN}{VN + FP} * 100 \tag{8}$$

$$\text{ACC} = \frac{VP + VN}{VP + VN + FN + FP} * 100 \tag{9}$$

To avoid overfitting, the cross-validation technique that evaluates the generalization capacity of a model will be used. Cross-validation partitions the dataset into mutually exclusive subsets, and later uses some of these subsets to estimation model parameters (training data), the remaining subsets (validation or test data) being employed in model validation. As a cross-validation technique, the K-fold method was used, which K = 10 (Friedman et at., 2001).

RESULTS

Response

After the segmentation of the ECG signal and the visualization of the existing deformations in the cardiac cycles with pathologies, we applied the bi-level feature extraction based on statistical moments to measure the deformations that differentiate the healthy signals from the three types of pathological signals and also the deformations that differentiate the three pathologies from each other (Figure 1).

Figure 1. The healthy signals and signals with Arrhythmia in a 3D space characterized by Eq. 4 applied on the variance of cardiac cycles - C(v), Eq. 4 applied on the kurtosis of cardiac cycles - C(k) and Eq. 4 applied on cardiac cycles - C(c)

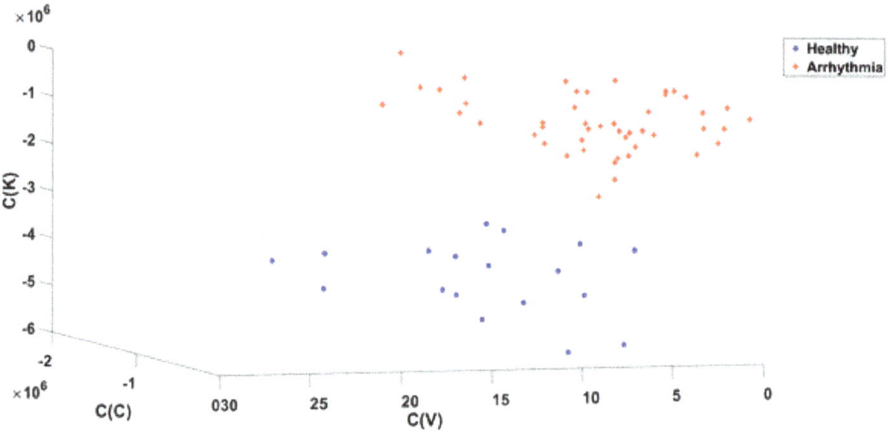

Then we obtained the matrices $\mathbf{M}_1$, $\mathbf{M}_2$, $\mathbf{M}_3$, $\mathbf{M}_4$ (Eq. 5) and **C** (Eq. 6), for healthy cardiac cycles, cardiac cycles with arrhythmia, cardiac cycles with atrial fibrillation, and cardiac cycles with supraventricular arrhythmia generated at the second-level of feature extraction.

At the end of the bi-level feature extraction using statistics moments, we performed some simulations to check between the matrices $\mathbf{M}_1$, $\mathbf{M}_2$, $\mathbf{M}_3$, $\mathbf{M}_4$ (Eq. 5) and **C** (Eq. 6) obtained, which of them allows the maximum reparability between the four classes in analysis. For this purpose, we verify the performance of the Multilayer Perceptron neural network, through the metrics of sensitivity, specificity, and accuracy, when applied to the MLP neural network inputs the lines of each of the matrices. We select the values of the MLP neural network parameters based on the best results presented during the simulations. These values have been specified according to 2.4. These simulations were performed using the K-fold cross-validation method (K= 10).

The results of the classification of the cardiac cycles by the MLP neural network for each of the matrices $\mathbf{M}_1$, $\mathbf{M}_2$, $\mathbf{M}_3$, $\mathbf{M}_4$ (Eq. 5) and **C** (Eq. 6) are presented in the Table 1.

From the results obtained, shown in Table 1, the second-level Eq. 4 of second order matrix **C** presented the best performance for the analyzed metrics in relation to the other matrices. We use factorial analysis to select the three most relevant attributes of the C matrix that evidence discrimination between healthy ECG signals and pathological ECG signals and also discrimination between the signals of pathological ECG and each other.

Finally, based on the results presented in Table 1 and the Figures 2 to 5, we verify the classification performance by the MLP neural network when we use all the attributes C *(C(m), C(v), C(s), C(k)* and *C(c))* and when we use only the three attributes resulting from factor analysis: C(v), C(k) and C(c). The result of the comparison of the classification performance verified by the metrics accuracy, sensitivity, and specificity for this simulation is presented in Table 2.

Table 1. Performance comparison between metrics M1, M2, M3, M4, and C

			Healthy	Arrhythmia	Supraventricular Arrhythmia	Atrial Fibrillation
		$\mathbf{M}_1$	92.55	92.75	91.42	97
		$\mathbf{M}_2$	95.65	95.29	89.44	92.49
SENS		$\mathbf{M}_3$	96.77	87	92.0	90.32
		$\mathbf{M}_4$	90	97.23	88.88	100
		C	100	97.18	98	100
		$\mathbf{M}_1$	91.63	95	91.08	93.15
		$\mathbf{M}_2$	97.77	92.89	94.28	91.11
SPEC		$\mathbf{M}_3$	82	90	81	81.52
		$\mathbf{M}_4$	95.29	89.48	84.28	91.25
		C	100	97	98.45	94.73
		$\mathbf{M}_1$	93	94.45	97.85	92.65
		$\mathbf{M}_2$	97.57	93.17	97.77	96.32
ACC		$\mathbf{M}_3$	87.11	88.87	80	88.50
		$\mathbf{M}_4$	94.30	93.60	84.48	93.17
		C	100	98.26	97.28	98.52

Note: Sensitivity (SENS); Specificity (SPEC); Accuracy (ACC); Second-level mean matrix - $\mathbf{M}_1$; Second-level variance matrix - $\mathbf{M}_2$; Second-level skewness matrix - $\mathbf{M}_3$; Second-level kurtosis matrix - $\mathbf{M}_4$; Second-level Eq. 4 of *k*-th order matrix - **C**.

Table 2. Comparison of performance between classification using five features C(m), C(v), C(s), C(k), and C(c) and three features C(v), C(k), and C(c) more relevant according to factor analysis (variance, kurtosis and Eq. 4)

	Number Features	Healthy	Arrhythmia	Supraventricular Arrhythmia	Atrial Fibrillation
	5	100	97.18	98	100
SENS	3	100	97.21	99.19	100
	Δ	0	-0.03	-1.19	0
	5	100	97	98.45	94.73
SPEC	3	100	97.12	98.90	95
	Δ	0	-0.12	-0.45	-0,27
	5	100	98.26	97.28	98.52
ACC	3	100	98.79	98.21	99.43
	Δ	0	-0.53	-0.93	-0.91

Note: Delta is the difference between a five and a three-feature classification. Sensitivity (SENS); Specificity (SPEC); Accuracy (ACC).

DISCUSSION AND CONCLUSIONS

A well-known method for exploring the repetitive characteristic of the cardiac muscle is the heart rate variability (HRV), which is a measure of the time between two consecutive heartbeats, i.e., the time between two consecutive R waves. This signal has been extensively explored as it shows the sympathetic or parasympathetic modulation of the autonomous nervous system. However, when it comes to diagnosing diseases, this method has shown limited capacity. Thus, we suggested here, rather than using the phase changing modulation occurring on the cardiac muscle, to explore the every-second information and the amplitude of the ECG.

Thus, unlike the HRV signal, in which each cardiac cycle is associated with a single real number, we associated each cardiac cycle with a set of points, that is, with a vector and thus build a new family of functions. For example, the sequence of functions generated by the statistical moments, that is, the average of variance, skewness and kurtosis of the moments and also families of functions generated by other metrics such as Eq. 4 of the moments.

In addition to frequency modulation, i.e., HRV, we believe that an additional idea is to use amplitude modulation from the ECG signal (Mazaheri & Khodadadi, 2020; Briceño et al., 2020; Qurraie & Afkhami, 2017). In fact, frequency and amplitude modulation are important concepts for combining two signals. In particular, frequency modulation occurs when a signal (modulating) is encoded in another (carrier) with larger frequency, varying the frequency of the carrier according to the variation in amplitude of the modulating signal. For example, the autonomic nervous system regulates changes in frequency in the ECG signal, thus generating heart rate variability (Victor et al., 2003). Similarly, amplitude modulation occurs when the modulating signal changes the amplitude of the ECG signal. Here, frequency and amplitude modulation information were used to assess the ECG signal either globally and locally (beat to beat). In this context, we used from the first to the fourth statistical moments to assess each cardiac cycle. Unlike other authors such as Martis et al. (2013), where they used the moments directly from the input data in the classifiers, we built a statistical moment signal, taken from every beat, to characterize either healthy or pathological ECG signals. Then, from this new beat-by-beat signal, we evaluated their moments: mean, variance, skewness, kurtosis and the Eq. 4.

After obtaining the matrices based on statistical moments in the bi-level feature extraction stage, we performed a multiclass classification of the ECG signals under analysis. This point of the proposed methodology differs from many studies that use dichotomous classifications in their approaches. For this, we use a Multilayer Perceptron neural network. The results obtained demonstrate the effectiveness of the methodology. As shown in the Table 1, average accuracy was achieved for the $\mathbf{M}_1$ matrix from 92.65% to 97.85%, for the $\mathbf{M}_2$ matrix from 93.17% to 97.77%, for the $\mathbf{M}_3$ from 80% to 88.87%, for the $\mathbf{M}_4$ matrix from 84.48% to 94.30% and for the $\mathbf{C}$ matrix from 97.28% to 100%. We verified that the best results were achieved by the new metric proposed in this study, Eq. 4, which was able to extract the information intrinsic to healthy signals and each cardiac pathology.

Due to the best classification result achieved with second-level analysis of Eq. 4, we used factor analysis to verify which attributes that make up this matrix are most relevant for the distinction between the classes of cardiac signals analyzed. Through factor analysis, we found that the second-level of the variance C(v), the second-level of the kurtosis C(k) and the second-level of the Eq. 2 C(c) are the most relevant features. These features were presented in a 3D scatter plot: healthy signals and signals with arrhythmia in Figure 1, healthy signals and signals with supraventricular arrhythmia in Figure 2, healthy signals and signals with atrial fibrillation in Figure 3, and healthy and pathological signals in Figure 4.

Figure 2. Healthy signals and signals with Supraventricular Arrhythmia in a 3D space characterized by Eq. 4 applied on the variance of cardiac cycles - C(v), Eq. 4 applied on the kurtosis of cardiac cycles - C(k) and Eq. 4 applied on cardiac cycles - C(c)

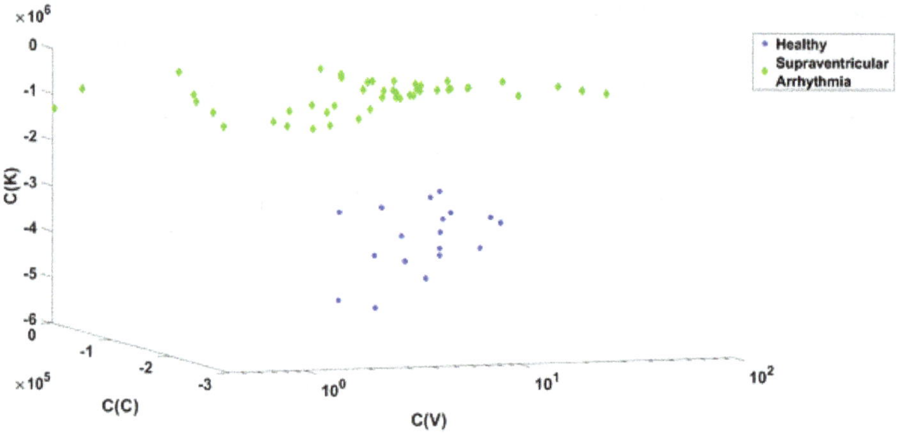

Figure 3. Healthy signals and signals with Atrial Fibrillation in a 3D space characterized by Eq. 4 applied on the variance of cardiac cycles - C(v), Eq. 4 applied on the kurtosis of cardiac cycles - C(k) and Eq. 4 applied on cardiac cycles - C(c)

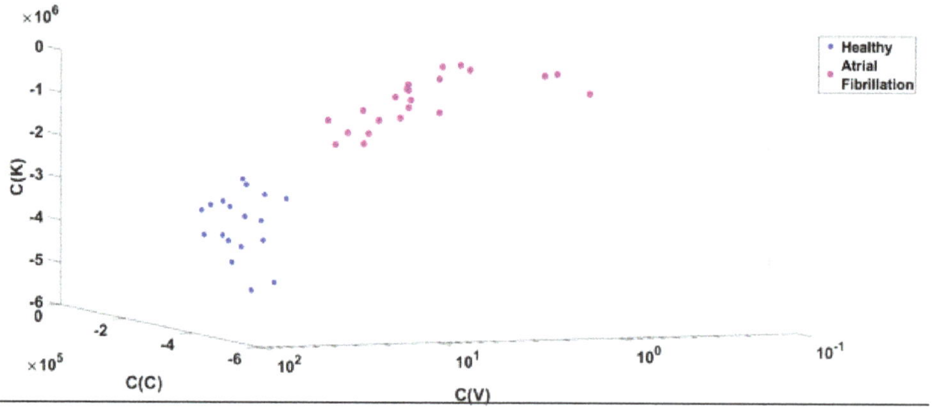

Figure 4. The signals healthy, arrhythmia, Supraventricular Arrhythmia, atrial fibrillation are presented in a 3D space characterized by Eq. 4 applied on the variance of cardiac cycles - C(v), Eq. 4 applied on the kurtosis of cardiac cycles - C(k) and Eq. 4 applied on cardiac cycles - C(c)

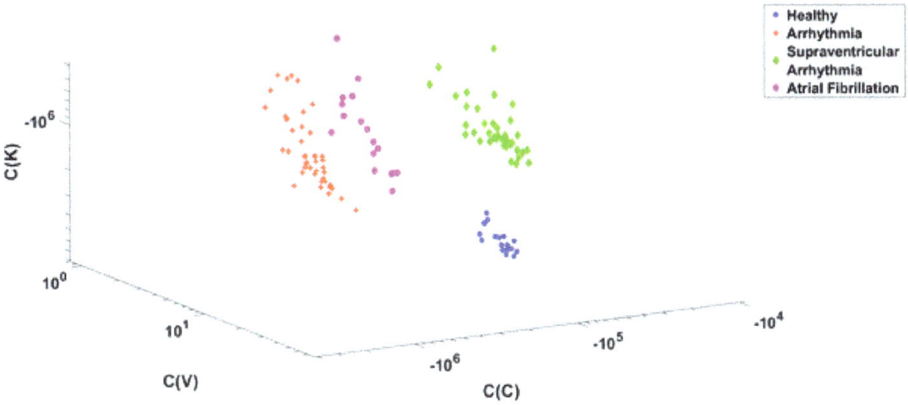

Looking at these plots, we can verify that classes do not overlap even using only 3 characteristics (C(v), C(k), C(c)). Therefore, we performed a new simulation using only these 3 features as the input classifier and the results were presented in Table 2. We found that results obtained by the MLP neural network using only three input features were higher than those obtained by the MLP using a total of five features. Therefore, it becomes more efficient, in addition to reducing computational cost, classifying healthy signals and pathological signals (atrial fibrillation; arrhythmia; supraventricular arrhythmia) using C(v), C(k) and C(c) as input MLP neural network.

To evaluate the contribution of the methodology proposed in this study in relation to the specialized literature, we present in Table 3 a comparison of performance in relation to sensitivity, specificity, and accuracy.

We observed in the papers presented in Table 3 that several authors, such as Raj et al. (2016) proposed classifications of up to 5 pathologies with accuracy of 99.18% using the SVM. However, these classifications are carried out using the one-against-all strategy. This strategy does not reflect the practical reality of the specialist, since each new sample should be submitted to one-against-all strategy for all classes evaluated and only then be classified in the class with the highest success rate. Rodriguez et al (2005) also performs dichotomous classification and proposes decision thresholds for the classification of different pathologies. Although the adaptation thresholds are simple and practical, they do not have a wide generalization capacity when compared to machine learning methods. The works proposed by

Chen et. Al. (2020), or Dokur & Ömez (2001) that use artificial neural networks as a classification algorithm are works with accuracy of 97% and 96% respectively. The accuracy rates proposed by these two authors are lower than those proposed by Raj Raj et al. (2016), on the other hand, the methods proposed by Chen et. Al (2020) and Dokur & Ömez (2001) simulates the non- dichotomous ability of the specialist to diagnose the healthy or pathological condition (one or more pathology) of a given signal. Our proposal also used MLP, but the difference is in the feature extraction stage, because we propose a simple and efficient extraction based on Eq. 4 which obtained average accuracy of 99.10%.

The main implication of this study is the complete analysis of the ECG signal, which allows the non-dicotomic classification of 3 classes of cardiac abnormalities, as well as healthy signals. The simplicity of this method may allow the application in systems similar to the 24-hour Holter. However, this system will be able not only to record the electrical activities of the heart and their variations but will also provide a prognosis for various cardiac pathologies. Our next challenge is to implement and evaluate the proposed method in real time.

Limitations of the Proposed Method

In practice, developing a diagnostic aid method has at least two major challenges to be overcome. One is the computational cost and the other is the infeasibility of obtaining the ECG signal over long periods of time so that the algorithms can be trained and tested. Therefore, a limitation to be overcome by the proposed method is that the moments can be calculated online, however, the calculation of the Eqs. 3 and 4 of the moments should be performed offline. The proposed data extraction is measured by statistical moments. The statistical moments are dependent on the average of the samples; actually, a large number of outliers can complicate the analysis of the data.

Table 3. Performance comparison of proposed methodology with the literature with respect to sensitivity, specificity, and accuracy. The symbol (-) represent values of sensibility, specificity, and accuracy are not specified in the works

Literature	Feature	Classifier	Classes	Accuracy
Proposed	**Frequ^encia, Amplitude,HOS**	**RNA**	**4**	**99,10**
Qiao et al. (2014)	Features	SVM	5	98.60
Martis et al. (2013)	PCA	LS-SVM	5	93.48
Kim et al. (2016)	SDRR, RMSSD, SDSD, pRR20,	SVM	2	98.91
	and pRR50,			
Raj et al. (2016)	DOST,PSO	SVM	5	99.18

Note: particle swarm optimization (PSO) discrete orthogonal Stockwell transform (DOST); SOM: self-organizing map, BPNN: back propagation neural network, LS-SVM: least square support vector machines, PCA: principal com- ponent analysis, WT: wavelet transform, MLP-NN: multi-layer perceptron neural network, GA: genetic algorithm, SDRR, standard deviation of RR-intervals; RMSSD, square root of the mean of squares of the successive differences between adjacent RR-intervals; SDSD, standard deviation of the successive differences between adjacent RR-intervals greater than 20 ms; pRR20, proportion of intervals presenting time duration differences between adjacent RR-intervals greater than 20ms; pRR50, proportion of intervals presenting time duration differences between adjacent RR-intervals greater than 50ms.

REFERENCES

Afkhami, R. G., Azarnia, G., & Tinati, M. A. (2016). Cardiac arrhythmia classification using statistical and mixture modeling features of ecg signals. *Pattern Recognition Letters*, *70*, 45–51. doi:10.1016/j.patrec.2015.11.018

Briceño, D. F., Santangeli, P., Frankel, D. S., Liang, J. J., Shirai, Y., Markman, T., Enriquez, A., Walsh, K., Riley, M. P., Nazarian, S., Lin, D., Kumareswaran, R., Arkles, J. S., Hyman, M. C., Deo, R., Supple, G. E., Garcia, F. C., Dixit, S., Epstein, A. E., . . . Schaller, R. D. (2020). Qrs morphology in lead v1 for the rapid localization of idiopathic ventricular arrhythmias originating from the left ventricular papillary muscles: A novel electrocardiographic criterion. *Heart Rhythm, 17*(10), 1711 – 1718. https://www.sciencedirect.com/science/article/pii/S1547527120304550

Chen, T.-M., Huang, C.-H., Shih, E. S., Hu, Y.-F., & Hwang, M.-J. (2020). Detection and classification of cardiac arrhyth- mias by a challenge-best deep learning neural network model. *iScience*, *23*(3), 100886. doi:10.1016/j.isci.2020.100886 PMID:32062420

Dokur, Z., & Ömez, T. (2001). Ecg beat classification by a novel hybrid neural network. *Computer Methods and Programs in Biomedicine, 66*(2), 167–181. doi:10.1016/S0169-2607(00)00133-4 PMID:11551391

Ebrahimzadeh, E., Kalantari, M., Joulani, M., Shahraki, R. S., Fayaz, F., & Ahmadi, F. (2018). Prediction of paroxysmal atrial fibrillation: A machine learning based approach using combined feature vector and mixture of expert classification on hrv signal. *Computer Methods and Programs in Biomedicine, 165*, 53–67. doi:10.1016/j.cmpb.2018.07.014 PMID:30337081

Elhaj, F. A., Salim, N., Harris, A. R., Swee, T. T., & Ahmed, T. (2016). Arrhythmia recognition and classification using combined linear and nonlinear features of ecg signals. *Computer Methods and Programs in Biomedicine, 127*, 52–63. doi:10.1016/j.cmpb.2015.12.024 PMID:27000289

Friedman, J., Hastie, T., & Tibshirani, R. (2001). The elements of statistical learning (Vol. 1). *Springer series in statistics, 1*(10).

Goldberger, A. L., Amaral, L. A., Glass, L., Hausdorff, J. M., Ivanov, P. C., Mark, R. G., Mietus, J. E., Moody, G. B., Peng, C.-K., & Stanley, H. E. (2000a). Physiobank, physiotoolkit, and physionet components of a new research resource for complex physiologic signals (Vol. 101). *Circulation, 101*(23), 215–220. doi:10.1161/01.CIR.101.23.e215 PMID:10851218

Goldberger, A. L., Amaral, L. A., Glass, L., Hausdorff, J. M., Ivanov, P. C., Mark, R. G., Mietus, J. E., Moody, G. B., Peng, C.-K., & Stanley, H. E. (2000b). Physiobank, physiotoolkit, and physionet. *Circulation, 101*(23), e215–e220. doi:10.1161/01.CIR.101.23.e215 PMID:10851218

Homaeinezhad, M., Atyabi, S., Tavakkoli, E., Toosi, H., Ghaffari, A., & Ebrahimpour, R. (2012). Ecg arrhythmia recognition via a neuro-svm–knn hybrid classifier with virtual qrs image- based geometrical features. *Expert Systems with Applications, 39*(2), 2047–2058. doi:10.1016/j.eswa.2011.08.025

Huang, C., Ye, S., Chen, H., Li, D., He, F., & Tu, Y. (2011). A novel method for detection of the transition between atrial fibrillation and sinus rhythm. *IEEE Transactions on Biomedical Engineering, 58*(4), 1113–1119. doi:10.1109/TBME.2010.2096506 PMID:21134807

Jovic, A., & Jovic, F. (2017). Classification of cardiac arrhythmias based on alphabet entropy of heart rate variability time series. *Biomedical Signal Processing and Control, 31*, 217–230. doi:10.1016/j.bspc.2016.08.010

Kim, Y. J., Heo, J., Park, K. S., & Kim, S. (2016). Proposition of novel classification approach and features for improved real-time arrhythmia monitoring. *Computers in Biology and Medicine*, *75*, 190–202. doi:10.1016/j.compbiomed.2016.06.009 PMID:27318329

Kuhtz-Buschbeck, J. P., Schaefer, J., Wilder, N., & Wolze, W. T. (2020). The origin of the heartbeat and theories of muscle contraction. Physiological concepts and conflicts in the 19th century. *Progress in Biophysics and Molecular Biology*, *159*, 3–9. doi:10.1016/j.pbiomolbio.2020.05.009 PMID:32622835

Li, Q., Rajagopalan, C., & Clifford, G. D. (2014). A machine learning approach to multi-level ecg signal quality classification. *Computer Methods and Programs in Biomedicine*, *117*(3), 435–447. doi:10.1016/j.cmpb.2014.09.002 PMID:25306242

Martis, R. J., Acharya, U. R., Mandana, K., Ray, A., & Chakraborty, C. (2013). Cardiac decision making using higher order spectra. *Biomedical Signal Processing and Control*, *8*(2), 193–203. doi:10.1016/j.bspc.2012.08.004

Mazaheri, V., & Khodadadi, H. (2020). Heart arrhythmia diagnosis based on the combination of morphological, frequency and nonlinear features of ecg signals and metaheuristic feature selection algorithm. *Expert Systems with Applications*, *161*, 113697. doi:10.1016/j.eswa.2020.113697

Moody, G. B., & Mark, R. G. (1983). *A new method for detecting atrial fibrillation using R-R intervals*. Academic Press.

Pan, J., & Tompkins, W. J. (1985). A real-time QRS detection algorithm (Vol. 32). *IEEE Transactions on Biomedical Engineering*, *32*(3), 230–236. doi:10.1109/TBME.1985.325532 PMID:3997178

Parvaneh, S., Rubin, J., Babaeizadeh, S., & Xu-Wilson, M. (2019). Cardiac arrhythmia detection using deep learning: A review. *Journal of Electrocardiology*, *57*, S70–S74. doi:10.1016/j.jelectrocard.2019.08.004 PMID:31416598

Petrėnas, L. S. A., & Vaidotas, M. (2015). Low-complexity detection of atrial fibrillation in continuous long-term monitoring. *Computers in Biology and Medicine*, 1 – 8. . doi:10.1016/j.compbiomed.2015.01.019

Poddar, M., Birajdar, A. C., & Virmani, J. (2019). Chapter 5 - automated classification of hypertension and coronary artery disease patients by pnn, knn, and svm classifiers using hrv analysis. In *Machine Learning in Bio-Signal Analysis and Diagnostic Imaging*. Academic Press. https://www.sciencedirect.com/science/article/pii/B9780128160862000059

Queiroz, J. A., Azoubel, L., & Barros, A. K. (2019). Support system for classification of beat-to-beat arrhythmia based on variability and morphology of electrocardiogram. *EURASIP Journal on Advances in Signal Processing*, *2019*(1), 16. Advance online publication. doi:10.1186/s13634-019-0613-9

Queiroz, J. A., Junior, A., Lucena, F., & Barros, A. K. (2017). Diagnostic decision support systems for atrial fibrillation based on a novel electrocardiogram approach. *Journal of Electrocardiology*. https://www.sciencedirect.com/science/article/pii/S0022073617304259

Qurraie, S. S., & Afkhami, R. G. (2017). Ecg arrhythmia classification using time frequency distribution techniques. *Biomedical Engineering Letters*, *7*(4), 325–332. Advance online publication. doi:10.1007/s13534-017-0043-2 PMID:30603183

Raj, S., Ray, K. C., & Shankar, O. (2016). Cardiac arrhythmia beat classification using dost and psotuned svm. *Computer Methods and Programs in Biomedicine*, *136*, 163–177. doi:10.1016/j.cmpb.2016.08.016 PMID:27686713

Rodriguez, J., Goni, A., & Illarramendi, A. (2005). Real-time classification of ecgs on a pda. *IEEE Transactions on Information Technology in Biomedicine*, *9*(1), 23–34. doi:10.1109/TITB.2004.838369 PMID:15787004

Sahoo, S., Dash, M., Behera, S., & Sabut, S. (2020). Machine learning approach to detect cardiac arrhythmias in ecg signals: A survey (Vol. 41). *Ingénierie et Recherche Biomédicale : IRBM = Biomedical Engineering and Research*, *41*(4), 185–194. doi:10.1016/j.irbm.2019.12.001

Sameni, R., Shamsollahi, M. B., Jutten, C., & Babaie-Zade, M. (2005). Filtering noisy ecg signals using the extended kalman filter based on a modified dynamic ecg model. Computers in Cardiology, 1017–1020. doi:10.1109/CIC.2005.1588283

Singh, S., Pandey, S. K., Pawar, U., & Janghel, R. R. (2018). Classification of ecg arrhythmia using recurrent neural networks (Vol. 132). *Procedia Computer Science, 132,* 1290 – 1297. https://www.sciencedirect.com/science/article/pii/S1877050918307774

Tsipouras, M. G., & Fotiadis, D. I. (2004). Automatic arrhythmia detection based on time and time–frequency analysis of heart rate variability. *Computer Methods and Programs in Biomedicine*, *74*(2), 95–108. doi:10.1016/S0169-2607(03)00079-8 PMID:15013592

Victor, S., & Nayak, V. M. (2003). Every heart beat is under neural command: An hypothesis relating to the cardiac rhythm. *Heart Lung and Circulation*, *12*(1), 11–17. doi:10.1046/j.1444-2892.2003.00178.x PMID:16352101

Yao, Q., Wang, R., Fan, X., Liu, J., & Li, Y. (2020). Multiclass arrhythmia detection from 12-lead varied-length ecg using attention-based time-incremental convolutional neural network. *Information Fusion*, *53*, 174–182. doi:10.1016/j.inffus.2019.06.024

Zarei, A., & Asl, B. M. (2020). Automatic classification of apnea and normal subjects using new features extracted from hrv and ecg-derived respiration signals. *Biomedical Signal Processing and Control*, *59*, 101927. doi:10.1016/j.bspc.2020.101927

Chapter 3
Digital Health, Smartphone-Based Motor Assessment, and Public Health

Givago Silva Souza
https://orcid.org/0000-0002-4525-3971
Federal University of Pará, Brazil

Brena Karoline Ataíde Furtado
Federal University of Pará, Brazil

Edilson Brabo Almeida
Federal University of Pará, Brazil

Bianca Callegari
Federal University of Pará, Brazil

Maria da Conceição Nascimento Pinheiro
Federal University of Pará, Brazil

ABSTRACT

The availability of technologies for assessing people's health is a limiting factor in many countries, especially the poorest ones. Smartphones offer a range of tools that can be useful for extracting biological signals that may be related to individuals' health conditions or diseases. Among these tools, inertial sensors and touchscreens enable the performance of motor tests that scientific literature has shown to be valid for the physical assessment of individuals. The integration of smartphones into public policies aimed at increasing health monitoring of individuals would allow for the expansion of the scope of quality assessment and preventive actions against functional declines.

DOI: 10.4018/979-8-3693-0851-6.ch003

DIGITAL HEALTH E mHEATLH

The World Health Organization proposed in 2020 the Global Digital Health Strategy 2020-2025 (World Health Organization, 2021), which outlines a comprehensive framework for leveraging digital technologies to enhance healthcare worldwide. The fundamental idea behind this strategy is to harness digital innovations to (i) expand access to healthcare services and information using digital tools, (ii) enhance the quality of healthcare through digital tools, (iii) strengthen healthcare systems by integrating technological solutions into existing healthcare infrastructure and policies, (iv) promote digital technology innovation for health monitoring and its responsible application, (v) empower individuals and enable them to control their own health monitoring through access to accurate health information, telemedicine services, and smartphone applications, (vi) ensure data privacy in the digital environment, and (vii) enhance governance and regulation to guide the responsible adoption and use of health technologies at both national and transnational levels. Essentially, the Global Digital Health Strategy 2020-2025 envisions a future where digital innovations play a pivotal role in improving access, quality, and efficiency of healthcare while prioritizing equity, privacy, and security in healthcare delivery. Its aim is to guide governments and stakeholders in utilizing digital technologies to enhance healthcare systems and outcomes worldwide over a five-year period. Some developing countries such as Brazil and India have their own planning for digital health implementation (Brazilian Ministry of Health, 2020; Ministry of Health and Family Welfare of India, 2020).

The use of digital technologies in the field of healthcare dates back to the second half of the 20th century (Mesko et al., 2017) when the first computers were used for medical research and for hospital purposes, such as storing and retrieving medical records in digital formats. With the advancement and development of more powerful computers, a broader range of digital tools began to emerge in various subfields of healthcare. A second wave of advancement and the development of new digital tools occurred with the popularization of the internet, including telehealth, in which services are provided remotely (Martínez-Caro et al., 2018). At the turn of the 21st century, the use of mobile phones and the creation of large databases made health monitoring a reality, leading to a proliferation of digital tools for health monitoring, often associated with artificial intelligence during the COVID-19 pandemic (Alhasan & Hasaneen, 2021; Dicianno et al., 2015). Thus, the term "Digital Health," also known as eHealth or electronic health, emerged, signifying the application of information and communication technologies in the healthcare sector to enhance service delivery, health monitoring, health information management, and the promotion of more effective and accessible health practices (Chan, 2021). In this field of knowledge, there is a convergence of health, informatics, and technology to transform the way healthcare is delivered, managed, and experienced, through solutions such as telemedicine, connected medical devices, health applications, and data analysis, all aimed at promoting the quality of life and well-being of individuals.

A branch of the Digital Health field is mHealth, which stands for the use of portable digital devices, especially smartphones, as a tool for health monitoring in various areas (McCool et al., 2022). The smartphone inherently incorporates a set of instruments that have come to be used for various purposes in the healthcare domain. Light sensors, touch screens, inertial sensors, microphone, speakers, temperature sensors, and video cameras are some of these instruments present in smartphones that enable user interaction with the device and thus facilitate the extraction of health-related information (Majumder & Deen, 2019).

Some important milestones for the implementation of mHealth were the appearance of the first mobile phones in the 1970s, which only began to gain popularity in the 1990s (Harris et al., 2019). During this period, the first studies on the use of mobile phones in health began to emerge and some regulation of health-oriented mobile phone applications began (Onodera & Sengoku, 2018). The term mHealth was first coined in 2003 (Istepanian et al., 2003), and then the first journals and books on this specific topic began to appear. The World Health Organization began discussing mHealth policies in various countries in 2009 and published a report on the subject in 2011 (World Health Organization, 2011). In 2007, with the launch of the iPhone, the concept of a smartphone was introduced, in which the presence of sensors made it possible to record a range of information and facilitated the start of the expansion of the creation of health applications, which continues up to date.

It is considered that smartphone applications aimed at people's health can be classified into supporting diagnosis and decision-making, supporting behavioral changes, supporting digital therapies and offering educational content (Rowland et al., 2020). Even considering the potential benefits that mHealth can bring, some challenges must be considered for these benefits to become real, such as structural barriers, especially in poor countries, the user's ability to use the technology and even the emergence of low-cost, highly durable technologies (Becker et al., 2014). Despite the immense supply of health apps, it is questionable how much they can actually improve patients' health conditions. There is a limited number of clinical scenarios supported by scientific evidence in which they are successful (Rowland et al., 2020), although there is a growing interest in studying mHealth, reflected by the increase in scientific production in the field (Fiordelli et al., 2013).

Today, mHealth already drives about dozens billion-dollar market with an estimated growth potential to exceed hundred billion dollars by 2030 (Bettencourt et al., 2023). The application of mHealth extends to all subdomains of healthcare, and one of the areas that has been at the forefront of technological solutions is the analysis of human movement[10]. Ultimately, assessing human movement is tantamount to evaluating an individual's overall health status. Ill individuals tend to exhibit reduced mobility. Therefore, the quantity and quality of movement can serve as clues, or biomarkers, of a person's health status. Below, we will describe some of the key smartphone technologies that have been widely employed for the non-invasive assessment of human movement.

SMARTPHONE SENSORS AND HUMAN MOVEMENT ANALYSIS

Due to the vast array of applications that can be performed with smartphones, including the use of forms for inquiries related to an individual's health, this discussion will focus primarily on applications that require interactions with the built-in sensors of the smartphone.

The fundamental concept behind using internal smartphone sensors in the assessment of human movement is that each of these sensors is sensitive to specific aspects related to human motion and can take readings at very short time intervals, enabling the recording of a time series of the desired parameters. From this time series, computational programs known as applications (apps) can extract information that can be associated with an individual's health condition (Abou et al., 2021). Among the sensors most applicable to the evaluation of human movement are the touch screen, inertial sensors, video camera, and microphone. In this chapter, we will explore how touchscreens and inertial sensors of smartphones can be utilized to assess certain motor functions and their applications. To learn more about the use of cameras and microphones in motor assessment, one can refer to other sources.

Application of Motor Tests Using Smartphone Touchscreens

Touchscreens, as the name suggests, are sensitive to pressure and allow for the identification of user interactions with the smartphone screen, such as tapping, long-pressing in one place, swiping, or releasing the finger from the screen (Nam et al., 2021). Furthermore, it is possible to determine the spatial coordinates where the touch occurred. Thanks to this functionality, various applications have been developed to assess user performance in different motor tasks.

The finger tapping test is a classic assessment of motor coordination in which the participant is asked to tap as quickly as possible on a surface for a specified duration. The more taps achieved, the better the individual's motor performance (Volkmar & Schmitt, 2013). Initially, in the 19th century, this test was conducted using telegraph keys, and with the advent of new technologies, it could be performed on computer keyboards and through video capture, among other methods (Akram et al., 2022; Guo et al., 2022; Hubel et al., 2013). Over the last decade, several studies have applied the finger tapping test on the touchscreens of smartphones (Lee et al., 2016). The use of smartphones has added an extra dimension to the assessment of the finger tapping test. Traditionally, the test's performance was entirely based on temporal parameters, but since touch screens allow for the evaluation of where the touch occurred, it is now possible to assess spatial patterns of taps made by the participant. The use of this test on smartphones has been successfully applied in diseases, especially Parkinson's disease (Lee et al., 2016). Factors such as hand dominance and participant gender appear to influence participant performance and should be considered when interpreting test results (Brito et al., 2023).

The reaction time test is another classic motor assessment test that is conducted through interaction with a smartphone's touchscreen (Burke et al., 2017). In the reaction time test, a sensory stimulus is presented, and the participant is instructed to perform a motor task on the smartphone screen, such as moving their finger from one place to another. Previously, this test was conducted using a computer and a cursor or response box. Adaptations on smartphones offer different sensory modalities (visual or auditory), and the response is given by touching the smartphone's touchscreen. Programmatically, it is possible to record the moment when the stimulus was presented and the interval until the participant reacts to the stimulus and performs the predetermined task. Reaction time and response time can be related to the neural information processing time.

Another test that has been frequently applied is an adaptation for smartphones of the spiral analysis (Ishii et al., 2020). The spiral analysis test was initially conducted on paper with the printed drawing of an Archimedean spiral. Participants are required to cover the spiral on the paper as perfectly as possible with a pen or pencil, and the number of traces deviating from the original path is quantified. On a smartphone, the spiral is displayed on the screen, and participants must trace it with their finger. Afterward, it is possible to verify if the pixels of the spiral on the screen coincide with the pixels traced by the participant on the spiral. This test initially aims to assess hand motor coordination; however, in a variety of articles, it has been used as a quantifier of hand tremors in patients with Parkinson's disease.

Application of Motor Tests Using Smartphone's Inertial Sensors

Smartphones are equipped with what are known as Inertial Measurement Units (IMUs), which consist of two inertial sensors: the accelerometer and the gyroscope. These sensors are tiny marvels of microelectronics, and their basic principle is to identify and quantify changes in the state of inertia of the smartphone (Picerno et al., 2021). Measurements of changes in inertia are indirect and are achieved by measuring acceleration using accelerometers and angular velocity using gyroscopes.

In general, accelerometers are better suited for describing linear displacements, while gyroscopes excel at evaluating rotational movements. The inertial sensors found in smartphones are typically tri-axial, meaning they sense linear and rotational movements in three axes (anteroposterior, mediolateral, and superoinferior). The choice of using time series data obtained from one or more axes depends on the specific study objectives.

A significant number of motor assessment protocols using built-in or external inertial sensors in smartphones have been proposed, including assessments of static balance, mobility, and tremor (Gawronska et al., 2020; Mellone et al., 2012; Sahin et al., 2022). The basic idea for conducting each of these assessments is to securely attach the smartphone to the body of the person being tested and instruct them to perform a specific task related to the function under analysis. The choice of where to attach the smartphone also depends on what is intended to be recorded. It is not uncommon to have studies that use different smartphone placements to assess the same function. Advances in future research should help determine the best smartphone positioning configurations for each activity.

The assessment of static balance is typically performed with the smartphone attached to the waist, at the level of the lower lumbar spine (Rodrigues et al., 2022). The task usually requested is to maintain an upright position, with the arms at the sides and as much immobility as possible. Since the smartphone is affixed to the body of the person being tested, body and smartphone oscillations will be the same. Generally, accelerometer time series data in the anteroposterior and mediolateral axes are used. If the smartphone is positioned upright (or in landscape mode), the anteroposterior axis corresponds to the Z-axis on the smartphone, which is the axis that runs perpendicular to the device's screen, and the mediolateral axis is related to the width of the smartphone (the device's X-axis). Consequently, oscillations related to movements in each axis over time can be studied through graphs known as stabilograms, and correlations between oscillations in both axes can be examined on a graph known as statokinesiogram. This nomenclature for static balance analysis is inherited from analyses conducted using the gold standard method, which is the force platform. There are no standardized names yet for these graphs using acceleration data. The more oscillations observed in both graphs, the worse the balance control of the person being tested is likely to be. Several studies have already provided results that support the validation of built-in accelerometers in smartphones for assessing static balance.

Another motor function that has been extensively investigated using smartphones is mobility through a test called the Timed Up and Go test (TUG) (Gerhardy et al., 2019). The TUG test was initially proposed to assess the mobility of elderly patients and over time has been strongly associated with the risk of falls. However, it can provide more information than just dynamic balance-related data. During the execution of this test, the participant is asked to sit in an armchair, then stand up, walk in a straight line for 3 meters, turn around, return along the same path to reach the front of the chair, turn around again, and sit down in the chair. The participant is required to perform this task as quickly as possible without running. Knowing the steps of the test, it can be inferred that there are components that allow for the evaluation of (i) the participant's muscle strength, as seen when they stand up from or sit down in the chair, (ii) mobility itself, as seen during the walking and turning portions of the test, and (iii) dynamic balance throughout the test, during which the participant must not fall at any point. Traditionally, the performance of this test was quantified only by the total duration of the task, which was measured by the experimenter using a stopwatch. However, after the application of inertial sensors during this test, it is possible to record changes in acceleration and angular velocity, thus obtaining representative waveforms of different stages of test execution. There is a wide variety of methods for analyzing inertial time series data, which involve evaluating time series data in each sensor axis or fusing time series data from different axes or sensors. Regardless of the mode of analysis, it is possible to extract information from these inertial time series regarding the total time for task execution, as well as the time and peak acceleration for standing up from the chair, the time and cadence of the outward walk, the time and amplitude of the turn at the 3-meter mark, or when returning to the front of the chair. In fact, new analyses are still being proposed, adding even more parameters for TUG analysis. Studies involving elderly individuals, diabetics, and patients with neurodegenerative diseases have shown that these additional parameters, beyond the test completion time, can provide insights into motor impairments resulting from different diseases.

For a long time, inertial sensors have also been used to describe and quantify physiological or pathological body tremors (Elble et al., 2016; Elble & McNames, 2016). With the presence of inertial sensors in smartphones, smartwatches, and other portable digital devices, a series of studies began to assess the validity of using these instruments for investigating body tremors[30]. In general, these studies showed that the sensors in these portable devices could differentiate tremors in healthy individuals from those with neurodegenerative diseases where tremors are a prominent symptom. Several apps are available for recording and analyzing tremors for use on smartphones. Given the frequent use of smartphones for tremor assessment, it is important to examine whether the mass or dimensions of the smartphone could influence the characteristics of the recorded tremor. It has been known since the 1960s-1970s that the mass attached to a body segment has an influence on the tremor recorded in that segment. Our group compared the results of tremor recorded using a smartphone weighing around a hundred grams and using a portable sensor weighing only a few grams (around 6 grams) in young and healthy individuals. It was found that the frequency spectrum of the tremor differed between the recordings obtained by the two devices and were not comparable (Santos et al., 2022). Body tremors involve at least three mechanisms contributing to their generation, and the mechanism that contributes the most energy is dependent on the elastic tension of tissues. Due to the differences in sensor masses, there are modifications in tissue tension and alterations in tremor characteristics.

APPLICATION OF SMARTPHONE-BASED MOTOR ASSESSMENT AND PUBLIC HEALTH

None of the smartphone-based motor assessment proposals described earlier have seen widespread use in communities thus far. In general, efforts have been focused on validating different assessment protocols and testing them in various populations living with different diseases while assessing the reliability of smartphones for self-testing.

However, governments could pay more attention to the possibility of using smartphones, especially in primary care. Motor tests, in general, serve as valuable biomarkers for the early detection of a range of health issues, particularly in the elderly. Nevertheless, these tests are not part of routine primary care because they require costly instruments to provide more diagnostic power for early detection.

Muscle weakness, decreased mobility, and falls are among the motor performance-related problems that could be monitored in primary healthcare using smartphones. Once a person is identified with some form of motor vulnerability, they could be directed to physical activity or motor rehabilitation programs. Several studies have already demonstrated the potential cost savings for individuals and governments if people could have their motor functions assessed and participate in physical activity programs.

The use of smartphones for motor assessment, as an easily accessible tool, brings benefits to both healthcare professionals and patients. This tool assists healthcare professionals in effective diagnosis and, consequently, in improving the quality of life of individuals. mHealth is already a reality and is present in various specialties within the healthcare field. Exploring the potential uses of smartphone sensors will allow for the expansion of monitoring, tracking, and diagnosis of significant diseases, especially those addressed in public health.

PRACTICAL APPLICATIONS OF mHEALTH IN PREVENTIVE MEDICINE

Among the many practical measures that a healthcare professional could use mHealth as a preventive measure in primary care, assessing the balance and mobility of the elderly could represent two initial actions. The elderly suffer from a number of physiological losses due to aging that predispose them to falls.

Older people who fall are less mobile, have a poorer quality of life, spend more time in hospital, consume more public and private healthcare dollars, and falls can lead to a range of morbidities and even death. Assessing static balance and mobility using smartphones can provide a great deal of information about the physical condition of the elderly in order to prevent falls.

To avoid creating a new monitoring activity, the elderly person's routine visit to the health service for another reason could be used. The tests don't take 10 minutes, and if the service has normative data available, it could provide the results of the assessments at the same time, indicating whether the elderly person has a loss of static balance or mobility. Over the years, the elderly person or even the service could create a time series of the elderly person's performance on these two tests, and whenever a loss of function is detected, the patient could be immediately referred to the rehabilitation sector or to a physical activity program and instructed to make environmental changes in his or her home or work environment to prevent falls. Studies have shown that multidimensional fall prevention strategies are more likely to have an impact on reducing the number of falls (Chang et al., 2004).

Another practical approach with great potential for success in primary care is to monitor the motor status of older people, with a focus on looking for the cardinal signs of Parkinson's disease, which are postural instability, bradykinesia, rigidity, and tremor. The idea is not to diagnose Parkinson's disease, but to monitor people in the community who may be showing one or more of these signs, so that they can be referred to a neurologist as soon as one of these symptoms is detected with quantitative parameters that are outside the population norms for their age. Here, static balance and mobility could also be assessed using the iTUG, as well as hand tremor and the finger tapping test to check the patient's degree of bradykinesia. The same tests could be continued after the clinical diagnosis, so that they could be used as functional indicators to evaluate the drug therapy applied to the patient.

ETHICAL CONSIDERATIONS FOR USING SMARTPHONES TO ASSESS MOTOR FUNCTION IN PRIMARY CARE

There is still no clear literature discussing the ethical issues surrounding the use of smartphones for motor assessment in primary health care. However, some points can be made to inform the debate. Not every application should be used by the patient. Many of the motor tests are challenging for patients and it would be safe to perform the test with some supervision, preferably by a healthcare professional. The use of smartphones as a public policy tool requires that the device on which the assessment is performed be owned by the service. Although smartphones are widely used by the population, the responsibility for monitoring people's health requires a certain level of quality and integrity of the device known to the service. In addition, the data must be stored in the service's database so that a historical series can be compiled for each patient. The presence of data on patients' smartphones would make it difficult for the service to access the data for proper planning of the service's actions. The same goes for tests performed on a healthcare professional's smartphone; the data would remain the property of the professional with little guarantee that it could be accessed by patients or the service when needed. A better strategy would be for the data to be automatically stored in the cloud, and for only registered service professionals to have access to it.

APPLICATIONS WITH POTENTIAL FOR USE IN PRIMARY CARE

It is thought that a motor assessment application for use in a primary health care service should (1) carry out an automated assessment of a given motor function, (2) compare the data with population norms and (3) save the data in a digital database for later access. It is thought that most of the validated movement assessment apps can fulfill the first requirement of doing what they set out to do. However, many applications described in the literature are still in the experimental phase of their development and are not yet one-button solutions. Some of them record the signal, but the user still needs to create programs for offline analysis of the data exported by the app. Many of the apps do not include a normality database for populations of different ages. If the user is interested, they will have to create their own population norms. As mHealth is still a growing area, it is not known exactly whether the population norms generated on a smartphone can be adapted to records obtained on other smartphone models. Our group has described some motor activities that may be influenced by the weight of the devices, thus implying comparability of data obtained from devices with very different physical characteristics.

Finally, there are applications that contain only one motor test, while others contain several of these tests, and the potential user should consider which test would be more important for the service, the importance of the motor results obtained from the app, and how to store the data from patients to choose many apps testing different motor tasks or few apps containing a variety of tests. In summary, a manager of a health service interested in introducing smartphones for health monitoring in primary care should include in the service costs the smartphones, licenses of apps and probably training in mHealth for the health professional staff.

REFERENCES

Abou, L., Peters, J., Wong, E., Akers, R., Dossou, M. S., Sosnoff, J. J., & Rice, L. A. (2021). Gait and Balance Assessments using Smartphone Applications in Parkinson's Disease: A Systematic Review. *Journal of Medical Systems*, *45*(9), 87. doi:10.1007/s10916-021-01760-5 PMID:34392429

Adams, J. L., Kangarloo, T., Tracey, B., O'Donnell, P., Volfson, D., Latzman, R. D., Zach, N., Alexander, R., Bergethon, P., Cosman, J., Anderson, D., Best, A., Severson, J., Kostrzebski, M. A., Auinger, P., Wilmot, P., Pohlson, Y., Waddell, E., Jensen-Roberts, S., ... Carroll, T. (2023). Using a smartwatch and smartphone to assess early Parkinson's disease in the WATCH-PD study. *NPJ Parkinson's Disease*, *9*(1), 64. doi:10.1038/s41531-023-00497-x PMID:37069193

Aghanavesi, S., Memedi, M., Dougherty, M., Nyholm, D., & Westin, J. (2017). Verification of a Method for Measuring Parkinson's Disease Related Temporal Irregularity in Spiral Drawings. *Sensors (Basel)*, *17*(10), 2341. doi:10.3390/s17102341 PMID:29027941

Akram, N., Li, H., Ben-Joseph, A., Budu, C., Gallagher, D. A., Bestwick, J. P., Schrag, A., Noyce, A. J., & Simonet, C. (2022). Developing and assessing a new web-based tapping test for measuring distal movement in Parkinson's disease: A Distal Finger Tapping test. *Scientific Reports*, *12*(1), 386. doi:10.1038/s41598-021-03563-7 PMID:35013372

Alhasan, M., & Hasaneen, M. (2021). Digital imaging, technologies and artificial intelligence applications during COVID-19 pandemic. *Computerized Medical Imaging and Graphics: The Official Journal of the Computerized Medical Imaging Society*, *91*, 101933. doi:10.1016/j.compmedimag.2021.101933

Becker, S., Miron-Shatz, T., Schumacher, N., Krocza, J., Diamantidis, C., & Albrecht, U.-V. (2014). mHealth 2.0: Experiences, possibilities, and perspectives. *JMIR mHealth and uHealth*, *2*(2), e3328.

Bettencourt, N., Wilson, C. J., Johnson, P. J., & D'Souza, F. (2023). A Rebalancing of Financial Valuations and Expectations Moving Forward in the Telehealth Sector as the United States Moves Toward a Post-COVID-19 Reality. *Journal of Medical Internet Research*, *25*, e35857. doi:10.2196/35857 PMID:37523216

Brito, F. A. C., Monteiro, L. C. P., Rocha Santos, E. G., de Lima, R. C., Santos-Lobato, B. L., Cabral, A. S., Callegari, B., Costa, E., Silva, A. A., & Souza, G. S. (2023). The role of sex and handedness in the performance of the smartphone-based Finger-Tapping Test. *PLOS Digital Health*, *2*(8), e0000304. doi:10.1371/journal.pdig.0000304 PMID:37585430

Burke, D., Linder, S., Hirsch, J., Dey, T., Kana, D., Ringenbach, S., Schindler, D., & Alberts, J. (2017). Characterizing Information Processing With a Mobile Device: Measurement of Simple and Choice Reaction Time. *Assessment*, *24*(7), 885–895. doi:10.1177/1073191116633752 PMID:26933140

Chan, J. (2021). Exploring digital health care: eHealth, mHealth, and librarian opportunities. *Journal of the Medical Library Association: JMLA*, *109*(3), 376–381. doi:10.5195/jmla.2021.1180 PMID:34629965

Chang, J. T., Morton, S. C., Rubenstein, L. Z., & ... . (2004). Interventions for the prevention of falls in older adults: Systematic review and meta-analysis of randomized clinical trials. *BMJ (Clinical Research Ed.)*, *328*, 680–683.

del Rosario, M. B., Redmond, S. J., & Lovell, N. H. (2015). Tracking the Evolution of Smartphone Sensing for Monitoring Human Movement. *Sensors (Basel)*, *15*(8), 18901–18933. doi:10.3390/s150818901 PMID:26263998

Dicianno, B. E., Parmanto, B., Fairman, A. D., Crytzer, T. M., Yu, D. X., Pramana, G., Coughenour, D., & Petrazzi, A. A. (2015). Perspectives on the evolution of mobile (mHealth) technologies and application to rehabilitation. *Physical Therapy*, *95*(3), 397–405. doi:10.2522/ptj.20130534 PMID:24925075

Elble, R. J., Hellriegel, H., Raethjen, J., & Deuschl, G. (2016). Assessment of Head Tremor with Accelerometers Versus Gyroscopic Transducers. *Movement Disorders Clinical Practice*, *4*(2), 205–211. doi:10.1002/mdc3.12379 PMID:30363428

Elble, R. J., & McNames, J. (2016). Using Portable Transducers to Measure Tremor Severity. *Tremor and Other Hyperkinetic Movements (New York, N.Y.)*, *6*(0), 375. doi:10.5334/tohm.320 PMID:27257514

Fiordelli, M., Diviani, N., & Schulz, P. J. (2013). Mapping mHealth research: A decade of evolution. *Journal of Medical Internet Research, 15*(5), e95.

Gawronska, A., Pajor, A., Zamyslowska-Szmytke, E., Rosiak, O., & Jozefowicz-Korczynska, M. (2020). Usefulness of Mobile Devices in the Diagnosis and Rehabilitation of Patients with Dizziness and Balance Disorders: A State of the Art Review. *Clinical Interventions in Aging, 15*, 2397–2406. doi:10.2147/CIA.S289861 PMID:33376315

Gerhardy, T., Gordt, K., Jansen, C. P., & Schwenk, M. (2019). Towards Using the Instrumented Timed Up-and-Go Test for Screening of Sensory System Performance for Balance Control in Older Adults. *Sensors (Basel), 19*(3), 622. doi:10.3390/s19030622 PMID:30717202

Guo, Z., Zeng, W., Yu, T., Xu, Y., Xiao, Y., Cao, X., & Cao, Z. (2022). Vision-Based Finger Tapping Test in Patients With Parkinson's Disease via Spatial-Temporal 3D Hand Pose Estimation. *IEEE Journal of Biomedical and Health Informatics, 26*(8), 3848–3859. doi:10.1109/JBHI.2022.3162386 PMID:35349459

Harris, A., & Cooper, M. (2019). Mobile phones: Impacts, challenges, and predictions. *Human Behavior and Emerging Technologies, 1*, 15–17. doi:10.1002/hbe2.112

Hubel, K. A., Reed, B., Yund, E. W., Herron, T. J., & Woods, D. L. (2013). Computerized measures of finger tapping: Effects of hand dominance, age, and sex. *Perceptual and Motor Skills, 116*(3), 929–952. doi:10.2466/25.29.PMS.116.3.929-952 PMID:24175464

Ishii, N., Mochizuki, Y., Shiomi, K., Nakazato, M., & Mochizuki, H. (2020). Spiral drawing: Quantitative analysis and artificial-intelligence-based diagnosis using a smartphone. *Journal of the Neurological Sciences, 411*, 116723. doi:10.1016/j.jns.2020.116723 PMID:32050132

Istepanian, R. S. H., & Lacal, J. Emerging Mobile Communication Technologies for Health: Some Imperative Notes on m-Health. In Proceedings of the 25th Annual International Conference of the IEEE Engineering in Medicine and Biology, EMBS, Cancun, Mexico, 17–21 September 2003; pp. 1414–1416.

Lee, C. Y., Kang, S. J., Hong, S. K., Ma, H. I., Lee, U., & Kim, Y. J. (2016). A Validation Study of a Smartphone-Based Finger Tapping Application for Quantitative Assessment of Bradykinesia in Parkinson's Disease. *PLoS One, 11*(7), e0158852. doi:10.1371/journal.pone.0158852 PMID:27467066

Majumder, S., & Deen, M. J. (2019). Smartphone Sensors for Health Monitoring and Diagnosis. *Sensors (Basel), 19*(9), 2164. doi:10.3390/s19092164 PMID:31075985

Martínez-Caro, E., Cagarra-Navarro, J. G., García-Perez, A., & Fait, M. (2018). Healthcare service evolution towards the Internet of Things: An end-user perspective. *Technological Forecasting and Social Change*, *136*, 268–276. doi:10.1016/j.techfore.2018.03.025

McCool, J., Dobson, R., Whittaker, R., & Paton, C. (2022). Mobile Health (mHealth) in Low- and Middle-Income Countries. *Annual Review of Public Health*, *43*(1), 525–539. doi:10.1146/annurev-publhealth-052620-093850 PMID:34648368

Mellone, S., Tacconi, C., & Chiari, L. (2012). Validity of a Smartphone-based instrumented Timed Up and Go. *Gait & Posture*, *36*(1), 163–165. doi:10.1016/j.gaitpost.2012.02.006 PMID:22421189

Meskó, B., Drobni, Z., Bényei, É., Gergely, B., & Győrffy, Z. (2017). Digital health is a cultural transformation of traditional healthcare. *mHealth*, *3*, 38. doi:10.21037/mhealth.2017.08.07 PMID:29184890

Nam, H., Seol, K. H., Lee, J., Cho, H., & Jung, S. W. (2021). Review of Capacitive Touchscreen Technologies: Overview, Research Trends, and Machine Learning Approaches. *Sensors (Basel)*, *21*(14), 4776. doi:10.3390/s21144776 PMID:34300514

Onodera, R., & Sengoku, S. (2018). Innovation process of mHealth: An overview of FDA-approved mobile medical applications. *International Journal of Medical Informatics*, *118*, 65–71. doi:10.1016/j.ijmedinf.2018.07.004

Picerno, P., Iosa, M., D'Souza, C., Benedetti, M. G., Paolucci, S., & Morone, G. (2021). Wearable inertial sensors for human movement analysis: a five-year update. *Expert Review of Medical Devices, 18*(sup1), 79–94. doi:10.1080/17434440.2021.1988849

Rodrigues, L. A., Santos, E. G. R., Santos, P. S. A., Igarashi, Y., Oliveira, L. K. R., Pinto, G. H. L., Santos Lobato, B. L., Cabral, A. S., Belgamo, A., Costa, E., Silva, A. A., Callegari, B., & Souza, G. S. (2022). Wearable Devices and Smartphone Inertial Sensors for Static Balance Assessment: A Concurrent Validity Study in Young Adult Population. *Journal of Personalized Medicine*, *12*(7), 1019. doi:10.3390/jpm12071019 PMID:35887516

Rowland, S. P. (2020). J. Edward Fitzgerald, Thomas Holme, John Powell, and Alison McGregor. "What is the clinical value of mHealth for patients?". *NPJ Digital Medicine*, *3*(1).

Sahin, G., Halje, P., Uzun, S., Jakobsson, A., & Petersson, P. (2022). Tremor evaluation using smartphone accelerometry in standardized settings. *Frontiers in Neuroscience*, *16*, 861668. doi:10.3389/fnins.2022.861668 PMID:35979340

Santos, P. S. A., Santos, E. G. R., Monteiro, L. C. P., Santos-Lobato, B. L., Pinto, G. H. L., Belgamo, A., Cabral, A. S., de Athayde Costa, E., Silva, A., Callegari, B., & Souza, G. S. (2022). The hand tremor spectrum is modified by the inertial sensor mass during lightweight wearable and smartphone-based assessment in healthy young subjects. *Scientific Reports*, *12*(1), 16808. doi:10.1038/s41598-022-21310-4 PMID:36207392

Schmitt, L. (2013). Finger-Tapping Test. In F. R. Volkmar (Ed.), *Encyclopedia of Autism Spectrum Disorders*. Springer.

World Health Organization. (2021). Global strategy on digital health 2020-2025. Geneva: World Health Organization.

World Health Organization (WHO). (2011). *M-Health: New Horizons for Health through Mobile Technologies; Global Observatory for eHealth Series: 3: 5–6*. WHO.

Chapter 4
Machine Learning-Based Application for Long-Term Electrocardiogram Analysis

Jonathan Araujo Queiroz
Federal University of Maranhão, Brazil

Juliana M. Silva
Federal University of Maranhão, Brazil

Yonara Costa Magalhães
Federal University of Maranhão, Brazil

Will Ribamar Mendes Almeida
Emil Brunner World University, Brazil

Bárbara Barbosa Correia
Equatorial Energia, Brazil

José Ricardo Santo de Lima
Emil Brunner World University, Brazil

Edilson Carlos Silva Lima
Universidade Estadual do Maranhão, Brazil

Marcos Jose Dos Passos Sa
Universidade Estadual do Maranhão, Brazil

Allan Kardec Barros Filho
Federal University of Maranhão, Brazil

ABSTRACT

Electrocardiogram (ECG) analyses can only be performed by health professionals whose demand for care is often greater than the availability. In this context, this work consists of the development of an application capable of processing long-lasting ECG signals to assist health professionals in making decisions. The application has an interactive interface that allows view the entire ECG signal in a single image generated by all overlapping cardiac cycles. The proposed application still has email communication between users with the objective of facilitating patient

DOI: 10.4018/979-8-3693-0851-6.ch004

follow-up. The application was tested on three different ECG signals, one artificial and two real. The first signal was an artificial signal generated in software Matlab. The second ECG signal has normal sinus rhythm, available in the MIT-BIH normal sinus rhythm database. The third ECG sign diagnosed with arrhythmia can be found in the MIT-BIH arrhythmia database. The results obtained by the proposed method can be used to support decision-making in clinical practice.

INTRODUCTION

The heart is an organ that has cells with its own rhythm, capable of generating action potentials of which are recorded by the ECG, that is, the ECG is an exam that registers the variation of the electrical potentials of the cardiac muscle and is composed of the P wave that corresponds to the depolarization of the atria, the QRS complex that corresponds to the depolarization of the ventricles, and the T wave that records the repolarization of the ventricles (Hall, 2015).

Changes in heart rate patterns provide a sensitive and anticipated indicator of an individual's health compromises. High heart rate and signal good adaptation, characterizing a healthy individual with efficient autonomic mechanisms. On the other hand, low heart rate and often an indicator abnormal or insufficient adaptation, which may indicate the presence of physiological malfunction in the individual, needing further investigations in order to find a specific diagnosis. This additional investigation can be performed based on the morphology of the cardiac cycles, deformations in the waves of the ECG signal, may be an indication of cardiovascular diseases.

Data released by the World Health Organization (WHO) indicate that 17.3 million people die each year worldwide victims of cardiovascular diseases, and 80% of these deaths are registered in low and middle income countries (World Health Organization, 2004). The number of deaths in the world is forecast to rise from 16.7 million in 2002 to 23.3 million deaths in 2030, which will put heart disease in a group responsible for 70% of all deaths in the world in 2030 (Mathers & Loncar, 2006). In this context, this work proposes the development of an application that can assist in making decisions medical decision, processing long-lasting ECG signals.

For the development of an application that analyzes signals long-term ECG, we propose a generalization of the method presented in our previous study, Queiroz et al. (2019). Here we will present a method to segment the P waves, QRS complex and T wave of the ECG signal and not just the cardiac cycle as previously proposed. In addition, a graphic interface was created, for the direct use of the health professional. The purpose of the graphical interface is assist in monitoring long-term ECG signals reducing the analysis of the long-term ECG signal in a single image to be analyzed. The image presented by Graphical interface consists of the overlapping of all available cardiac cycles of the ECG signal. The graphical interface still allows the sending of the processed signal and information added by the specialist, such as diagnostics via email.

BACKGROUND

Usually, cardiac variability is used to provide temporal information of the ECG signal. The cardiac variability is obtained by the R-R interval, which is defined like Eq.1

$$\text{interval } R\text{-}R = RR = R_m - R_{m-1}, \tag{1}$$

m being the time point of the m-th peak R.

Several authors use the R-R interval, which is the difference between two consecutive R waves, as a source of data extraction. However, the analysis of RR intervals does not is able to measure changes in other ECG signal waves, such as the P wave distortions for AF and the appearance of the F wave in the atrial flutter (Maan et al.,2014). The proposed method in our previous study uses the voltage variability in each heartbeat, unlike the R-R interval, in which each cardiac cycle is associated with a single real number, the proposed method associates each cardiac cycle to a set of points, that is, to a vector.

This method uses the variation of tension in each cardiac cycle is defined as

$$\mathbf{b} = (b_1, b_2, \cdots, b_m), \tag{2}$$

where **b** is a cardiac cycle, b_m the m-th sample in millivolt (mV) of **b** with $L_I(\mathbf{b}) \leq m \leq L_S(\mathbf{b})$, in such a way that $L_S(\mathbf{b})$ is the upper limit and $L_I(\mathbf{b})$ the lower limit of *m*. The upper limit $L_S(\mathbf{b})$ and lower $L_I(\mathbf{b})$ are given by Eq.3

$$L_S(\mathbf{b}) = P_R + F_s\lambda, \tag{3}$$

and

$$L_I(\mathbf{b}) = P_R - F_s\theta, \tag{4}$$

L_S and L_I being position limits, P_R the peak position R P_R is found using the Pan and Tompkins algorithm that has a 99.3% success rate of *QRS* complexes for the MIT-BIH arrhythmia database [0], F_s the frequency of sampling in seconds, λ and θ are heuristically assigned parameters and function as sliding windows in the cardiac cycle. Based on the duration of the *P* wave to Pico *R*(0.4s) and in the duration of peak *R* to *T* wave (0.6s), the values maximum values for healthy cardiac cycles should be $\theta = 0.4$ and $\lambda = 0.6$, that is, $\lambda + \theta \leq 1$ (Mansur & Favarato, 2016).

METHOD

Analysis

The previous section presents an analysis for the cycle cardiac by the usual analysis and the analysis proposed by. At the However, a cardiac cycle is composed of a set of ripples initiated by the sinoatrial module. Therefore, we propose use a generalization of the method proposed by Queiroz et al. (2019) for extracting data from each wave in the cardiac cycle. For this, it is necessary to rewrite the cardiac cycle (Eq. 2) as in Eq. 5 and Eq. 6

$$C = P <> QRS <> T =$$

$$= (p_1, p_2, \cdots, p_i) <> (qrs_1, qrs_2, \cdots, qrs_j) <> \tag{5}$$

$$(t1, t2, \cdots, tk), \tag{6}$$

i, *j* and *k* being the lengths and <> and the concatenation of the P, QRS and T waves, respectively, and $P \cap QRS \cap T = \{\varnothing\}$.

When you want to isolate all waves from the ECG signal, upper and lower limits are defined for each wave. The lower limit of the P wave ($L_I(P)$) is given by Eq. 4 and the upper limit ($L_S(P)$) is given by Eq.7

$$L_S(P) = \mathbf{t}_{\mathrm{P}} F_s, \tag{7}$$

being the duration time in seconds of the $\mathbf{t}_P$ wave.

The lower ($L_I(QRS)$) and upper ($L_S(QRS)$)) limit of the *QRS* complex are given by Eq. 8

$$L_I(QRS) = L_s(P) + 1, \tag{8}$$

and Eq. 9

$$L_S(QRS) = \mathbf{t}_{\mathrm{QRS}} F_s, \tag{9}$$

$\mathbf{t}_{QRS}$ being the duration time in seconds of the complex QRS.

The lower limit ($L_I(T)$) and upper limit ($L_S(T)$) of the T wave are given by Eq. 10

$$L_I(T) = L_s(QRS) + 1, \tag{10}$$

and Eq. 11

$$L_S(T) = \mathbf{t}_T F_s, \tag{11}$$

$\mathbf{t}_T$ being the duration time in seconds of the T wave.

Graphic Interface

The proposed methodology was implemented in the numerical computation Matlab (2020), and because they are algorithms implemented directly through a language programming, are not very easy to use for the end user. In order for the functions to have greater interactivity, it was decided to implement a graphical interface for the methodology proposed in the previous section.

The graphical interface is an application developed using the APP Designer available in the Matlab software version 2020b, App Designer is an interactive application development environment, where it is possible to design a layout for the application interface and program your behavior.

The platform is intuitive, offering the user an area to fill in information about the patient in question, and it is an environment where the end user, possible health professionals, such as cardiologists and professionals who need to perform ECG analysis, may have facilitated and streamlined the analysis of long-term ECG signals life in a simple way.

In the interface, the user uploads a file ECG signal in ".mat" format, and the application applies in this signal the proposed processing. Returning to the user at the interface the result of the ECG processing and the amount of Beats Per Minutes (BPM) of the signal analyzed, allowing the user an agility in the analysis long-term ECG's, making service of these professionals, as well as e-mail communication between health professionals or between professional and patient. Through of the interface, the user can save the graphics for analysis and send information you want to share via email.

Through App Designer, the implemented application can be shared as an installer file. In that case, it is necessary that the desktop on which the application will be used has installed the Matlab software, or sharing it by creating a Web application, where in this case the end user needs to have MATLAB Compiler, a compiler Matlab application that allows both sharing regarding the installation of web applications, installed on your desktop https://pt.overleaf.com/project/602a8aaf35265f0978fa48a8

App Designer makes it possible to implement the application DAASINC. The interface was implemented to load and process an ECG signal previously saved by the professional of health, that when loaded in the Application is processed by the proposed method and the result of this processing

RESULTS

Responses and Discussions

The 14.575 cardiac cycles were obtained by applying methodology proposed in ECG signals obtained in banks Normal Sinus Rhythm MIT-BIH data (Goldberger et al, 2000a), from where a signal with 4678 cardiac cycles was used, and arrhythmia MIT-BIH (Goldberger et al, 2000b), from where a signal with 6297 was used cardiac cycles, and by an artificial signal produced by a function implemented in Matlab with 3600 cardiac cycles. The classifiers used in this study were:

Goldberger

Linear Discriminant Analysis (LDA)) with linear version; nearest k-neighbors (from english, k-Nearest Neighbors (k-NN)) with a range of 2-5 closest neighbors; support vectors machine (do English, Support Vector Machine (SVM)) with the polynomial nucleus. The 10-fold cross-validation technique was used to test classifiers. Overall performance of classifiers was assessed by taking the average accuracy of cross-validations. In addition, the parameters for the windows of the ECG signal are shown in Table 1.

Table 1. Parameters used to window the ECG signal

Localization	Terminology	Time (ms)		Limit	
In the heart	ECG	Healthy	Arrhythmia	Higher	Bottom
Cycle cardiac	B	1000	500 1500	80	0
Atrium	P	70-110	60 120	20	0
Node atrioventricular	PQ	120-200	210 110	25	20
Depolarization Ventricular	QRS	60-100	50 120	60	25
Repolarization Ventricular	T	90-130	70 150	80	65

Table 2 shows the average accuracy of the classifiers (LDA, k-NN, SVM) for 5 different windows in the ECG signal (Cardiac cycles, *P* wave, *QRS* complex, *T* wave, segment *PQ*).

Table 2. ECG signal window using the proposed method in 5 different cases (cardiac cycles, P wave, QRS complex, T wave, PQ segment).

	Average Accuracy %							
	Training	Test	Classifiers	Cycles Cardiac	P	QRS	T	PQ
			LDA	99.99	99.98	99.98	99.98	99.90
Healthy	18h	4h	k-NN	99.99	99.99	99.99	99.98	99.92
			SVM	99.99	99.98	99.98	99.98	99.92
			LDA	99.97	99.52	99.57	99.03	99.02
Arrhythmia	18h	4h	k-NN	99.97	99.97	99.97	99.95	99.80
			SVM	99.97	99.80	99.87	99.57	99.43

The classifiers were trained to recognize the different windows in the ECG signal (cardiac cycle, P wave, QRS complex, T wave, PQ segment). For patients healthy, the average accuracy was greater than or equal to 99.90%. This behavior is a consequence of the uniformity of the signal, in fact, there is little or no difference in signal morphology. The ECG signal with arrhythmia had average accuracy of up to 99.97% in heart rate classification cardiac wave P and QRS complexes; on the other hand, the lowest average accuracy was in segment segmentation P Q using

LDA (99.02%). This is due to the fact that in some cardiopathies the contraction of the atria and ventricles occurs in a non-synchronized way, impairing the filling and ejection phases of the cardiac chambers, making it difficult thus, the distinction between atrial and ventricular electrical activity in the ECG signal (World Health Organization, 2004). The implementation of the method The proposed strategy proves to be relevant, above all, in low and medium income, where 80% of deaths from arrhythmias occur (World Health Organization, 2004), or even in countries of great territorial extension, such as Brazil, in which the concentration of the number of doctors and 2.09 doctors per 1000 inhabitants. In this context the application is presented to assist the health professionals in the analysis of long-term ECG signs duration, the (DAASINC). The aim of DAASINC is to assist the assessment of cardiac condition using the methodology proposed in (Queiroz et al., 2019). The test application is available at the link: https://1drv.ms/u/s!AsvmLwSVT8QIaucopiyXefyTKjk?e=fNkbaX and for its execution it is necessary that the Matlab software installed on the desktop that the user uses, also if makes it necessary that the signal to be processed is in the same application workbook.

In the implementation of the graphical interface so that the user can process the signal directly through the application, the interface shown in Figure 1.

Figure 1. The interface contains three areas, contains instructions for use, patient filling data and the amount of BPM. In the second, central area, we see the displays for plotting wave graphs of the processed ECG signal. In the third area, on the left, we see the email sending area.

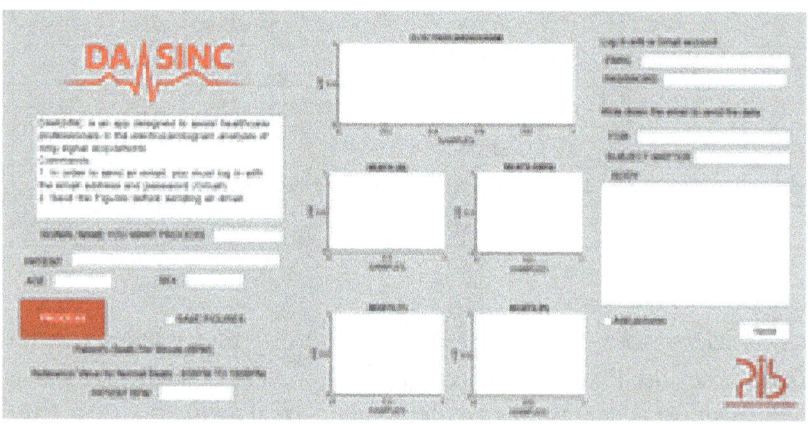

In Figure 1, we can see, in general, the interface graphics of the application, where we note the interactivity allowing the user a skillful manipulation. Initially, right we observe the instructions area of the application, where we see the area for inserting the data on the ECG that will be analyzed. After requesting the ECG to be processed by the user select in the check box if you want to save the figures resulting from processing. Just below, after processing the signal, the amount of BPM will be shown on the application screen, which can be compared with the limit values and can be used by the professional to help classify the signal. In the central area of the interface we observe the plotting areas graphs related to the ECG, where in the first figure and the signal shown as previously collected by the user, and in the following figures the processed signal of the cycles P waves, QRS complexes, and T waves. All the figures show a set of suspended controls that through them the user can increase or decrease the visualization of the ECG signal, as desired.

You can also see in the interface the area for sending e-mails, on the left of Figure 1, where the user after log in to a google server account, you can send an email to any user with analysis and comments. In this area, we note that by definition, the subject of the e-mail contains the patient's name, and the body of the email attached in its structure to the end of the typed text by the user the information of age, sex and quantity of User Beats Per Minute (BPM). Still, in the area of sending e-mails we observe the check box destined to the choice of the user about the attachment of the images of the processing of the ECG in email.

When submitting the application to tests, it was decided to process three distinct signals with the same duration, 1 hour of duration each signal. The first one, and an artificial signal, and the processing of this signal in the application is shown in Figure 2.

Figure 2. Artificial ECG signal processed in DAASINC - ECG signal from 3600 cardiac cycles processed by the application, in the first graph the electrocardiogram in its entirety. In the processed graphs of B, P, QRS and T, there are few lines as it is an artificial, clean signal, without noise.

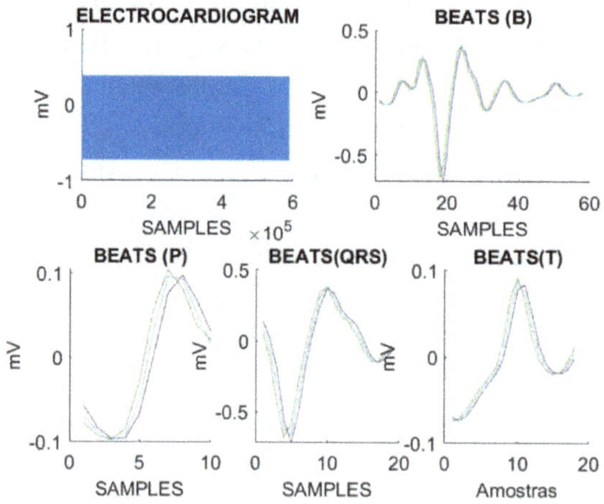

Then a normal ECG signal was processed, where the the result of this processing is shown in Figure 3.

Figure 3. Normal ECG signal processed in DAASINC - In the first graph the complete ECG, the user can enlarge the desired areas in the application. Next we see the processed graphs of B, P, QRS and T, we can see many overlapping lines indicating a normal signal.

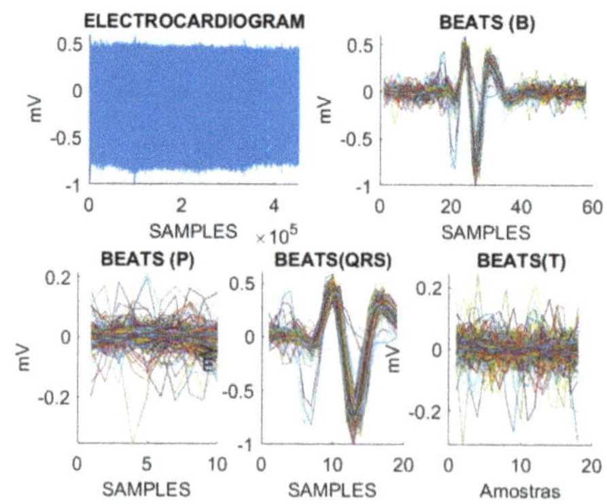

Soon after, in Figure 4, we observe the third sign processed, a signal with cardiac arrhythmia was chosen.

Figure 4. ECG signal with arrhythmia processed in DAASINC - In the first graph, the complete ECG, the user can enlarge the desired areas in the application. Next, we see the processed graphs of B, P, QRS and T, we can observe a disorder in the processed lines, indicating the arrhythmia.

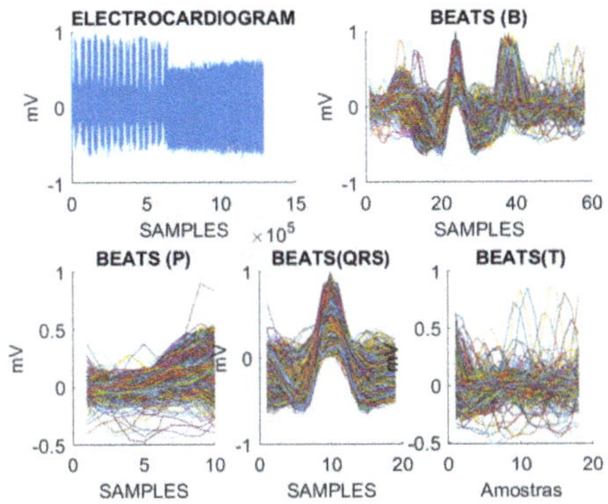

When we look at Figure 2, we notice the occurrence of few lines in the image, which was to be expected, since that an artificial signal always has the same beat, without presenting the variations normally found in a real ECG signal. In Figure 3, we observe the variations intrinsic to the ECG, with variations in amplitudes and lags. In Figure 4, when we observe an ECG signal with arrhythmia, we notice the characteristic disorder of this signal. Comparing three signs, we observed a progression in the disorder starting of the artificial ECG signal as the most ordered, passing by the healthy signal until the signal with arrhythmia that presents greater disorder, indicating the progression in the alterations of the standard characteristics of the ECG signal.

Also, at the end of the first application area, the amount of BPM of the analyzed signal is observed after processing. In Figure 5, we note the amount of BPM for the signal Normal database used in the tests.

In Figure 5, we can see the limits for a BPM to be considered normal, and we observe the processing result for the normal ECG signal used for testing the app. Through this result shown in the interface, the user may have more information at his disposal for medical decision making.

Figure 5. Amount of BPM for a normal ECG processed in DAASINC - Amount of BPM for a patient with a normal ECG signal processed by the application, the displayed value can be compared with the limits shown in the application.

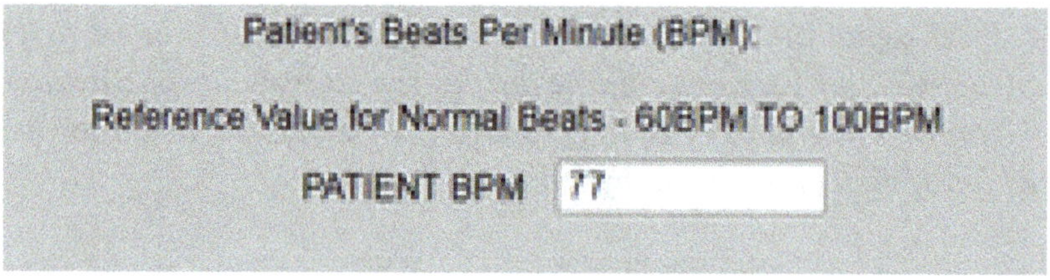

In the section for sending email, the user can make login with a Gmail account and password, and just below insert the recipient and the subject of the email. In the field immediately below, the user can write information about the signal analyzed, be it information for patients or information to be shared with other health professionals.

Then the user decides on the attachment of the figures of ECG processing and chooses to send the email. In Figure 6 we can notice the result of an email received through application.

Figure 6. E-mail sent using the DAASINC interface. An e-mail was sent with an ECG from a 23-year-old female patient whose beats were from a normal ECG signal, the data from the ECG signal shown are fictitious.

When sending the email, the user will be able to communicate remotely with other users of the application, being able to be patients or other health professionals, which would facilitate distance service, in places with difficult access type of iteration is important for monitoring cardiac patients.

CONCLUSION

In practice, develop a computational method to aid diagnosis for cardiac pathologies using the ECG has at least two major challenges to be overcome. One is the computational cost and the other is the infeasibility of obtaining the ECG signal for long periods of time to that the algorithms can be trained and tested. In order to overcome this challenge in parts, a method was presented able to analyze the ECG signal in a systematic and simplified way. The simplicity of the proposed method in segmenting cardiac cycles and subsequently each wave of the same cardiac cycle, made it possible to implement an interface graph that can be used to aid diagnosis applications in systems similar to the 24-hour Holter. The DAASINC application facilitates the care and monitoring of patients whose physical distances make it difficult to access by health professionals, and the patient's ECG analysis can be streamlined through the interface and shared with more Health professionals. The proposed methodology has been tested on long-term ECGs, however, it can be used to assist in the evaluation of any ECG signal. So our next challenge and evaluate the proposed methodology in real time.

REFERENCES

Goldberger, R. A. L., Amaral, L. A., Glass, L., Hausdoff, J. M., Ivanov, P. C., Mark, R. G., Mietus, J. E., Moody, G. B., Peng, C.-K., & Stanley, H. E. (2000a). Physiobank, physiotoolkit, and physionet components of a new research resource for complex physiologic signals. *Circulation*, *101*(23), 215–220. doi:10.1161/01.CIR.101.23.e215 PMID:10851218

Hall, J. E. (2015). *Guyton and Hall Textbook of Medical Physiology* (13th ed.). Elsevier Health Sciences. https://books.google.com.br/books?id=krLSCQAAQBAJhttp://circ.ahajournals.org/content/101/23/e215http://www.scielo.br/scielo.phpscript=sciarttextpid=S0066-782X2016004000020nrm=isohttps://apps.who.int/iris/bitstream/handle/10665/43007/9786009317059-fas.pdf

Maan, A., Mansour, M., Ruskin, J. N., & Heist, E. K. (2014). Impact of catheter ablation on p-wave parameters on 12-lead electrocardiogram in patients with atrial fibrillation. *Journal of Electrocardiology*, *47*(5), 725–733. doi:10.1016/j.jelectrocard.2014.04.010 PMID:24850319

Mansur, A. d. P., & Favarato, D. A. (2016). Trends in Mortality Rate from Cardiovascular Disease in Brazil, 1980-2012. *Arquivos Brasileiros de Cardiologia, 107*, 20 – 25.

Mathers, C. D., & Loncar, D. (2006). Projections of global mortality and burden of disease from 2002 to 2030. *PLOS Medicine, 3*(11), 1–20. doi:10.1371/journal.pmed.0030442

Moody, G. B., Peng, C.-K., & Stanley, H. E. (2000b). Physiobank, physiotoolkit, and physionet. *Circulation, 101*(23), e215–e220. PMID:10851218

Queiroz, J., Azoubel, L., & Barros, A. (2019). Support system for classification of beat-to-beat arrhythmia based on variability and morphology of Electrocardiogram. *EURASIP Journal on Advances in Signal Processing, 16*(1), 16. Advance online publication. doi:10.1186/s13634-019-0613-9

World Health Organization. (2004). *The Atlas of Heart Disease and Stroke*. Author.

Chapter 5
First Episode of Psychosis: From Psychosis Continuum to Specialized Treatment

Walter Barbalho Soares
Onofre Lopes University Hospital, Brazil

Amannda Melo de Oliveira Lima
Federal University of Rio Grande do Norte, Brazil

ABSTRACT

The first episode of psychosis is the interval between the onset of the first positive psychotic symptom above the threshold for clinical psychosis for at least 1 week until the first 5 years of starting treatment. The at-risk mental state is subdivided into three ultra high-risk populations for psychosis: brief intermittent psychotic symptoms; attenuated positive symptoms syndrome; genetic risk and deterioration syndrome. The incidence of psychotic disorders varies between 15-34/100,000 person-years at risk, most of the specific diagnoses are schizophrenia. The duration of untreated psychosis is the most studied variable and closely related to the assessment of the impact of early treatment on the patient's prognosis. Psychosis can be divided into primary (affective and non-affective) or secondary causes. Intervening early in the course of psychotic illness is important as centers specialized in FEP aim reduce DUP, achieve remission of the psychotic condition, reduce recurrence, and reduce the number of hospital admissions.

DOI: 10.4018/979-8-3693-0851-6.ch005

INTRODUCTION AND DEFINITION

In 1845, Austrian psychiatrist Ernst V. Feuchtersleben developed the term "psychosis" to differentiate the psychic symptoms of mental illness from the neural symptoms (Gaebel & Zielasek, 2015). Over time, the term was modified, and currently psychosis is understood as: distortion of understanding and interpretation of reality, characterized by symptoms such as hallucinations (alteration of sensorial perception), delusions (alteration of critical judgment of reality) and disorganization of speech (change in thinking) (Galletly et al., 2016).

In general, psychotic episodes appear in adolescence or early adulthood (Cosgrave et al., 2021), with 80% of them between 16 and 30 years of age, with a higher incidence around the age of 20 (Healy et al., 2019). This is a period marked by the search for vocational achievements, relationships and identity definition, which increases the financial and emotional risks faced by affected individuals and their families (Malla et al., 2005).

We understand the first episode of psychosis (FEP) as the period between the onset of the first psychotic event, when the patient continues to present positive symptoms (hallucinations, delusions, disordered thoughts) above the threshold for clinical psychosis for at least 1 week, until the first 5 years of starting treatment for this presentation (Galletly et al., 2016).

Despite this, important references in the study of psychotic disorders still do not consider a specific definition for this condition (Healy et al., 2019). Part of the difficulty in this definition is the multivariable clinical presentation of the acute psychotic event, regarding its duration, symptoms, prodromal period and impact in functionality; which also makes diagnostic subcategorization difficult (Galletly et al., 2016). In view of this, it is understood that psychotic manifestations are part of a psychotic continuum, a syndromic diagnosis, which varies from brief and self-limited psychotic experiences, through prodromal states, to psychotic illness itself. Therefore, it is observed that for a specific diagnosis, longitudinal monitoring of the patient's evolution is necessary (Malla et al., 2005).

Despite the difficulties in making a specific diagnosis in the initial presentation of FEP, it is necessary to consider whether psychotic symptoms occur exclusively associated with depressive or mild symptoms in order to consider the addition of a mood stabilizer to the patient's therapeutic regimen (Galletly et al., 2016). Studies show that cases of non-Affective Psychosis have a higher incidence rate than those of Affective Psychosis. The AESOP Study - carried out in three large centers in England, revealed that 67% of cases were non-affective FEP, while 28% of the sample corresponded to affective FEP (Kirkbride et al., 2006).

Data from a meta-analysis of DSM/ICD diagnostic results in high-risk individuals who later developed a frank psychotic episode (N = 560) show that approximately

73% of individuals were diagnosed with schizophrenia spectrum psychoses and only 11% with affective psychoses. The proportion of each psychotic diagnosis: 59% schizophrenia, 13% psychosis not otherwise specified, 8% schizoaffective, 6% schizophreniform disorder, 6% bipolar affective disorder, 5% psychotic depression, 2% brief psychotic disorder, 1% delirium (deFusar-Poli et al, 2014). Delusional perception is the most common symptom in non-affective PEP (Malinowski FR, et al., 2020).2016).

Psychotic Continuum and Their Psychotic Experiences

It is generally recognized that psychotic phenomena do not occur on an all-or-nothing basis. They are expressed in the population on a continuum of varying degrees of intensity and severity, ranging from normality to full-blown psychotic disorders. A meta-analytic study reported an average prevalence of 5–8% of psychotic experiences in the general population, while only 3% of individuals were diagnosed with a psychotic disorder (Sieradzka D, et al, 2014). Consequently, psychotic symptoms are ubiquitously distributed in the general population, not restricted to diagnoses of psychotic disorders and not even restricted to psychiatric diagnoses in general (Gaebel & Zielasek, 2015). Although most of these phenomena found in the community do not persist (Galletly et al., 2016), they are considered risk factors for future psychotic disorders and general psychiatric diagnoses (Healy et al., 2019) and a marker of disease severity (Malla et al., 2005).

Psychotic experiences (PE) are defined by: symptoms considered comparable to the positive symptoms of psychosis (hallucinations/delusions/speech disorganization) but without severe clinical presentation or significant functional impairment to diagnose and may occur outside of a psychotic disorder (Cosgrave et al., 2021). PE are common in childhood, with an estimated prevalence of 17% between 9 and 12 years old, decreasing to around 7% in adolescence and adulthood. Although PE in childhood is common and can be normal and transient, it can also share genetic and environmental vulnerabilities with psychotic disorders, corresponding to an increased risk of psychotic disorders throughout life with worse global functioning, both in children with or without mental disorders. non-psychotic (Rimvall et al., 2020).

Individuals who report PEs have a 4-fold increased risk of developing a psychotic disorder and a 3-fold increased risk for any mental disorder (Healy et al., 2019). Therefore, PE is a severity and transdiagnostic marker for the development of mental disorders, especially persistent cases (Staines et al., 2022) and is associated with greater use of mental health services (up to a 2-fold increase (Bhavsar et al., 2018)), higher healthcare costs and use of psychotropic drugs, worse quality of life (Rimvall et al., 2021) and global functioning (Staines et al., 2022) and average life expectancy, reduced by up to 5 years, mainly due to neoplasms and suicide (Sharifi et al., 2015). Systematic reviews and meta-analyses have indicated that those with PE were at a 2–3 times greater risk of suicidal behaviors (Yates et al., 2019), with recurrent PE being associated with an even greater risk of suicidal behavior relative to transient PE (Connell et al., 2016). Risk factors for persistence of psychotic experience: family psychiatric history and lower socioeconomic status, lower level of education, childhood trauma and other adverse life events, in addition to emotional and behavioral problems. The paranoid symptom cluster has a greater tendency to persist (Havers et al., 2023). Risk factors for recurrence of psychotic experience: early developmental problems, including birth complications, maternal smoking and infections, in addition to female gender and a greater number of cumulative environmental risk factors, such as: frequent cannabis use, stressful life events, trauma, history of abuse, victimization and bullying. Cognitive deficits currently show no association with recurrent PE, nor family history of psychosis (Havers et al., 2023; Staines et al., 2022, 2023).

The results of large general population studies show that for the vast majority of people, estimated between 75% and 90% (Van Os et al., 2009), psychotic experiences are transient and not indicative of a need for care (Hanssen et al., 2005). A minority of individuals, on the other hand, will transition to clinical psychosis, with an estimated prevalence rate of around 4% (Van Os et al., 2009). Transition from psychotic experiences to clinical psychosis is preceded by the persistence of psychotic experiences and is accompanied by increasing impairment, depending on additional risk factors (Cougnard et al., 2007, Van Os et al., 2009). Previous research has indeed demonstrated that the persistence of psychotic experiences is associated with the transition to clinical psychosis and/or need for care (Cougnard et al., 2007, Dominguez et al., 2011), as well as functional impairment (Dominguez et al., 2011, Kaymaz et al., 2012, De Loore et al., 2011). There are indications that the negative symptom dimension of psychotic disorders, consisting of symptoms such as blunted affect and avolition, is also present in the general population in attenuated forms (Stefanis et al., 2002).

PEs can be subcategorized into six dimensions: perceptual anomalies, persecutory ideation, bizarre ideas, cognitive disorganization, delusional mood, and negative symptoms. The Oxford Wellbeing Life and Sleep Survey (OWLS)(Cosgrave et al., 2021) was designed with multivariate logistic regression models to examine how environmental risks are associated with specific dimensions of PE. The predictive risk factors that occurred in all dimensions of PE were anxiety and social withdrawal in childhood. Cannabis and depression predicted bizarre ideas and persecutory ideation. Psychological abuse and bullying predicted persecutory ideation, age predicted cognitive disorganization, physical abuse predicted bizarre ideas, and circadian phase predicted delusional mood (Cosgrave et al., 2021). The dimensions also vary in terms of severity, degree of associated suffering and risk of further mental health difficulties; in addition to also differing in terms of allelic variants and genetic heritability: varying from 33 to 57% depending on the size of the PE (Sieradzka et al., 2014).

High Risk Population for Psychosis

The At Risk Mental State (ARMS) construct was introduced in 1996 with the aim of identifying individuals at high clinical risk of psychosis before symptoms worsen (Yung et al., 1996, 2021). It is currently subdivided into 3 ultra-high risk populations for psychosis (UHR) (Fusar-Poli et al., 2018): 1) Brief intermittent psychotic symptoms (BIPS): the psychotic symptom must be present at least a few minutes a day, for at least 1x/month, as long as the person has never reported it before, it must not be caused by another disorder; despite representing only 10% of the high-risk population, it is the subgroup that has the highest rate of conversion to psychosis (about 50%) (Fusar-Poli, Cappucciati, et al., 2017); 2) Attenuated positive symptoms syndrome (APSS): these are positive symptoms of clinical relevance, but do not constitute psychosis. It must occur at least once a week and must also not be caused by another disorder; APSS are the most common and when compared to BLIPS, they present a greater chance of conversion to a non-psychotic disorder (Fusar-Poli, Rutigliano, et al., 2017) and a lower risk of conversion to psychosis (Oliver et al., 2023); 3) Genetic risk and deterioration syndrome (GRD): refers to the presumed genetic risk (schizoid personality and/or first-degree family member diagnosed with psychosis) associated with functional deterioration (30% drop in functioning scale); It has low conversion to psychosis and low prevalence in the general population (Fusar-Poli et al., 2018).

A recent meta-analysis estimated that the prevalence of the high-risk population is about 1.7% in community samples and 19.2% in clinical samples (Salazar de Pablo, Woods, et al., 2021). Overall, high-risk individuals have a 20% chance of transitioning to a first episode of psychosis in two years and 35% in 10 years, which only stabilizes after the 4th year (Salazar de Pablo, Davies, et al., 2021). The majority of those who do not develop psychosis on follow-up have persistent comorbid mental disorders (Rutigliano et al., 2016).

Psychotic symptoms can also arise from clinical risk syndromes for non-psychotic disorders, although with a much lower incidence when compared to the ultra-high-risk population, but with a pluripotential risk of developing several other types of mental disorders (Lee et al ., 2018). Incidence rates for emerging psychotic disorders were 4.4 per 1,000 person-years (95% CI: 2.0 to 9.8) for depression risk syndrome, 24.1 per 1,000 person-years (95% CI: 2.0 to 9.8). 95%: 15.1 to 36.5) for bipolar risk syndrome and 33.7. per 1,000 person-years (95% CI: 21.6 to 50.2) for obsessive-compulsive risk syndrome. There have been no longitudinal studies on panic risk syndrome. When the prospective studies were pooled (n = 4, sample = 1,051), the incidence of psychotic disorders emerging from clinical risk syndromes to non-psychotic disorders was 12.9 per 1,000 person-years (95% CI: 4.3 to 38.6) (Lee et al., 2018).

Natural History of Psychotic Illness

Naturalistic follow-up studies have found highly divergent results in PEP. Although an episodic course is most common (Morgan et al., 2014) and most patients initially achieve remission (Lally et al., 2017), a minority experience early and sustained recovery (Lappin et al., 2018) or has a treatment-resistant antipsychotic illness since the onset of the illness (Demjaha et al., 2017). The challenge for the physician treating patients with PEP is to predict these different disease trajectories and make the best treatment choices for each individual patient (Suvisaari et al., 2018).

Some clinical features - such as longer duration of untreated psychosis, premorbid poor adjustment, insidious mode of onset, greater severity of negative symptoms, comorbid substance use disorders, history of suicide attempts and suicidal ideation, and having non-affective psychosis – are associated with a worse outcome. Of the social and demographic factors, male gender, social disadvantage, neighborhood deprivation, dysfunctional family environment, and ethnicity may be relevant. Non-adherence to treatment is a substantial risk factor for relapse, but a small proportion of patients with acute onset of PEP and early remission may benefit from antipsychotic discontinuation (Suvisaari et al., 2018).

The clinical deterioration that occurs in psychosis may actually begin in the prepsychotic phase (J. A. Lieberman et al., 2001). In 75% of cases, the first admission is preceded by a pre-psychotic prodromal phase lasting an average of 5 years and a psychotic prodromal phase lasting an average of 1 year (Hafner, 1998; Häfner, 2019). Prodromal symptoms, especially negative ones, are often associated with social and cognitive deficits already at this stage. On average, women become ill 3 to 4 years later than men and have a second peak of onset close to menopause (Hafner, 1998). After the end of the first episode, average social and work disability values remain quite stable over a period of at least three years (Hafner, 1998). Until now, treatment and rehabilitation measures are only initiated after the first admission, that is, after several years of disease duration, when most of the damage has already manifested itself (Hafner, 1998). Lieberman, et al (J. Lieberman, 1993) classified it as follows: pre-morbid phase (including genetic, environmental and personality factors), 2) Critical phase - subdivided into a) prodromal - early phase with non-specific manifestations of the disease. At this stage, the disease is already ongoing, but without clinically significant symptoms. and b) progressive - when the first positive psychotic symptoms occur and 3) Chronic phase, which would encompass cognitive and functional impairment in a more stable manner over time.

Interventions for the Prodrome Stage of Psychosis

Research on intervention measures to prevent the transition to psychosis is numerous. Some deal with omega-3 fatty acids, low-dose antipsychotics, cognitive behavioral therapy, family intervention, social skills training, N-acetylcysteine, glycine, among others, but until the current review there is no robust evidence to support the indication (Bechdolf et al., 2023; Bosnjak Kuharic et al., 2019; McGorry et al., 2013; Yung, 2017).

Psychotic Disorders Epidemiology

The incidence of psychotic disorders varies between 15 and 34/100,000 person-years at risk (Correa-Oliveira et al, 2021; Del-Ben et al, 2019; Menezes et al, 2007; National Centre of Excellence in Youth Mental Health, 2016; Starzer et al, 2023). Incidences around 15 are samples of "schizophrenia-like" disorders, cohorts of FEP services have younger individuals with a greater variety of diagnoses and are therefore at the other end of the range (Starzer et al, 2023). The global prevalence of psychotic disorders is 4.6 per 1.000 people and the lifetime prevalence is 7,49 per 1.000 people (Moreno-Küstner et al, 2018). The highest percentage of specific diagnoses from FEP patient samples is schizophrenia, almost 40% (Del-Ben et al, 2019) (Menezes et al, 2007). Around 3% of FEPs have "organic" origins (National Centre of Excellence in Youth Mental Health, 2016).

The duration of untreated psychosis (DUP) is the time between the onset of psychosis (positive symptoms: delirium, hallucinations) and the beginning of treatment (National Centre of Excellence in Youth Mental Health, 2016). It is an important concept within the study of FEP, as it is the most studied variable and closely related to the assessment of the impact of early treatment on the patient's prognosis. The American average is 74 weeks, 68% have more than 6 months (Addington et al, 2015). In Brazil, it is between 14 and 15 weeks, being shorter for affective PEP (Del-Ben et al, 2019). A systematic review from 2005 demonstrated a range of 4 to 50 weeks for the global average (Freudenreich, 2020; Marshall et al, 2005), currently we have more centers specialized in the early treatment of FEP, leading to a tendency towards a reduction in DUP (Del-Ben et al., 2019).

Higher DUP were associated with: early onset, older age at entry into the programs; worse premorbid functioning; be an outpatient; previous use of psychoactive substances (PAS); poor prognosis; treatment resistance; worse severity of symptoms (Addington et al., 2015; Cavalcante et al., 2020; Correa-Oliveira et al., 2021; Del-

Ben et al., 2019; National Centre of Excellence in Youth Mental Health., 2016). Better prognosis is linked to lower DUP; less intense negative symptoms; not living alone; good pre-morbid social functioning. Poor prognosis is related to: diagnosis of schizophrenia; premorbid group A personality characteristics; low social and occupational functioning (Alvarez-Jimenez et al., 2016).

Clinical Assessment

The main symptoms of psychotic disorders in the acute (or progressive) phase are delusions, hallucinations and formal thought alterations (positive symptoms). In addition to them, we can find negative symptoms: blunted affect, impoverished contact, avolition, alogia and social withdrawal, characteristically present before the emergence of positive symptoms in schizophrenia spectrum disorders. In addition to these symptoms, disorganized behavior and catatonia can also be included in broader concepts of psychosis (Freudenreich, 2020; Lieberman et al., 2001; National Centre of Excellence in Youth Mental Health, 2016).

As previously stated, the critical phase of the disease can be subdivided into prodromal and progressive. The prodromal phase is marked by attenuated positive symptoms, a lack of extraordinary conviction, or a shorter period of frank psychosis, usually resolving spontaneously. More attention should be paid to individuals with first-degree relatives with schizophrenia, in addition to the presence of group A personality disorders in the patient, especially schizoid and schizotypal. Negative symptoms are also common problems in individuals in this prodromal state and cognitive symptoms may already be present, mainly disturbances in attention, concentration, thought blockage, reduced abstraction capacity, which can lead to impaired learning (Freudenreich, 2020; National Centre of Excellence in Youth Mental Health, 2016). Currently, the prodrome is established retrospectively, when the patient in the first episode is adequately evaluated, but although most patients in this state do not transition to psychotic disorder, changing to a prospective perspective could further reduce this conversion. Furthermore, even if they do not make the transition, they are teenagers or young adults with significant mental changes, that is, they need some care in their approach (National Centre of Excellence in Youth Mental Health, 2016). Between 15 and 51% of individuals at risk for psychosis develop an anxiety disorder, 41 to 46% develop depression, 66% have suicidal ideation; 49% cause self-harm; 33% use psychoactive substances (Fusar-Poli et al, 2014; Fusar-Poli et al, 2020; Lieberman et al., 2001).

Affective episodes with psychotic symptoms can be divided into mania or depression. The first is marked by a hyperthymic, irritable, euphoric mood, decreased need for sleep, excitement, increased energy, flow of words, disinhibition, grandiosity delusions, involving supernatural powers, for example and erythomania (Freudenreich, 2020)

In the case of psychotic depression, in addition to symptoms of depression (generally severe), patients may present delusions of ruin, guilt, misery, death, denial, as well as ideas of reference. Thoughts of death and suicidal ideation are more frequent and generally involve more motor changes, whether slowness or agitation (Freudenreich, 2020).

In such an acute disease, it is difficult to establish a clear, delimited symptom profile, or even the duration of such symptoms. Furthermore, distinguishing diagnostic categories can be difficult in FEP, as they are fluid symptoms and difficult to predict. The presence of psychotic symptoms during periods of euthymia suggests schizophrenia spectrum disorders. Even affective FEPs can progress to a diagnosis of schizophrenia, which happens to 10% of them. Therefore, longitudinal follow-up is essential to understand the patient's diagnosis and prognosis (National Centre of Excellence in Youth Mental Health, 2016).

Instruments: The Anchored Brief Psychiatric Rating Scale (BPRS-A) and Positive and Negative Syndrome Scale (PANSS) are scales for assessing the severity of psychotic symptoms in individuals over 18 years of age with psychotic disorders, mainly diagnosed with schizophrenia. They are widely used in research contexts and to evaluate the response of symptoms to antipsychotic treatment. The BPRS-A has 18 items and is anchored. The PANSS is a more extensive scale, taking 40 to 60 minutes to apply. Both require adequate training for application (Gorestein et al, 2016).

Young Mania Rating Scale (YOUNG), Calgary Depression Rating Scale for Schizophrenia (CDRS) and Patient Health Questionnaire-9 (PHQ-9 help) can be responsible for the assessment of mood symptoms (the first one for manic episodes with psychotic symptoms and the last one for the assessment of depressive symptoms, the CDRS used only for patients with schizophrenia) primary or secondary (Gorestein et al, 2016).

FEP and Cognition

Cognitive symptoms, as well as positive and negative symptoms occur as a continuum in nature. Cognitive deficits can already be observed in childhood, they are robust in those at UHR, stable and severe in those psychotic since FEP (Sheffield et al., 2018). UHR individuals have intermediate cognitive performance between FEPs and healthy controls, both globally and specifically. Within the UHR, those that convert are closer to FEPs and those that do not convert to normal controls (Bora & Murray, 2014; de Mello Ayres et al., 2010).

Currently, it is possible to verify in the literature involving individuals undergoing FEP evidence of general and specific cognitive impairment. In general, it is demonstrated through a reduction in IQ tests and academic performance. The cognitive functions that are most replicated as specific impairment are executive functions, brain processing speed and verbal memory. However, it is also possible to find deficits in working memory, verbal fluency, attention and motor skills (Bora & Murray, 2014; de Mello Ayres et al., 2010; Sánchez-Torres et al., 2018; Sheffield et al, 2018).

The approach to these compromised cognitive functions is based mainly on the prevention of cognitive impairment, therefore it must focus on premorbid states, focusing on early intervention, since most of the damage happens in the FEP or even before. This cognitive state is linked to the psychotic state, so improving one will help the other. Domains that are more easily trained/practiced have greater gains, especially memory (Bora & Murray, 2014; Sánchez-Torres et al, 2018).

Remission of the clinical condition and no relapse during the first-year post-pep is related to a better cognitive prognosis (Bergh et al, 2016; Rund et al, 2016). Furthermore, patients on FEP followed for 10 years demonstrated improvement in cognitive performance until the sixth year of follow-up, from that year until the tenth there was stabilization; global functional improvement may be linked to patients' cognitive improvement. Given this, the first years after FEP are a critical period in which patients can make the greatest advances in social functioning and a relevant opportunity to seek full recovery (Torgalsbøen et al., 2023).

FEP and Sleep

Insomnia is the most common, severe, important and treatable sleep disorder in patients with FEP (Reeve et al, 2019). Furthermore, early-stage PEP patients present significant impairments in the continuity, duration and onset of subjective sleep quality (Reeve et al, 2019). The literature has shown that systematic sleep assessments in early intervention environments can allow early identification and treatment of sleep disorders in this population, which would lead to an improvement in cognitive and functional changes. This is because the results of neuropsychological tests that measure executive functions, learning and memory support the idea that there is a global cognitive deterioration from the initial course of PEP in association with sleep disorders (Reeve et al., 2019).

Differential Diagnosis

Differential diagnosis is a mandatory exercise within the study of FEP, after all it is a transitory diagnosis. Psychosis can be divided into primary or secondary causes. The primary ones are related to primary mental disorders and are divided into affective and non-affective, the first ones are due to mania or depression with psychotic symptoms, while the non-affective ones are: brief psychotic disorder, schizophreniform disorder, schizophrenia, schizoaffective disorder and delusional disorder. The secondary ones are related to clinical conditions that cause psychotic symptoms, such as the use of psychoactive substances, dementia syndromes, other medical conditions, use of medications and *delirium* (National Centre of Excellence in Youth Mental Health, 2016).

Physical assessment is essential in the comprehensive assessment, in order to search for possible non-psychiatric causes. It is essential to properly examine the patient, not forgetting to check important parameters, such as vital signs (blood pressure, heart rate, temperature, weight, waist circumference, BMI) and the basic neurological examination, electrocardiogram (if there is a cardiac risk). Laboratory tests (including blood count, sodium, potassium, magnesium, calcium, liver and kidney function, VHS, FAN, blood glucose, prolactin, sorology for hepatitis, syphilis and HIV, thyroid function, dosage of vitamin B12 and folate) and imaging (when indicated) are part of the screening (Freudenreich et al., 2009).

It is common for patients on FEP to experience depressive episodes during psychosis, or even when they are showing signs of improvement. At this moment of recovery, patients experience shame, isolation; decreased self-control, self-esteem and expectations (Dillinger & Kersun, 2020).

Another important differential diagnosis is the use of psychoactive substances, almost 50% have problems with their use. Nicotine is the most used substance (72%), followed by alcohol (10 to 33%) (National Centre of Excellence in Youth Mental Health, 2016). Cannabis is used more than samples without the diagnosis and is related to the induction of psychotic conditions, as well as negative factors after the onset of psychosis, such as more intense positive symptoms, worse engagement in services, more hospitalizations, ideation and attempted suicide (Brunette et al, 2018; National Centre of Excellence in Youth Mental Health, 2016).

Interventions

Intervening early in the course of psychotic illness is important as it is associated with reduction in the time of untreated psychosis, better overall outcome in the short term; greater safety, acceptability and efficacy; and lower cost of treatment (Albert et al., 2017; Hastrup et al., 2013; National Centre of Excellence in Youth Mental Health, 2016). The benefit is evident in the first two years of follow-up, with a reduction in psychotic symptoms in their various dimensions: positive, negative and disorganization, in addition to greater adherence, shorter duration of hospitalizations, better expression of emotions and functionality (Secher et al., 2015). There is no evidence to support an ideal length of follow-up in a specialized center to prevent relapse or improve symptom outcomes. Beneficial effects of the first 2 years of treatment at the specialized center can be sustained without the patient necessarily remaining in the program (Robinson et al., 2022; Secher et al, 2015).

Pharmacotherapy is considered the first line of treatment for FEP, as positive symptoms usually have a good rate of reduction in severity and even resolution with the use of antipsychotics (Jackson & McGorry, 2009). 50% of treated patients respond within the first 2 weeks, 80% in 4 weeks and 10% can respond within 8 weeks (Freudenreich, 2020). Second-generation antipsychotics are used in preference to first-generation antipsychotics, due to better tolerability, side effect profile, reducing the chances of extrapyramidal symptoms, and the possibility of having better cognitive effects (National Centre of Excellence in Youth Mental Health, 2016; Vita et al., 2022). In case of mania with psychotic symptoms, a mood stabilizer, preferably lithium and valproate, are the first-line indications, however, the use of antipsychotics alone (National Centre of Excellence in Youth Mental Health, 2016) can be considered. In the face of psychotic depression, the use of antidepressants is necessary. There is no scientific support the suspension of antipsychotic, as it is not possible to predict the second episode, continuous treatment is associated with better functioning and less hospitalization, fewer relapses (National Centre of Excellence in Youth Mental Health, 2016).

This is a challenging period for family members, as patients have strange, unusual behaviors. Generating uncertainty, mourning, a feeling of loss, guilt for the genetic load transmitted, for the mistake of not having noticed it before, fear, panic, apprehension, anger, sadness, anxiety. 80% report stress more than two-thirds have clinically significant depressive symptoms, almost a quarter of them have a depressive disorder (Dillinger & Kersun, 2020; Verdolini et al, 2021). Approaching family members reduced the risk by 2 times, as well as preventing relapse in general, serious or not; it also reduces the chance of having psychotic episodes severe enough to be hospitalized, increasing functionality and severity of symptoms (Camacho-Gomez & Castellvi, 2020; Onwumere et al., 2011). From the beginning it is necessary to psychoeducate and provide support. Family members need to be heard and considered partners of the treatment team.

Specialized centers in FEP present evidence of more assertive operational policies and are associated with patients with better global functionality, emotional well-being, social recovery, greater adherence to treatment, lower rates of relapses, hospitalizations, mortality and suicide rates (Albert et al., 2017; Hastrup et al., 2013; National Centre of Excellence in Youth Mental Health, 2016; Posselt et al., 2021). These services aim primarily to reduce DUP, achieve remission of the psychotic condition, reduce recurrence and reduce the number of hospital admissions (Bird et al., 2010; Dillinger & Kersun, 2020; National Centre of Excellence in Youth

Mental Health, 2016). The multidisciplinary approach acts directly on potentially modifiable factors that influence the recurrence of psychotic episodes: medication adherence, use of psychoactive substances and the family environment (Bird et al., 2010; Hastrup et al., 2013; National Centre of Excellence in Youth Mental Health, 2016; Posselt et al., 2021). It is an environment with the potential to deconstruct stigmas among patients, families and healthcare professionals. If the team does not believe it is possible to change the somber prognosis posed by society and the history of the psychotic illness, it will not be able to convince the patient and his family of the possibility of having an independent life, setting and pursuing goals, aiming for a full functional recovery.

CONCLUSION

The first psychotic episode exists and needs to be better tracked. Psychiatrists, doctors, family members and even teachers need to be aware of prodromal signs and symptoms. Many patients experience long periods of attenuated positive and negative symptoms even before the onset of psychosis without any follow-up. Longitudinal monitoring is essential to establish the specific diagnosis of the mental disorder; we must have caution during this process. Early intervention and adequate treatment of FEP can change the patient's prognosis. The pharmacological treatment is characterized by lower doses of antipsychotics and the non-pharmacological approach is characterized by psychosocial interventions for the patient and their family, initially based on psychoeducation and later on intervention for manage psychotic symptoms.

REFERENCES

Addington, J., Heinssen, R.K., Robinson, D.G., Schooler, N.R., Marcy, P., Brunette, M.F., Correll, C.U., Estroff, S., Mueser, K.T., Penn, D., Robinson, J.A., Rosenheck, R.A., Azrin, S.T., Goldstein, A.B., Severe, J., & Kane, J.M. (2015). Duration of Untreated Psychosis in Community Treatment Settings in the United States. *Psychiatr Serv., 66*(7), 753-6. doi:10.1176/appi.ps.201400124

Albert, N., Melau, M., Jensen, H., Emborg, C., Jepsen, J.R., Fagerlund, B., Gluud, C., Mors, O., Hjorthøj, C., & Nordentoft, M. (2017). Five years of specialised early intervention versus two years of specialised early intervention followed by three years of standard treatment for patients with a first episode psychosis: randomised, superiority, parallel group trial in Denmark (OPUS II). *BMJ, 12*(356).

Alvarez-Jimenez, M., O'Donoghue, B., Thompson, A., Gleeson, J.F., Bendall, S., Gonzalez-Blanch, C., Killackey, E., Wunderink, L., & McGorry, P.D. (2016). Beyond Clinical Remission in First Episode Psychosis: Thoughts on Antipsychotic Maintenance vs. Guided Discontinuation in the Functional Recovery Era. *CNS Drugs, 30*(5), 357-68. . doi:10.1007/s40263-016-0331-x

Bechdolf, A., Müller, H., Hellmich, M., de Millas, W., Falkai, P., Gaebel, W., Gallinat, J., Hasan, A., Heinz, A., Janssen, B., Juckel, G., Karow, A., Krüger-Özgürdal, S., Lambert, M., Maier, W., Meyer-Lindenberg, A., Pützfeld, V., Rausch, F., Schneider, F., ... Klosterkötter, J. (2023). Prevention of First-Episode Psychosis in People at Clinical High Risk: A Randomized Controlled, Multicentre Trial Comparing Cognitive-Behavioral Therapy and Clinical Management Plus Low-Dose Aripiprazole or Placebo (PREVENT). *Schizophrenia Bulletin, 49*(4), 1055–1066. doi:10.1093/schbul/sbad029 PMID:37021666

Bergh, S., Hjorthøj C., Sørensen H.J., Fagerlund B., Austin S., Secher R.G., Jepsen J.R., & Nordentoft M. (2016). Predictors and longitudinal course of cognitive functioning in schizophrenia spectrumdisorders, 10 years after baseline: The OPUS Study. *Schizophr. Res., 175*(1-3), 57-63. . doi:10.1016/j.schres.2016.03.025

Bhavsar, V., McGuire, P., MacCabe, J., Oliver, D., & Fusar-Poli, P. (2018). A systematic review and meta-analysis of mental health service use in people who report psychotic experiences. *Early Intervention in Psychiatry, 12*(3), 275–285. doi:10.1111/eip.12464 PMID:28805304

Bird, V., Premkumar, P., Kendall, T., Whittington, C., Mitchell, J., & Kuipers, E. (2010). Early intervention services, cognitive-behavioural therapy and family intervention in early psychosis: Systematic review. *Br J Psychiatry, 197*(5), 350-6. . doi:10.1192/bjp.bp.109.074526

Bora, E., & Murray, R.M. (2014). Meta-analysis of cognitive deficits in ultra-high risk to psychosis and first-episode psychosis: do the cognitive deficits progress over, or after, the onset of psychosis? *Schizophr Bull., 40*(4), 744-55. . doi:10.1093/schbul/sbt085

Bosnjak Kuharic, D., Kekin, I., Hew, J., Rojnic Kuzman, M., & Puljak, L. (2019). Interventions for prodromal stage of psychosis. *Cochrane Database of Systematic Reviews, 2019*(11). doi:10.1002/14651858.CD012236.pub2

Brunette, M.F., Mueser, K.T., Babbin, S., Meyer-Kalos, P., Rosenheck, R., Correll, C.U., Cather, C., Robinson, D.G., Schooler, N.R., Penn, D.L., Addington, J., Estroff, S.E., Gottlieb, J., Glynn, S.M., Marcy, P., Robinson, J., & Kane, J.M. (2018). Demographic and clinical correlates of substance use disorders in first episode psychosis. *Schizophr Res., 194,* 4-12. doi:10.1016/j.schres.2017.06.039

Camacho-Gomez, M., & Castellvi, P. (2020). Effectiveness of Family Intervention for Preventing Relapse in First-Episode Psychosis Until 24 Months of Follow-up: A Systematic Review With Meta-analysis of Randomized Controlled Trials. *Schizophr Bull., 46*(1), 98-109.

Cavalcante, D.A, Coutinho, L.S., Ortiz, B.B., Noto, M.N., Cordeiro, Q., Ota, V.K., Belangeiro, S.I., Bressan, R.A., Gadelha, A., & Noto, C. (2020). Impact of duration of untreated psychosis in short-term response to treatment and outcome in antipsychotic naïve first-episode psychosis. *Early Interv Psychiatry., 14*(6), 677-683. . doi:10.1111/eip.12889

Connell, M., Betts, K., McGrath, J. J., Alati, R., Najman, J., Clavarino, A., Mamun, A., Williams, G., & Scott, J. G. (2016). Hallucinations in adolescents and risk for mental disorders and suicidal behaviour in adulthood: Prospective evidence from the MUSP birth cohort study. *Schizophrenia Research, 176*(2), 546–551. doi:10.1016/j.schres.2016.06.009 PMID:27374323

Corrêa-Oliveira, G. E., Scarabelot, L. F., Araújo, J. M., Boin, A. C., de Paula Pessoa, R. M., Leal, L. R., & Del-Ben, C. M. (2021). Early intervention in psychosis in emerging countries: Findings from a first-episode psychosis programme in the Ribeirão Preto catchment area, southeastern Brazil. *Early Intervention in Psychiatry, 16*(7), 800–807. doi:10.1111/eip.13252 PMID:34794209

Cosgrave, J., Purple, R. J., Haines, R., Porcheret, K., van Heugten-van der Kloet, D., Johns, L., Alexander, I., Goodwin, G. M., Foster, R. G., & Wulff, K. (2021). Do environmental risk factors for the development of psychosis distribute differently across dimensionally assessed psychotic experiences? *Translational Psychiatry, 1*(1), 226. Advance online publication. doi:10.1038/s41398-021-01265-2 PMID:33875641

De Loore, E., Gunther, N., Drukker, M., Feron, F., Sabbe, B., Deboutte, D., & van Os, J. (2011). Myin-Germeys I. Persistence and outcome of auditory hallucinations in adolescence: a longitudinal general population study of 1800 individuals. *Schizophr Res., 127*(1-3), 252-6. doi:10.1016/j.schres.2011.01.015

de Mello Ayres, A., Scazufca, M., Menezes, P.R., Nakano, E.Y., Regina, A.C., Schaufelberger, M.S., Murray, R.M., McGuire, P.K., Rushe, T., & Busatto, G.F. (2010). Cognitive functioning in subjects with recent-onset psychosis from a low-middle-income environment: multiple-domain deficits and longitudinal evaluation. *Psychiatry Res., 179*(2), 157-64. doi:10.1016/j.psychres.2009.11.001

Del-Ben, C. M., Shuhama, R., Loureiro, C. M., Ragazzi, T. C., Zanatta, D. P., Tenan, S. H., Ferreira Santos, J. L., Louzada-Junior, P., dos Santos, A. C., Morgan, C., & Menezes, P. R. (2019). Urbanicity and risk of first-episode psychosis: Incidence Study in Brazil. *The British Journal of Psychiatry*, *215*(6), 726–729. doi:10.1192/bjp.2019.110 PMID:31099323

Demjaha, A., Lappin, J.M., Stahl, D., Patel, M.X., MacCabe, J.H., & Howes, O.D. (2017). Antipsychotic treatment resistance in first-episode psychosis: prevalence, subtypes and predictors. *Psychol Med.*, *47*(11), 1981–9.

Dillinger, R.L., & Kersun, J.M. (2020). Caring for caregivers: Understanding and meeting their needs in coping with first episode psychosis. *Early Interv Psychiatry, 14*(5), 528-534. doi:10.1111/eip.12870

Dominguez, M.D., Wichers, M., Lieb, R., Wittchen, H.U., & van Os, J. (2011). Evidence that onset of clinical psychosis is an outcome of progressively more persistent subclinical psychotic experiences: an 8-year cohort study. *Schizophr Bull.*, *37*(1), 84-93. doi:10.1093/schbul/sbp022

Emsley, R., Oosthuizen, P.P., Koen, L., Niehaus, D.J., & Martinez, G. (2012). Symptom recurrence following intermittent treatment in first-episode schizophrenia successfully treated for 2 years: a 3-year open-label clinical study. *J Clin Psychiatry, 73*(4), e541-7. doi:10.4088/JCP.11m07138

Freudenreich, O. (2020). *Psychotic disorders: A practical guideline* (2nd ed.). Springer International Publishing. doi:10.1007/978-3-030-29450-2

Freudenreich, O., Schulz, S.C., & Goff, D.C. (2009). Initial medical work-up of first-episode psychosis: a conceptual review. *Early Interv Psychiatry, 3*(1), 10-8. . doi:10.1111/j.1751-7893.2008.00105.x

From, H. H. (2019). Onset and Prodromal Stage to a Life-Long Course of Schizophrenia and Its Symptom Dimensions: How Sex, Age, and Other Risk Factors Influence Incidence and Course of Illness. *Psychiatry Journal, 2019*, 9804836. PMID:31139639

Fusar-Poli, P., Cappucciati, M., De Micheli, A., Rutigliano, G., Bonoldi, I., Tognin, S., Ramella-Cravaro, V., Castagnini, A., & McGuire, P. (2017). Diagnostic and Prognostic Significance of Brief Limited Intermittent Psychotic Symptoms (BLIPS) in Individuals at Ultra High Risk. *Schizophrenia Bulletin*, *43*(1), 48–56. doi:10.1093/schbul/sbw151 PMID:28053130

Fusar-Poli, P., De Micheli, A., Cappucciati, M., Rutigliano, G., Davies, C., Ramella-Cravaro, V., Oliver, D., Bonoldi, I., Rocchetti, M., Gavaghan, L., Patel, R., & McGuire, P. (2018). Diagnostic and Prognostic Significance of DSM-5 Attenuated Psychosis Syndrome in Services for Individuals at Ultra High Risk for Psychosis. *Schizophrenia Bulletin*, *44*(2), 264–275. doi:10.1093/schbul/sbx055 PMID:28521060

Fusar-Poli, P., Nelson, B., Valmaggia, L., Yung, A.R., & McGuire, P.K. (2014). Comorbid depressive and anxiety disorders in 509 individuals with an at-risk mental state: impact on psychopathology and transition to psychosis. *Schizophr Bull*, *40*(1), 120-131.

Fusar-Poli, P., Rutigliano, G., Stahl, D., Davies, C., De Micheli, A., Ramella-Cravaro, V., Bonoldi, I., & McGuire, P. (2017). Long-term validity of the At Risk Mental State (ARMS) for predicting psychotic and non-psychotic mental disorders. *European Psychiatry*, *42*, 49–54. doi:10.1016/j.eurpsy.2016.11.010 PMID:28212505

Fusar-Poli, P., Salazar de Pablo, G., Correll, C.U., Meyer-Lindenberg, A., Millan, M.J., Borgwardt, S., Galderisi, S., Bechdolf, A., Pfennig, A., Kessing, L.V., van Amelsvoort, T., Nieman, D.H., Domschke, K., Krebs, M.O., Koutsouleris, N., McGuire, P., Do, K.Q., & Arango, C. (2020). Prevention of Psychosis: Advances in Detection, Prognosis, and Intervention. *JAMA Psychiatry*, *77*(7), 755-765.

Gaebel, W., & Zielasek, J. (2015). Focus on psychosis. *Dialogues in Clinical Neuroscience*, *17*(1), 9–18. doi:10.31887/DCNS.2015.17.1/wgaebel PMID:25987859

Galletly, C., Castle, D., Dark, F., Humberstone, V., Jablensky, A., & Killackey, E. (2016). Royal Australian and New Zealand College of Psychiatrists clinical practice guidelines for the management of schizophrenia and related disorders. *Aust N Z J Psychiatry*, *50*(5), 410–72.

Gorestein, C., Wang, Y., & Hungerbuhler, I. (2016). Instrumentos de avaliação em saúde mental. *The Art of Medication*.

Hafner, H. (1998). Onset and course of the first schizophrenic episode. *Kaohsiung J Med Sci.*, *14*(7), 413–31.

Hanssen, M., Bak, M., Bijl, R., Vollebergh, W., & van Os, J. (2005). The incidence and outcome of subclinical psychotic experiences in the general population. *Br J Clin Psychol., 44*(Pt 2), 181-91. doi:10.1348/014466505X29611

Hastrup, L.H., Kronborg, C., Bertelsen, M., Jeppesen, P., Jorgensen, P., Petersen, L., Thorup, A., Simonsen, E., & Nordentoft, M. (2013). Cost-effectiveness of early intervention in first-episode psychosis: economic evaluation of a randomised controlled trial (the OPUS study). *Br J Psychiatry, 202*(1), 35-41.

Havers, L., Stumm, S. von, Cardno, A.G., Freeman, D., & Ronald, A. (2023). Psychotic experiences and negative symptoms from adolescence to emerging adulthood: Developmental trajectories and associations with polygenic scores and childhood characteristics. *Psychol Med., 53*(12), 5685–97.

Healy, C., Brannigan, R., Dooley, N., Coughlan, H., Clarke, M., Kelleher, I., & Cannon, M. (2019). Childhood and adolescent psychotic experiences and risk of mental disorder: A systematic review and meta-analysis. *Psychological Medicine, 49*(10), 1589–1599. doi:10.1017/S0033291719000485 PMID:31088578

Jackson, H. J., & McGorry, P. D. (2009). *The recognition and management of early psychosis a preventive approach*. Cambridge University Press. doi:10.1017/CBO9780511576287

Kaymaz, N., Drukker, M., Lieb, R., Wittchen, H.U., Werbeloff, N., Weiser, M., Lataster, T., & van Os, J. (2012). Do subthreshold psychotic experiences predict clinical outcomes in unselected non-help-seeking population-based samples? A systematic review and meta-analysis, enriched with new results. *Psychol Med., 42*(11), 2239-53. doi:10.1017/S0033291711002911

Kirkbride, J.B., Fearon, P., Morgan, C., Dazzan, P., Morgan, K., & Tarrant, J. (2006). Heterogeneity in incidence rates of schizophrenia and other psychotic syndromes: findings from the 3-center AeSOP study. *Arch Gen Psychiatry, 63*(3), 250–8.

Lally, J., Ajnakina, O., Stubbs, B., Cullinane, M., Murphy, K. C., & Gaughran, F. (2017). Remission and recovery from first-episode psychosis in adults: Systematic review and meta-analysis of long-term outcome studies. *Br J Psychiatry J Ment Sci. Dez, 211*(6), 350–358. PMID:28982659

Lappin, J.M., Heslin, M., Lomas, B., Jones, P.B., Doody, G.A., & Reininghaus, U.A. (2019). Early sustained recovery following first episode psychosis: Evidence from the AESOP10 follow-up study. *Schizophr Res., 199*, 341–5.

Lee, T.Y., Lee, J., Kim, M., Choe, E., & Kwon, J.S. (2018). Can We Predict Psychosis Outside the Clinical High-Risk State? A Systematic Review of Non-Psychotic Risk Syndromes for Mental Disorders. *Schizophr Bull.*, *44*(2), 276–85.

Lieberman, J. (1993). Time Course and Biologic Correlates of Treatment Response in First-Episode Schizophrenia. *Arch Gen Psychiatry*, *50*(5), 369.

Lieberman, J. A., Perkins, D., Belger, A., Chakos, M., Jarskog, F., Boteva, K., & Gilmore, J. (2001). The early stages of schizophrenia: Speculations on pathogenesis, pathophysiology, and therapeutic approaches. *Biological Psychiatry*, *50*(11), 884–897. doi:10.1016/S0006-3223(01)01303-8 PMID:11743943

Lo Buglio, G., Pontillo, M., Cerasti, E., Polari, A., Schiano Lomoriello, A., Vicari, S., Lingiardi, V., Boldrini, T., & Solmi, M. (2022). A network analysis of anxiety, depressive, and psychotic symptoms and functioning in children and adolescents at clinical high risk for psychosis. *Front Psychiatry, 28.* . doi:10.3389/fpsyt.2022.1016154

Malinowski, F.R., Tasso, B. de C., Ortiz, B.B., Higuchi, C.H., Noto, C., & Belangero, S.I. (1999). Schneider's first-rank symptoms as predictors of remission in antipsychotic-naive first episode psychosis. *Rev Bras Psiquiatria Sao Paulo Braz*, *42*(1), 22–6.

Malla, A. K., Norman, R. M. G., & Joober, R. (2005). First-episode psychosis, early intervention, and outcome: What have we learned? *Canadian Journal of Psychiatry*, *50*(14), 881–891. doi:10.1177/070674370505001402 PMID:16494257

Marshall, M., Lewis, S., Lockwood, A., Drake, R., Jones, P., & Croudace, T. (2005). Association Between Duration of Untreated Psychosis and Outcome in Cohorts of First-Episode Patients: A Systematic Review. *Archives of General Psychiatry*, *62*(9), 975–983. doi:10.1001/archpsyc.62.9.975 PMID:16143729

Marshall, M., Lewis, S., Lockwood, A., Drake, R., Jones, P., & Croudace, T. (2005). Association Between Duration of Untreated Psychosis and Outcome in Cohorts of First-Episode Patients: A Systematic Review. *Archives of General Psychiatry*, *62*(9), 975–983. doi:10.1001/archpsyc.62.9.975 PMID:16143729

McGorry, P. D., Nelson, B., Phillips, L. J., Yuen, H. P., Francey, S. M., Thampi, A., Berger, G. E., Amminger, G. P., Simmons, M. B., Kelly, D., Dip, G., Thompson, A. D., & Yung, A. R. (2013). Randomized controlled trial of interventions for young people at ultra-high risk of psychosis: Twelve-month outcome. *The Journal of Clinical Psychiatry*, *74*(4), 349–356. doi:10.4088/JCP.12m07785 PMID:23218022

Menezes, P. R., Scazufca, M., Busatto, G. F., Coutinho, L. M., Mcguire, P. K., & Murray, R. M. (2007). Incidence of first-contact psychosis in São Paulo, Brazil. *The British Journal of Psychiatry, 191*(S51), s102–s106. Advance online publication. doi:10.1192/bjp.191.51.s102 PMID:18055924

Moreno-Küstner, B., Martín, C., & Pastor, L. (2018) Prevalence of psychotic disorders and its association with methodological issues. A systematic review and meta-analyses. *PLoS One, 13*(4), e0195687. . doi:10.1371/journal.pone.0195687

Morgan, C., Lappin, J., Heslin, M., Donoghue, K., Lomas, B., & Reininghaus, U. (2014). Reappraising the long-term course and outcome of psychotic disorders: The AESOP-10 study. *Psychol Med., 44*(13), 2713–26.

National Centre of Excellence in Youth Mental Health. (2016). *Early Psychosis Guidelines Writing Group and EPPIC National Support Program, Australian Clinical Guidelines for Early Psychosis* (2nd ed.). Orygen.

Oliver, D., Davies, C., Zelaya, F., Selvaggi, P., De Micheli, A., Catalan, A., Baldwin, H., Arribas, M., Modinos, G., Crossley, N. A., Allen, P., Egerton, A., Jauhar, S., Howes, O. D., McGuire, P., & Fusar-Poli, P. (2023). Parsing neurobiological heterogeneity of the clinical high-risk state for psychosis: A pseudo-continuous arterial spin labelling study. *Frontiers in Psychiatry, 14*, 1092213. doi:10.3389/fpsyt.2023.1092213 PMID:36970257

Onwumere, J., Bebbington, P., & Kuipers, E. (2011). Family interventions in early psychosis: Specificity and effectiveness. *Epidemiology and Psychiatric Sciences, 20*(2), 113–119. doi:10.1017/S2045796011000187 PMID:21714356

Posselt, C.M., Albert, N., Nordentoft, M., & Hjorthøj, C. (2021). The Danish OPUS Early Intervention Services for First-Episode Psychosis: A Phase 4 Prospective Cohort Study With Comparison of Randomized Trial and Real-World Data. *Am J Psychiatry, 178*(10), 941-951. doi:10.1176/appi.ajp.2021.20111596

Reeve, S., Sheaves, B., & Freeman, D. (2019). Sleep Disorders in Early Psychosis: Incidence, Severity, and Association With Clinical Symptoms. *Schizophr Bull., 45*(2), 287-295. doi:10.1093/schbul/sby129

Rimvall, M. K., van Os, J., Verhulst, F., Wolf, R. T., Larsen, J. T., Clemmensen, L., Skovgaard, A. M., Rask, C. U., & Jeppesen, P. (2020). Mental Health Service Use and Psychopharmacological Treatment Following Psychotic Experiences in Preadolescence. *The American Journal of Psychiatry, 177*(4), 318–326. doi:10.1176/appi.ajp.2019.19070724 PMID:32098486

Rimvall, M. K., Wolf, R. T., Olsen, E. M., Skovgaard, A. M., Clemmensen, L., Oxholm, A. S., Verhulst, F., Rask, C. U., van Os, J., & Jeppesen, P. (2021). Healthcare Costs, School Performance, and Health-related Quality of Life in Adolescence Following Psychotic Experiences in Preadolescence: A Longitudinal Cohort Study. *Schizophrenia Bulletin*, *47*(3), 682–691. doi:10.1093/schbul/sbaa175 PMID:33345286

Robinson, D.G., Schooler, N.R., Marcy, P., Gibbons, R.D., Hendricks Brown, C., John, M., Mueser, K.T., Penn, D.L., Rosenheck, R.A., Addington, J., Brunette, M.F., Correll, C.U., Estroff, S.E., Mayer-Kalos, P.S., Gottlieb, J.D., Glynn, S.M., Lynde, D.W., Gingerich, S., Pipes, R., Miller, A.L., Severe, J.B., & Kane, J.M. (2022). Outcomes During and After Early Intervention Services for First-Episode Psychosis: Results Over 5 Years From the RAISE-ETP Site-Randomized Trial. *Schizophr Bull.*, *48*(5), 1021-1031. . doi:10.1093/schbul/sbac053

Rund, B.R., Barder, H.E., Evensen, J., Haahr, U., ten Velden Hegelstad, W., Joa, I., Johannessen, J.O., Langeveld, J., Larsen, T.K., Melle, I., Opjordsmoen, S., Røssberg, J.I., Simonsen, E., Sundet, K., Vaglum, P., McGlashan, T., & Friis, S. (2016). Neurocognition and Duration of Psychosis: A 10-year Follow-up of First-Episode Patients. *Schizophr Bull.*, *42*(1), 87-95. doi:10.1093/schbul/sbv083

Rutigliano, G., Valmaggia, L., Landi, P., Frascarelli, M., Cappucciati, M., Sear, V., Rocchetti, M., De Micheli, A., Jones, C., Palombini, E., McGuire, P., & Fusar-Poli, P. (2016). Persistence or recurrence of non-psychotic comorbid mental disorders associated with 6-year poor functional outcomes in patients at ultra high risk for psychosis. *Journal of Affective Disorders*, *203*, 101–110. doi:10.1016/j.jad.2016.05.053 PMID:27285723

Salazar de Pablo, G., Davies, C., de Diego, H., Solmi, M., Shin, J. I., Carvalho, A. F., Radua, J., & Fusar-Poli, P. (2021). Transition to psychosis in randomized clinical trials of individuals at clinical high risk of psychosis compared to observational cohorts: A systematic review and meta-analysis. *European Psychiatry*, *64*(1), e51. doi:10.1192/j.eurpsy.2021.2222 PMID:34315555

Salazar de Pablo, G., Woods, S. W., Drymonitou, G., de Diego, H., & Fusar-Poli, P. (2021). Prevalence of Individuals at Clinical High-Risk of Psychosis in the General Population and Clinical Samples: Systematic Review and Meta-Analysis. *Brain Sciences*, *11*(11), 1544. doi:10.3390/brainsci11111544 PMID:34827543

Sánchez-Torres, A.M., Moreno-Izco, L., Lorente-Omeñaca, R., Cabrera, B., Lobo, A., González-Pinto, A.M., Merchán-Naranjo, J., Corripio, I., Vieta, E., de la Serna, E., Butjosa, A., Contreras, F., Sarró, S., Mezquida, G., Ribeiro, M., Bernardo, M., & Cuesta, M.J. (2018). Individual trajectories of cognitive performance in first episode psychosis: a 2-year follow-up study. *Eur Arch Psychiatry Clin Neurosci., 268*(7), 699-711.

Secher, R.G., Hjorthøj, C.R., Austin, S.F., Thorup, A., Jeppesen, P., Mors, O., & Nordentoft, M. (2015). Ten-year follow-up of the OPUS specialized early intervention trial for patients with a first episode of psychosis. *Schizophr Bull., 41*(3), 617-26.

Sharifi, V., Eaton, W. W., Wu, L. T., Roth, K. B., Burchett, B. M., & Mojtabai, R. (2015). Psychotic experiences and risk of death in the general population: 24–27 year follow-up of the Epidemiologic Catchment Area study. *The British Journal of Psychiatry, 207*(1), 30–36. doi:10.1192/bjp.bp.113.143198 PMID:25953893

Sheffield, J.M., Karcher, N.R., & Barch, D.M. (2018). Cognitive Deficits in Psychotic Disorders: A Lifespan Perspective. *Neuropsychol Rev., 28*(4), 509-533. doi:10.1007/s11065-018-9388-2

Sieradzka, D., Power, R.A., Freeman, D., Cardno, A.G., McGuire, P., & Plomin, R. (2014). Are Genetic Risk Factors for Psychosis Also Associated with Dimension-Specific Psychotic Experiences in Adolescence? *PLOS ONE, 9*(4), e94398.

Staines, L., Healy, C., Coughlan, H., Clarke, M., Kelleher, I., Cotter, D., & Cannon, M. (2022). Psychotic experiences in the general population, a review; definition, risk factors, outcomes and interventions. *Psychological Medicine, 52*(15), 3297–3308. doi:10.1017/S0033291722002550 PMID:36004805

Starzer, M., Hansen, H.G., Hjorthøj, C., Albert, N., Nordentoft, M., & Madsen, T. (2023). 20-year trajectories of positive and negative symptoms after the first psychotic episode in patients with schizophrenia spectrum disorder: results from the OPUS study. *World Psychiatry, 22*(3), 424-432. doi:10.1002/wps.21121

Stefanis, N.C., Hanssen, M., Smirnis, N.K., Avramopoulos, D.A., Evdokimidis, I.K., Stefanis, C.N., Verdoux, H., & Van Os, J. (2002). Evidence that three dimensions of psychosis have a distribution in the general population. *Psychol Med., 32*(2), 347-58.

Suvisaari, J., Mantere, O., Keinänen, J., Mäntylä, T., Rikandi, E., & Lindgren, M. (2018). Is It Possible to Predict the Future in First-Episode Psychosis? *Front Psychiatry, 9*(580).

Torgalsbøen, A.K., Mohn, C., Larøi, F., Fu, S., & Czajkowski, N. (2023). A ten-year longitudinal repeated assessment study of cognitive improvement in patients with first-episode schizophrenia and healthy controls: The Oslo Schizophrenia Recovery (OSR) study. *Schizophr Res., 260,* 92-98.

Van Os, J., Linscott, R.J., Myin-Germeys, I., Delespaul, P., & Krabbendam, L. (2009). A systematic review and meta-analysis of the psychosis continuum: evidence for a psychosis proneness-persistence-impairment model of psychotic disorder. *Psychol Med., 39*(2), 179-95. doi:10.1017/S0033291708003814

Verdolini, N., Amoretti, S., Mezquida, G., Cuesta, M.J., Pina-Camacho, L., García-Rizo, C., Lobo, A., González-Pinto, A., Merchán-Naranjo, J., Corripio, I., Salagre, E., Baeza, I., Bergé, D., Garriga, M., Bioque, M., Vallespir, C., Serra, M., Vieta, E., & Bernardo, M. (2021). The effect of family environment and psychiatric family history on psychosocial functioning in first-episode psychosis at baseline and after 2 years. *Eur Neuropsychopharmacol., 49,* 54-68.

Vita, A., Gaebel, W., Mucci, A., Sachs, G., Barlati, S., Giordano, G.M., Nibbio, G., Nordentoft, M., Wykes, T., & Galderisi, S. (2022). European Psychiatric Association guidance on treatment of cognitive impairment in schizophrenia. *European Psychiatry, 65*(1), e57.

Yates, K., Lång, U., Cederlöf, M., Boland, F., Taylor, P., Cannon, M., McNicholas, F., DeVylder, J., & Kelleher, I. (2019). Association of Psychotic Experiences With Subsequent Risk of Suicidal Ideation, Suicide Attempts, and Suicide Deaths: A Systematic Review and Meta-analysis of Longitudinal Population Studies. *JAMA Psychiatry, 76*(2), 180–189. doi:10.1001/jamapsychiatry.2018.3514 PMID:30484818

Yung, A. R. (2017). Treatment of people at ultra-high risk for psychosis. *World Psychiatry; Official Journal of the World Psychiatric Association (WPA), 16*(2), 207–208. doi:10.1002/wps.20424 PMID:28498602

Yung, A. R., McGorry, P. D., McFarlane, C. A., Jackson, H. J., Patton, G. C., & Rakkar, A. (1996). Monitoring and care of young people at incipient risk of psychosis. *Schizophrenia Bulletin, 22*(2), 283–303. doi:10.1093/schbul/22.2.283 PMID:8782287

Chapter 6
First Episode Psychosis and Cognition

Gilberto Sousa Alves
https://orcid.org/0000-0002-0463-6183
Federal University of Rio de Janeiro, Brazil

Romulo Kunrath Pinto Silva
Nina Rodrigues Hospital, Brazil

Marielia Barbosa Freitas Leal
Federal University of Ceará, Brazil

Bianca de Melo Ferro
Federal University of Maranhao, Brazil

Leandro de Oliveira Trovão
Nina Rodrigues Hospital, Brazil

ABSTRACT

The experience of the psychosis can open a critical window into primary psychiatric conditions, particularly schizophrenia or bipolar disorder. More recently, the role of environment in early adolescence or late childhood, including psychological abuse or physical violence, has been associated with a greater likelihood of developing psychosis years later. Evidence has also shown, even in symptomatic remission, the occurrence of cognitive symptoms, particularly in complex attention, executive functioning, episodic memory, and verbal and procedural IQ. The current chapter discusses the epidemiology of first episode psychosis, neurobiological characteristics, and main risk factors, including early exposure to the trauma situation and early interventions.

DOI: 10.4018/979-8-3693-0851-6.ch006

INTRODUCTION

Psychosis can be conceptualized as a combination of symptoms resulting in a disconnection of reality and a severe impact on an individual's functionality and quality of life. The experience of the psychotic phenomenon can open a critical window into primary psychiatric conditions, particularly schizophrenia, schizoaffective disorder, or bipolar disorder. The use of psychoactive substances, such as marijuana, cocaine, and synthetic drugs (LSD, MDMA), as well as a variety of infectious, autoimmune, and vascular medical conditions, among others, may be related to psychotic episodes as etiological or predisposing factors. More recently, the role of environmental experiences in early adolescence or late childhood, including psychological abuse or physical violence, has been associated with a greater likelihood of developing psychosis years later. The emergence of first-episode psychosis (FEP) varies widely, usually occurring at the end of adolescence and the beginning of adult life. The operational concept of first-episode psychosis usually encompasses a wide range of categories: first treatment contact, duration of antipsychotic medication, and duration of psychotic episode (Breitborde et al., 2009). A determining factor for progression to severe symptoms is the intensity of psychotic symptoms, such as delusions and hallucinations. In addition to behavioral symptoms, psychotic conditions are often accompanied by negative symptoms (reduced emotional expression, sociability, avolition, and anhedonia). Furthermore, evidence has shown, even in individuals with symptomatic remission, the occurrence of cognitive symptoms, particularly in complex attention, executive functioning, episodic memory, and verbal and procedural IQ, among others. The current chapter discusses factors such as the epidemiology of FEP, neurobiological characteristics, and main risk factors, including early exposure to the trauma situation and early interventions.

1. Prevalence of Psychosis in Adolescents and Early Adulthood

Psychosis, from a broader perspective, affects approximately 3% of the global population, with schizophrenia representing its most severe and prevalent form, affecting approximately 1% of individuals throughout their lives (McCleery & Nuechterlein, 2019). Adolescence represents a significant sociocultural phenomenon associated with psychological reshaping regarding how individuals perceive themselves and the world. During this period, certain aspects of personality, such as self-concept and self-esteem, develop, influencing preferences and the quest for acceptance within peer groups (Lepre & De Oliveira, 2022).

The neurobiology of FEP may involve the disruption of crucial brain functions, notably affecting fronto-striatal and temporal circuits. Furthermore, its clinical onset occurs during adolescence or early adulthood due to the complex interplay of sociocultural and genetic predisposing risk factors during "key moments" of neurodevelopment. As a result, a prodromal period of these symptoms is observable in individuals with Clinical High-Risk Syndrome (CHR) (de Vos et al., 2019).

Evidence demonstrates a strong association between childhood trauma and psychosis. In one literature review, 83.8% of FEP individuals had experienced traumatic events, compared to 62.6% without psychotic symptoms. Trauma experiences before the age of 17 exhibited a 2.91 times greater likelihood of developing psychotic symptoms at age 18, and those who endured three or more types of childhood traumata were 4.7 times more likely to exhibit psychotic symptoms. Thus, the risk of psychosis increased 2.5 times for each additional trauma, with an overall risk increase of 5 times for those who experienced three or more traumatic events (Stanton, Denietolis, Goodwin, & Dvir, 2020).

Another study affirms that early exposure to moderate to severe traumas during childhood is related to the development of psychosis. It was observed that the population with Clinical High-Risk Syndrome (CHR) has an increased risk of 16-35% of developing full psychosis within the 2.5 years following the initial diagnosis. Therefore, it is crucial to assess the history of trauma before the onset of full psychosis to elucidate potential mechanisms linking psychosis to traumatic events in childhood (Loewy et al., 2019).

Furthermore, the connection between trauma and the development of psychiatric disorders has been demonstrated. Individuals aged 18 or older diagnosed with schizophrenia or schizoaffective disorder who experienced emotional neglect had a higher tendency to develop depression. In contrast, those who suffered physical neglect were more likely to develop posttraumatic stress disorder. The severity of trauma was also associated with higher rates of suicide and nonsuicidal self-harm, worse clinical outcomes, and poorer adaptive functioning (Stanton, Denietolis, Goodwin, & Dvir, 2020).

In addition to psychosis, childhood trauma can result in other consequences, such as anxiety, depression, and mood disorders. Additionally, assessments have indicated that high rates of childhood trauma are not related to a family history of psychosis or lower socioeconomic status, suggesting a specific connection with CHR syndrome (Isvoranu et al., 2017; Loewy et al., 2019).

It is also essential to differentiate potentially traumatic events from less severe stressful events. Possible traumatic events include exposure to physical, sexual, or emotional abuse and neglect during childhood. Moreover, patients at high clinical risk may have pronounced comorbid affective disorders associated with psychosis and decreased neuropsychological performance in attention, working memory, and cognitive flexibility (Stanton, Denietolis, Goodwin, & Dvir, 2020).

Additionally, there are still gaps in research regarding the relationship between the age of trauma exposure and its consequences and the increased likelihood of developing psychosis. Typically, the age range is between 12 and 30 years, and it is essential to note that trauma exposure tends to accumulate over time. Therefore, distinguishing between traumas experienced before and after the onset of CHR syndrome is challenging. However, there is evidence that, as exemplified by bullying, vulnerable youth face a higher risk of trauma (Heins et al., 2011).

It should be noted that an increased risk of psychosis has been observed in children under 12 years old, and such a finding does not imply that traumatic events after this age are incapable of triggering psychosis but rather that, at older ages, these events need to be more intense to produce similar effects (Loewy et al., 2019).

Furthermore, it has been observed that in women, the onset of symptoms occurs slightly later, with a higher prevalence of positive symptoms, fewer negative symptoms, and a less severe course of the disease. Gender can affect children's susceptibility to certain types of abuse, with women, for example, being more likely to endure physical or sexual abuse. It also affects how they respond to these traumas, ultimately altering the course of psychosis (Trovão et al., 2022).

2. Risk Factors for Psychosis

Epigenetics

Putative mechanisms linking emotional and social experiences with early neurobiological abnormalities in schizophrenia have been intensively debated (Figure). Early life experiences, combined with genetic factors, are thought to significantly influence behavioral patterns presented in adulthood. Hence, stressful events experienced early in life may influence brain development, as demonstrated by neuroimaging and neuropsychological studies (de Oliveira et al., 2010). Evidence shows that early exposure to traumatic experiences may influence the age of onset in schizophrenia. Epidemiological studies have suggested that psychotic-like experiences were more common by the age of 12 in children previously exposed to child trauma (Loewy et al., 2019). Child and adolescent abuse and neglect result in vulnerability to developing positive and negative symptoms. According to the epigenetic hypothesis (van Os et al., 2008), experiencing trauma leads to neurobiological and gene–environment changes, leading to affective disturbances and emotional distress (anxiety, tension, depression) and ultimately to clinical psychosis (Isvoranu et al., 2017). In a few studies, the breakdown of psychosis during adolescence (early-onset psychosis-EOP) has been associated with a history of family violence and sexual abuse (Fergusson et al., 2008; Thompson et al., 2014). Studies have shown that onset in adults (adult-onset psychosis-AOP) is less linked to social problems or a history of family violence. Furthermore, EOP patients were likely to have worse premorbid functioning and a significantly longer duration of untreated psychosis than those with later onset (Chen et al., 2018; Schimmelmann et al., 2007).

The experience of physical violence may be linked to the onset of psychosis at an earlier age (Croft et al., 2019). In a recent population cohort with 4,433 patients with an average age of 17.8 years old, exposure to any trauma up to 17 years of age was associated with an increased chance for psychotic experiences at the age of 18 (odds ratio adjusted, 2.91; 95% CI, 2.15-3.93). All types of traumata from early childhood to late adolescence were associated with an increased chance for psychotic experiences. The fraction attributable to the population for childhood and adolescence trauma in psychotic experiences at 18 was 45% (IC 95%, 25% -60%) (Croft et al., 2019).

In childhood and adolescence, functional and connectivity changes in several brain regions as part of the maturation process may occur, with significant associations between early trauma and higher levels of experiential anomalies, which can be considered disturbances of self-awareness in patients in the early stages of schizophrenia. Self-awareness disturbances seem to be triggered by emotional abuse, emotional neglect, and physical neglect, in addition to the frequent relation between sexual and physical abuse and the development of psychotic features, as well as prodromal symptomatology (Haug et al., 2015). Previous literature has demonstrated the correlation between the diagnosis of schizophrenia and early neglect, while nonpsychotic disorders were correlated with different types of abuse (Vogel et al., 2011).

Figure 1.

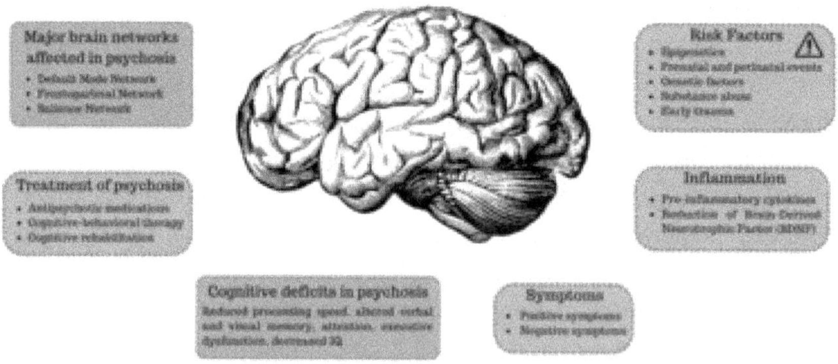

4. Other Risk Factors

Specific groups are at particular risk of psychosis, and various modifiable and nonmodifiable risk factors mentioned in the following topics may influence the clinical course of this disorder.

Prenatal and Perinatal Events

Although typically appearing in early adulthood, the neurodevelopment hypothesis is supported by substantial evidence (Jaaro-Peled & Sawa, 2020; Kahn & Sommer, 2015). Individuals who experience severe intrauterine or peripartum complications have an increased risk of developing psychotic disorders (Barch et al., 2022). There is also consistent evidence of a similar relationship involving atypical neurodevelopment, cognitive, motor, and attentional impairments in childhood, as well as altered cutaneous markers of ectodermal development (Erlenmeyer-Kimling et al., 2000; Jameson et al., 2023).

A meta-analysis of prospective population-based studies revealed that perinatal and intrauterine events are potentially associated with schizophreniform psychosis: 1) pregnancy complications (bleeding, diabetes, rhesus incompatibility, preeclampsia); 2) abnormal fetal growth and development (low birth weight, congenital malformations, reduced head circumference); and 3) birth complications (uterine atony, asphyxia, emergency cesarean section (Cannon et al., 2002; Jenkins, 2013).

Furthermore, people born between winter and spring are more represented among patients with schizophrenia (meaning an increase of 7–10%). This excess may be due to a greater likelihood of the fetal brain being exposed to maternal respiratory infections or maternal malnutrition, including folic acid or vitamin D deficiency, during winter. However, none of the explanations for this phenomenon have been firmly established. Presumably, these risk factors affect the neural connectivity of the developing brain.

Genetic Factors

Psychotic conditions may have significant genetic heritability. In schizophrenia, for example, twins and other studies have consistently shown a significant genetic component, with heritability estimated at approximately 80% (Sullivan et al., 2003). There are also some genetic variants involving the deletion or duplication of sections of DNA (copy number, variants) that are associated with a significantly increased risk of schizophrenia on their own but are only found in 2% to 3% of people with cases of schizophrenia. One of the best established is a deletion of several megabases of DNA at chromosome 22q11.2, which is associated with a 30% to 40% lifetime risk of developing schizophrenia (Bassett & Chow, 2008; Schneider et al., 2014). It should also be recognized that, as with environmental risk factors, many genetic variants associated with schizophrenia may also be associated with other psychiatric disorders, suggesting overlap in risk factors and potential mechanisms.

However, even among identical twins, pairwise concordance for schizophrenia is approximately 50% (Hilker et al., 2018). Interactions with genetic factors increase the risk of schizophrenia and other disorders.

Paternal Age

Evidence has shown a greater likelihood of developing schizophrenia among the offspring of older parents (Kaltsas et al., 2023). However, whether this risk is due to psychological or biological factors is unclear. For example, men with a schizotypal personality may be more likely to marry later in life, or older men may harbor more mutations that increase the risk due to repeated mitoses in sperm progenitors. The evidence currently favors the idea that the association between late and schizotypal fatherhood personality is the predominant driver of this effect (Petersen et al., 2011).

Urban Environment

Schizophrenia is most common in disadvantaged areas of inner cities, a finding first noted in Chicago in 1939 (Stilo & Murray, 2010). There is significant and independent variation in the incidence of schizophrenia and other psychoses regarding sex, age, ethnicity, and place, and such findings suggest that environmental effects or neighborhood level may interact with genetic factors in the etiology of psychosis (Kirkbride et al., 2006). One paper compiled all incident cases of ICD-10 psychoses from a large 2-year epidemiological study of first-episode psychoses in southeastern London. Neighborhood variation in the incidence of nonaffective psychoses could not be explained by individual-level risk, implicating neighborhood-level socioenvironmental factors in their etiology (Kirkbride et al., 2007).

Migration Status

A personal or family history of migration is a significant risk factor for developing psychotic conditions (Cantor-Graae & Selten, 2005). In Great Britain, for example, evidence suggests that the second-generation Afro-Caribbean population is particularly vulnerable to some environmental risk factors for schizophrenia or that some environmental factors act selectively on this population in Great Britain (Hutchinson et al., 1996).

Ethnic density also appears important, as the relative proportion of nonwhite minorities in a neighborhood increases, and the risk of psychotic conditions in the minority population decreases. Thus, a lack of social support or increased exposure to discrimination may work to increase the risk of developing the disease, especially in areas with only a small minority population. One fact that may support this is that schizophrenia in nonwhite ethnic minorities in London is higher when they comprise a smaller proportion of the local population (Boydell et al., 2001).

In contrast, a recent study in British Columbia (BC), Canada, revealed that the risk of a diagnosis of nonaffective psychotic disorder was lower among first- and second-generation migrants from some Asian regions than among nonmigrants. These results highlight the complexity of the association between migration and psychotic disorders. The differential risk pattern between subgroups suggests a role for psychosocial adversity in this etiology (Cantor-Graae & Selten, 2005). Future research should investigate why certain groups of migrants are less likely to be diagnosed and whether specific subgroups face elevated risk (Magee et al., 2023).

Drug Abuse

Persistent abuse of amphetamine, methamphetamine, and cocaine, as well as cathinone-derived ‘legal highs,’ can produce a state almost identical to paranoid schizophrenia (Murray et al., 2013). Furthermore, experimental administration of cannabis or its active ingredient, delta9-tetrahydrocannabinol (THC), can induce an acute and transient psychotic reaction in psychiatrically well individuals, and the extent of this psychotic reaction has not been related to the degree of anxiety or cognitive impairment (Morrison et al., 2009). In addition, a series of prospective studies have shown that young people who heavily use cannabis have an increased risk of subsequent schizophrenia and that this relationship is dose-dependent (Malaspina et al., 2001).

Continued cannabis use after the onset of psychosis predicts adverse outcomes, including higher recovery rates, more extended hospital stays, and more severe positive symptoms, than for individuals who have discontinued cannabis use and those who have not. These findings point to reductions in cannabis use as a crucial intervention target to improve outcomes in patients with psychosis (Schoeler et al., 2016). The risk is greater in those who start cannabis use in early adolescence than in those who start use later in life and in those using high-potency varieties of cannabis (Di Forti et al., 2019). In recent years, even more potent synthetic cannabinoids (spice) have become widely available over the Internet and have been linked to acute psychotic reactions (Murray et al., 2013).

On the other hand, a meta-analysis indicated that cannabis use is generally not associated with neurocognitive functioning in patients with FEP. However, it highlights the deleterious effect of low doses of cannabis on some patients. It also highlights the importance of the type of antipsychotic prescription and the dose of cannabis as moderating variables in the neurocognitive functioning of this population (Sánchez-Gutiérrez et al., 2020).

The table below lists some of the main risk factors and their respective increased risk of developing nonaffective psychosis:

Table 1. Association of environmental factors and risk of psychosis in schizophrenia [1]

Risk Factor	Odds Ratio (95% CI)
Obstetric complications	1.84 (1.25-2.70)
Winter birth in the northern hemisphere	1.04 (1.02-1.06)
Childhood trauma	2.87 (2.07-3.98)
Urban living	2.19 (1.55-3.09)
Migration (first generation)	2.10 (1.72-2.56) [2]
Cannabis use	5.17 (3.64-7.36)

[1]Odds ratios were taken from RADUA*et al.*, (2018)
[2]An incidence rate ratio is reported, rather than an odds ratio.
Adapted McCutcheon et al. (2020)

5. Brain Networks Affected by Psychosis

In recent decades, the hypothesis that psychoses may derive from the disconnection of the networks that integrate the brain has been strengthened, and studies with functional magnetic resonance imaging (fMRI) are helping to discover areas that undergo deactivation or hyperactivation during the process of psychosis, from the early stages of the condition to more advanced stages (Gong et al., 2017).

Functional connectivity assesses altered brain networks in a resting state or during the execution of a task, which helps study mental disorders and explore the healthy brain. Among the various neural networks studied, the triple network model stands out: the default mode network (DMN), the frontoparietal network (FPN), also called the central executive network (CEN), and the salience network (SN). The limbic and motor networks are also targets of research (Schimmelpfennig et al., 2023).

The DMN is a set of distinct brain regions active in passive tasks or a resting state. It is anatomically linked to the posterior cingulate cortex (PCC), precuneus, and ventromedial prefrontal cortex and is related to social cognition, recall of autobiographical data, past events, or prospective thoughts (Ribeiro da Costa et al., 2021). The frontoparietal network connects the lateral posterior parietal cortex (PPC) and the dorsolateral prefrontal cortex, becoming active during tasks that require more significant cognitive effort and are negatively related to the default network. Its function involves selecting activities, making decisions, and executing problems (Lindquist & Barrett, 2012; Seeley et al., 2007).

The salience network encompasses the dorsolateral cingulate cortex (dACC) and anterior insula (AI). It is related to the socio affective route and the integration of emotional, sensory, and cognitive information and is associated with processing reward, pain, and motivation. Furthermore, the SN activates the amygdala and creates an interconnection between the DNM and FPN, functioning as a switch that modulates internal and external stimuli originating from the networks. Studies reinforce a high negative correlation between the FPN and DNM networks and better performance in executive functions (Saviola et al., 2021; Uddin, 2015).

Recent evidence suggests that altered communication between brain networks is related to the symptoms of psychosis. A deep discriminant autoencoding network study demonstrated dysfunctional integration within the triple model networks and their connections. Less joint integration of the DMN, FPN, and sensory/somatomotor network is associated with the severity of positive symptoms in psychosis (Lee et al., 2018).

A meta-analysis carried out in 2018 investigated brain network activity in FEP; a reduction in functional connectivity in the DMN was evidenced within regions of the network itself and with regions of the SN and FPN, showing a general brain hypoconnectivity of the DMN in psychosis. Furthermore, the DMN was associated with greater severity of negative symptoms of psychosis. The FPN and SN presented mixed results in this research (O'Neill et al., 2019). Children and adolescents at high risk for psychosis were evaluated in two studies that showed hyperconnectivity of the DMN in the context of a working memory task and activation-deactivation mismatch between the DMN and FPN network in those activities where one would typically expect FPN activation (Satterthwaite & Baker, 2015).

Further studies are needed on brain networks, their functions, connections, and the activation-deactivation process of networks that are organized in a complex way, as a mismatch in this orchestra of signaling alters brain functioning illustrated through the clinical symptoms of mental disorders.

6. Main Cognitive Deficits in Early Psychosis

Cognitive function may often be present throughout the psychosis course and plays a crucial role as a positive predictor for the development of psychosis, with approximately 62% of patients exhibiting cognitive impairment at the onset of the illness. A wide range of difficulties are found in executive function, language, memory, and attention domains, shortly after the disease outbreak or as residual symptoms after psychotic stabilization. Medication-related cognitive impairment may also be noticed. Although antipsychotics act well on symptoms such as delusions and hallucinations, cognitive deficits can persist or progress throughout psychosis, impairing the individual's functionality (Ayres et al., 2007).

Bombassaro et al. (2023) carried out a systematic review on cognition and schizophrenia, highlighting more than seven cognitive deficits, such as processing speed, problem solving, visual and verbal memory, and logical reasoning. Research in which the Wechsler Adult Intelligence Scale (WAIS-III) was applied to healthy individuals and those with schizophrenia demonstrated a reduction in the total intelligence quotient (IQ). Although there are several neuropsychological tests, the WAIS is considered the gold standard.

Although most research finds broader cognitive deficits covering all cognitive domains, changes in processing speed, attention, and working memory stand out discreetly. Other cohorts showed a drop in IQ in the premorbid period and a worsening of IQ (14-15 points) after the manifestation of schizophrenia. In 2019, Zanelli *et al.* evaluated individuals in their first psychotic episode for ten years and found significant changes in IQ, verbal knowledge, and memory compared to control patients (Mesholam-Gately et al., 2009; Zanelli et al., 2019).

Sheffield et al. (2018) studied cognitive deficits throughout life and showed that children and adolescents who later developed schizophrenia had deficits in general cognitive abilities with poor school performance. The drop in IQ worsened after psychosis took hold. Verbal skills such as receptive language were also affected in children. The impairment of verbal memory and processing speed in the first psychotic episode was significant and remained stable, even without antipsychotics.

The impact of antipsychotic use on cognition in FEP has found mixed results. Studies show a slight improvement in general cognitive domains without specifications. The use of haloperidol and second-generation antipsychotics also resulted in cognitive improvement. Despite expecting a more significant cognitive improvement with second-generation antipsychotics, a meta-analysis with first- and second-generation medications found no difference in results, questioning the use of other alternative pharmacological treatments and cognitive rehabilitation therapies (Feber et al., 2023).

Functional impairment persists in more than two-thirds of individuals who recover from FEP, making it challenging to establish interpersonal relationships and independence in solving tasks. Data suggest that 80% of patients with primary psychotic disorders experience cognitive decline, but most longitudinal studies have not illustrated the long-term progression of cognitive changes. PMID 31749648. The complexity of assessing cognition is due to its heterogeneous manifestations in the different psychotic disorders, requiring more research to better understand the process of cognitive loss and the role of cognition in psychosis.

7. Prognosis of Early Psychosis

The challenges related to the prognosis of individuals developing psychosis remain significant. The progression of cognitive deficits is still under debate since a clear correlation between older age and the evolution of cognitive deficits in psychosis has not been established. The lack of longitudinal cohort studies and the predominance of cross-sectional research favored this context. Watson *et al.*, 2022 carried out a meta-analysis on the longitudinal assessment of cognitive deficits in the first five years of the condition, finding findings similar to those of Bora and Murray in 2014, weakening the hypothesis of neuroprogression and supporting the hypothesis of neurodevelopment of schizophrenia in which the deficits would occur from the period premorbid stage of psychosis (Bora & Murray, 2014; Steele et al., 2021).

8. Concluding Remarks

There is a need for further research on psychosis in adolescents and young adults. The stigma surrounding the condition remains substantial, and social prejudice limits and hinders its treatment. Knowledge about predictors of psychosis can not only reduce its occurrence but also prevent its development in individuals with CHR (Sönmez et al., 2020).

Additionally, purely medication-based treatment often fails due to therapeutic dropout, making cognitive–behavioral therapy (CBT) an essential and more effective method in treating and supporting individuals with CHR (Turkington et al., 2006). Thus, providing clear and objective information about the diagnosis to the patient and their family is crucial, avoiding and addressing any existing prejudices.

REFERENCES

Ayres, A. M., Busatto, G., Menezes, P., Schaufelberger, M., Coutinho, L., Murray, R., McGuire, P., Rushe, T., & Scazufca, M. (2007). Cognitive deficits in first-episode psychosis: A population-based study in São Paulo, Brazil. *Schizophrenia Research*, *90*(1-3), 338–343. doi:10.1016/j.schres.2006.09.026 PMID:17123787

Bassett, A. S., & Chow, E. W. C. (2008). Schizophrenia and 22q11.2 deletion syndrome. *Current Psychiatry Reports*, *10*(2), 148–157. doi:10.1007/s11920-008-0026-1 PMID:18474208

Bombassaro, T., Carrilho, C. G., Peixoto, C., Alves, G. S., Kahn, J. P., Nardi, A. E., & Veras, A. B. (2023). Cognition in Schizophrenia: A Systematic Review of Wechsler Adult Intelligence Scale Studies. *The Primary Care Companion for CNS Disorders*, *25*(5). doi:10.4088/PCC.22r03456 PMID:37857289

Bora, E., & Murray, R. M. (2014). Meta-analysis of cognitive deficits in ultrahigh risk to psychosis and first-episode psychosis: Do the cognitive deficits progress over, or after, the onset of psychosis? *Schizophrenia Bulletin*, *40*(4), 744–755. doi:10.1093/schbul/sbt085 PMID:23770934

Boydell, J., van Os, J., McKenzie, K., Allardyce, J., Goel, R., McCreadie, R. G., & Murray, R. M. (2001). Incidence of schizophrenia in ethnic minorities in London: Ecological study into interactions with environment. *BMJ (Clinical Research Ed.)*, *323*(7325), 1336–1336. doi:10.1136/bmj.323.7325.1336 PMID:11739218

Breitborde, N. J. K., Srihari, V. H., & Woods, S. W. (2009). Review of the operational definition for first-episode psychosis. *Early Intervention in Psychiatry*, *3*(4), 259–265. doi:10.1111/j.1751-7893.2009.00148.x PMID:22642728

Cannon, M., Jones, P. B., & Murray, R. M. (2002). Obstetric Complications and Schizophrenia: Historical and Meta-Analytic Review. *The American Journal of Psychiatry*, *159*(7), 1080–1092. doi:10.1176/appi.ajp.159.7.1080 PMID:12091183

Cantor-Graae, E., & Selten, J.-P. (2005). Schizophrenia and Migration: A Meta-Analysis and Review. *The American Journal of Psychiatry*, *162*(1), 12–24. doi:10.1176/appi.ajp.162.1.12 PMID:15625195

Chen, L., Selvendra, A., Stewart, A., & Castle, D. (2018). Risk factors in early and late onset schizophrenia. *Comprehensive Psychiatry*, *80*, 155–162. doi:10.1016/j.comppsych.2017.09.009 PMID:29096206

Croft, J., Heron, J., Teufel, C., Cannon, M., Wolke, D., Thompson, A., Houtepen, L., & Zammit, S. (2019). Association of Trauma Type, Age of Exposure, and Frequency in Childhood and Adolescence With Psychotic Experiences in Early Adulthood. *JAMA Psychiatry*, *76*(1), 79. doi:10.1001/jamapsychiatry.2018.3155 PMID:30477014

de Oliveira, P. A., Scivoletto, S., & Cunha, P. J. (2010). Estudos neuropsicológicos e de neuroimagem associados ao estresse emocional na infância e adolescência. *Archives of Clinical Psychiatry*, *37*(6), 271–279. doi:10.1590/S0101-60832010000600004 de Vos, C., Thompson, A., Amminger, P., Bendall, S., de Haan, L., Hartmann, J., Lavoie, S., Markulev, C., McGorry, P., Nieman, D., Phillips, L., Reininghaus, U., Yuen, H. P., Yung, A., & Nelson, B. (2019). The relationship between childhood trauma and clinical characteristics in ultrahigh risk for psychosis youth. *Psychosis*, *11*(1), 28–41. doi:10.1080/17522439.2019.1582686

Di Forti, M., Quattrone, D., Freeman, T. P., Tripoli, G., Gayer-Anderson, C., Quigley, H., Rodriguez, V., Jongsma, H. E., Ferraro, L., La Cascia, C., La Barbera, D., Tarricone, I., Berardi, D., Szöke, A., Arango, C., Tortelli, A., Velthorst, E., Bernardo, M., Del-Ben, C. M., ... van der Ven, E. (2019). The contribution of cannabis use to variation in the incidence of psychotic disorder across Europe (EU-GEI): A multicentre case-control study. *The Lancet. Psychiatry*, *6*(5), 427–436. doi:10.1016/S2215-0366(19)30048-3 PMID:30902669

Erlenmeyer-Kimling, L., Rock, D., Roberts, S. A., Janal, M., Kestenbaum, C., Cornblatt, B., Adamo, U. H., & Gottesman, I. I. (2000). Attention, Memory, and Motor Skills as Childhood Predictors of Schizophrenia-Related Psychoses: The New York High-Risk Project. *The American Journal of Psychiatry*, *157*(9), 1416–1422. doi:10.1176/appi.ajp.157.9.1416 PMID:10964857

Feber, L., Peter, N., Schneider-Thoma, J., Siafis, S., Bighelli, I., Hansen, W.-P., Prates Baldez, D., Salanti, G., Keefe, R. S. E., Engel, R. R., & Leucht, S. (2023). Antipsychotic drugs and their effects on cognitive function: Protocol for a systematic review, pairwise, and network meta-analysis. *Systematic Reviews*, *12*(1), 54. doi:10.1186/s13643-023-02213-5 PMID:36959619

Fergusson, D. M., Boden, J. M., & Horwood, L. J. (2008). Exposure to childhood sexual and physical abuse and adjustment in early adulthood. *Child Abuse & Neglect*, *32*(6), 607–619. doi:10.1016/j.chiabu.2006.12.018 PMID:18565580

Gong, Q., Hu, X., Pettersson-Yeo, W., Xu, X., Lui, S., Crossley, N., Wu, M., Zhu, H., & Mechelli, A. (2017). Network-Level Dysconnectivity in Drug-Naïve First-Episode Psychosis: Dissociating Transdiagnostic and Diagnosis-Specific Alterations. *Neuropsychopharmacology*, *42*(4), 933–940. doi:10.1038/npp.2016.247 PMID:27782128

Haug, E., Øie, M., Andreassen, O. A., Bratlien, U., Nelson, B., Aas, M., Møller, P., & Melle, I. (2015). Anomalous self-experience and childhood trauma in first-episode schizophrenia. *Comprehensive Psychiatry*, *56*, 35–41. doi:10.1016/j.comppsych.2014.10.005 PMID:25458477

Heins, M., Simons, C., Lataster, T., Pfeifer, S., Versmissen, D., Lardinois, M., Marcelis, M., Delespaul, P., Krabbendam, L., van Os, J., & Myin-Germeys, I. (2011). Childhood trauma and psychosis: A case–control and case-sibling comparison across different levels of genetic liability, psychopathology, and type of trauma. *The American Journal of Psychiatry, 168*(12), 1286–1294. doi:10.1176/appi.ajp.2011.10101531 PMID:21955935

Hilker, R., Helenius, D., Fagerlund, B., Skytthe, A., Christensen, K., Werge, T. M., Nordentoft, M., & Glenthøj, B. (2018). Heritability of Schizophrenia and Schizophrenia Spectrum Based on the Nationwide Danish Twin Register. *Biological Psychiatry, 83*(6), 492–498. doi:10.1016/j.biopsych.2017.08.017 PMID:28987712

Hutchinson, G., Takei, N., Fahy, T. A., Bhugra, D., Gilvarry, C., Moran, P., Mallett, R., Sham, P., Leff, J., & Murray, R. M. (1996). Morbid Risk of Schizophrenia in First-Degree Relatives of White and African–Caribbean Patients with Psychosis. *The British Journal of Psychiatry, 169*(6), 776–780. doi:10.1192/bjp.169.6.776 PMID:8968638

Isvoranu, A.-M., van Borkulo, C. D., Boyette, L.-L., Wigman, J. T. W., Vinkers, C. H., & Borsboom, D. (2017). A Network Approach to Psychosis: Pathways Between Childhood Trauma and Psychotic Symptoms. *Schizophrenia Bulletin, 43*(1), 187–196. doi:10.1093/schbul/sbw055 PMID:27165690

Jaaro-Peled, H., & Sawa, A. (2020). Neurodevelopmental Factors in Schizophrenia. *The Psychiatric Clinics of North America, 43*(2), 263–274. doi:10.1016/j.psc.2020.02.010 PMID:32439021

Jameson, C., Boulton, K. A., Silove, N., Nanan, R., & Guastella, A. J. (2023). Ectodermal origins of the skin-brain axis: A novel model for the developing brain, inflammation, and neurodevelopmental conditions. *Molecular Psychiatry, 28*(1), 108–117. doi:10.1038/s41380-022-01829-8 PMID:36284159

Jenkins, T. A. (2013). Perinatal complications and schizophrenia: Involvement of the immune system. *Frontiers in Neuroscience, 7*. doi:10.3389/fnins.2013.00110 PMID:23805069

Kahn, R. S., & Sommer, I. E. (2015). The neurobiology and treatment of first-episode schizophrenia. *Molecular Psychiatry, 20*(1), 84–97. doi:10.1038/mp.2014.66 PMID:25048005

Kaltsas, A., Moustakli, E., Zikopoulos, A., Georgiou, I., Dimitriadis, F., Symeonidis, E. N., Markou, E., Michaelidis, T. M., Tien, D. M. B., Giannakis, I., Ioannidou, E. M., Papatsoris, A., Tsounapi, P., Takenaka, A., Sofikitis, N., & Zachariou, A. (2023). Impact of Advanced Paternal Age on Fertility and Risks of Genetic Disorders in Offspring. *Genes*, *14*(2), 486. doi:10.3390/genes14020486 PMID:36833413

Kirkbride, J. B., Fearon, P., Morgan, C., Dazzan, P., Morgan, K., Murray, R. M., & Jones, P. B. (2007). Neighborhood variation in the incidence of psychotic disorders in Southeast London. *Social Psychiatry and Psychiatric Epidemiology*, *42*(6), 438–445. doi:10.1007/s00127-007-0193-0 PMID:17473901

Kirkbride, J. B., Fearon, P., Morgan, C., Dazzan, P., Morgan, K., Tarrant, J., Lloyd, T., Holloway, J., Hutchinson, G., Leff, J. P., Mallett, R. M., Harrison, G. L., Murray, R. M., & Jones, P. B. (2006). Heterogeneity in Incidence Rates of Schizophrenia and Other Psychotic Syndromes: Findings From the 3-Center ÆSOP Study. *Archives of General Psychiatry*, *63*(3), 250. doi:10.1001/archpsyc.63.3.250 PMID:16520429

Lee, W. H., Doucet, G. E., Leibu, E., & Frangou, S. (2018). Resting-state network connectivity and metastability predict clinical symptoms in schizophrenia. *Schizophrenia Research*, *201*, 208–216. doi:10.1016/j.schres.2018.04.029 PMID:29709491

Lepre, R. M. & De Oliveira, J. (2022). Adolescência e construção da personalidade moral. *Dialogia,* e21333 . doi:10.5585/41.2022.21333

Lindquist, K. A., & Barrett, L. F. (2012). A functional architecture of the human brain: Emerging insights from the science of emotion. *Trends in Cognitive Sciences*, *16*(11), 533–540. doi:10.1016/j.tics.2012.09.005 PMID:23036719

Lipner, E., O'Brien, K. J., Pike, M. R., Ered, A., & Ellman, L. M. (2022). Environmental Risk Factors and Cognitive Outcomes in Psychosis: Pre, Perinatal, and Early Life Adversity. In D. M. Barch & J. W. Young (Eds.), *Cognitive Functioning in Schizophrenia: Leveraging the RDoC Framework* (Vol. 63, pp. 205–240). Springer International Publishing. doi:10.1007/7854_2022_378

Loewy, R. L., Corey, S., Amirfathi, F., Dabit, S., Fulford, D., Pearson, R., Hua, J. P. Y., Schlosser, D., Stuart, B. K., Mathalon, D. H., & Vinogradov, S. (2019). Childhood trauma and clinical high risk for psychosis. *Schizophrenia Research*, *205*, 10–14. doi:10.1016/j.schres.2018.05.003 PMID:29779964

Magee, C., Oberle, E., Guhn, M., Gadermann, A., & Puyat, J. H. (2023). Risk of Diagnosed Adolescent-Onset Non-Affective Psychotic Disorder by Migration Background in British Columbia: A Retrospective Cohort Study. *Canadian Journal of Psychiatry*, *68*(1), 33–42. doi:10.1177/07067437221100351 PMID:35698751

Malaspina, D., Harlap, S., Fennig, S., Heiman, D., Nahon, D., Feldman, D., & Susser, E. S. (2001). Advancing Paternal Age and the Risk of Schizophrenia. *Archives of General Psychiatry*, *58*(4), 361. doi:10.1001/archpsyc.58.4.361 PMID:11296097

McCleery, A., & Nuechterlein, K. H. (2019). Cognitive impairment in psychotic illness: Prevalence, profile of impairment, developmental course, and treatment considerations. *Dialogues in Clinical Neuroscience*, *21*(3), 239–248. doi:10.31887/DCNS.2019.21.3/amccleery PMID:31749648

McCutcheon, R. A., Reis Marques, T., & Howes, O. D. (2020). Schizophrenia—An Overview. *JAMA Psychiatry*, *77*(2), 201. doi:10.1001/jamapsychiatry.2019.3360 PMID:31664453

Mesholam-Gately, R. I., Giuliano, A. J., Goff, K. P., Faraone, S. V., & Seidman, L. J. (2009). Neurocognition in first-episode schizophrenia: A meta-analytic review. *Neuropsychology*, *23*(3), 315–336. doi:10.1037/a0014708 PMID:19413446

Morrison, P. D., Zois, V., McKeown, D. A., Lee, T. D., Holt, D. W., Powell, J. F., Kapur, S., & Murray, R. M. (2009). The acute effects of synthetic intravenous Δ9-tetrahydrocannabinol on psychosis, mood and cognitive functioning. *Psychological Medicine*, *39*(10), 1607. doi:10.1017/S0033291709005522 PMID:19335936

Murray, R. M., Paparelli, A., Morrison, P. D., Marconi, A., & Di Forti, M. (2013). What can we learn about schizophrenia from studying the human model, drug-induced psychosis? *American Journal of Medical Genetics. Part B, Neuropsychiatric Genetics*, *162*(7), 661–670. doi:10.1002/ajmg.b.32177 PMID:24132898

O'Neill, A., Mechelli, A., & Bhattacharyya, S. (2019). Dysconnectivity of Large-Scale Functional Networks in Early Psychosis: A Meta-analysis. *Schizophrenia Bulletin*, *45*(3), 579–590. doi:10.1093/schbul/sby094 PMID:29982729

Petersen, L., Mortensen, P. B., & Pedersen, C. B. (2011). Paternal Age at Birth of First Child and Risk of Schizophrenia. *The American Journal of Psychiatry*, *168*(1), 82–88. doi:10.1176/appi.ajp.2010.10020252 PMID:20952457

Ribeiro da Costa, C., Soares, J. M., Oliveira-Silva, P., Sampaio, A., & Coutinho, J. F. (2021). Interplay Between the Salience and the Default Mode Network in a Social-Cognitive Task Toward a Close Other. *Frontiers in Psychiatry*, *12*, 718400. doi:10.3389/fpsyt.2021.718400 PMID:35197871

Sánchez-Gutiérrez, T., Fernandez-Castilla, B., Barbeito, S., González-Pinto, A., Becerra-García, J. A., & Calvo, A. (2020). Cannabis use and nonuse in patients with first-episode psychosis: A systematic review and meta-analysis of studies comparing neurocognitive functioning. *European Psychiatry, 63*(1), e6. doi:10.1192/j.eurpsy.2019.9 PMID:32093788

Satterthwaite, T. D., & Baker, J. T. (2015). How can studies of resting-state functional connectivity help us understand psychosis as a disorder of brain development? *Current Opinion in Neurobiology, 30*, 85–91. doi:10.1016/j.conb.2014.10.005 PMID:25464373

Saviola, F., Bellani, M., Perlini, C., Squarcina, L., Maggioni, E., Zacà, D., Lasalvia, A., Dusi, N., Bonetto, C., Cristofalo, D., Alessandrini, F., Zoccatelli, G., Ciceri, E., Mesiano, L., Semrov, E., Lo Parrino, R., Furlato, K., Pratelli, M., Ruggeri, M., ... Jovicich, J. (2021). First-episode psychosis: Structural covariance deficits in salience network correlate with symptoms severity. *Journal of Psychiatric Research, 136*, 409–420. doi:10.1016/j.jpsychires.2021.01.044 PMID:33647856

Schimmelmann, B. G., Conus, P., Cotton, S., McGorry, P. D., & Lambert, M. (2007). Pretreatment, baseline, and outcome differences between early-onset and adult-onset psychosis in an epidemiological cohort of 636 first-episode patients. *Schizophrenia Research, 95*(1-3), 1–8. doi:10.1016/j.schres.2007.06.004 PMID:17628441

Schimmelpfennig, J., Topczewski, J., Zajkowski, W., & Jankowiak-Siuda, K. (2023). The role of the salience network in cognitive and affective deficits. *Frontiers in Human Neuroscience, 17*, 1133367. doi:10.3389/fnhum.2023.1133367 PMID:37020493

Schneider, M., Debbané, M., Bassett, A. S., Chow, E. W. C., Fung, W. L. A., van den Bree, M. B. M., Owen, M., Murphy, K. C., Niarchou, M., Kates, W. R., Antshel, K. M., Fremont, W., McDonald-McGinn, D. M., Gur, R. E., Zackai, E. H., Vorstman, J., Duijff, S. N., Klaassen, P. W. J., Swillen, A., ... Eliez, S. (2014). Psychiatric Disorders From Childhood to Adulthood in 22q11.2 Deletion Syndrome: Results From the International Consortium on Brain and Behavior in 22q11.2 Deletion Syndrome. *The American Journal of Psychiatry, 171*(6), 627–639. doi:10.1176/appi.ajp.2013.13070864 PMID:24577245

Schoeler, T., Monk, A., Sami, M. B., Klamerus, E., Foglia, E., Brown, R., Camuri, G., Altamura, A. C., Murray, R., & Bhattacharyya, S. (2016). Continued versus discontinued cannabis use in patients with psychosis: A systematic review and meta-analysis. *The Lancet. Psychiatry, 3*(3), 215–225. doi:10.1016/S2215-0366(15)00363-6 PMID:26777297

Seeley, W. W., Menon, V., Schatzberg, A. F., Keller, J., Glover, G. H., Kenna, H., Reiss, A. L., & Greicius, M. D. (2007). Dissociable intrinsic connectivity networks for salience processing and executive control. *The Journal of Neuroscience : The Official Journal of the Society for Neuroscience*, *27*(9), 2349–2356. doi:10.1523/JNEUROSCI.5587-06.2007 PMID:17329432

Sheffield, J. M., Karcher, N. R., & Barch, D. M. (2018). Cognitive Deficits in Psychotic Disorders: A Lifespan Perspective. *Neuropsychology Review*, *28*(4), 509–533. doi:10.1007/s11065-018-9388-2 PMID:30343458

Sönmez, N., Romm, K. L., Østefjells, T., Grande, M., Jensen, L. H., Hummelen, B., Tesli, M., Melle, I., & Røssberg, J. I. (2020). Cognitive behavior therapy in early psychosis with a focus on depression and low self-esteem: A randomized controlled trial. *Comprehensive Psychiatry*, *97*, 152157. doi:10.1016/j.comppsych.2019.152157 PMID:31935529

Stanton, K. J., Denietolis, B., Goodwin, B. J., & Dvir, Y. (2020). Childhood Trauma and Psychosis: An Updated Review. *Child and Adolescent Psychiatric Clinics of North America*, *29*(1), 115–129. doi:10.1016/j.chc.2019.08.004 PMID:31708041

Steele, P., Cheng, N., Phillips, L. J., Bryce, S., Alvarez-Jimenez, M., & Allott, K. (2021). Cognitive strengths in first episode psychosis: A thematic analysis of clinicians' perspectives. *BMC Psychiatry*, *21*(1), 612. doi:10.1186/s12888-021-03627-y PMID:34876068

Stilo, S. A., & Murray, R. M. (2010). The epidemology of schizophrenia: Replacing dogma with knowledge. *Dialogues in Clinical Neuroscience*, *12*(3), 305–315. doi:10.31887/DCNS.2010.12.3/sstilo PMID:20954427

Sullivan, P. F., Kendler, K. S., & Neale, M. C. (2003). Schizophrenia as a Complex Trait: Evidence From a Meta-analysis of Twin Studies. *Archives of General Psychiatry*, *60*(12), 1187. doi:10.1001/archpsyc.60.12.1187 PMID:14662550

Thompson, A. D., Nelson, B., Yuen, H. P., Lin, A., Amminger, G. P., McGorry, P. D., Wood, S. J., & Yung, A. R. (2014). Sexual Trauma Increases the Risk of Developing Psychosis in an Ultra High-Risk "Prodromal" Population. *Schizophrenia Bulletin*, *40*(3), 697–706. doi:10.1093/schbul/sbt032 PMID:23455040

Trovão, L. D. O., Alves, G. S., Carrilho, C. G., Ricci, T. G., Bittencourt, L., Alves, C., Brito, N. C., Nardi, A. E., Malaspina, D., & Veras, A. B. (2022). Early trauma and schizophrenia onset: Preliminary results of an outpatient cohort in Brazil. *Trends in Psychiatry and Psychotherapy*. Advance online publication. doi:10.47626/2237-6089-2020-0024 PMID:33872478

Turkington, D., Kingdon, D., & Weiden, P. J. (2006). Cognitive Behavior Therapy for Schizophrenia. *The American Journal of Psychiatry*, *163*(3), 365–373. doi:10.1176/appi.ajp.163.3.365 PMID:16513854

Uddin, L. Q. (2015). Salience processing and insular cortical function and dysfunction. *Nature Reviews. Neuroscience*, *16*(1), 55–61. doi:10.1038/nrn3857 PMID:25406711

van Os, J., Rutten, B. P., & Poulton, R. (2008). Gene–Environment Interactions in Schizophrenia: Review of Epidemiological Findings and Future Directions. *Schizophrenia Bulletin*, *34*(6), 1066–1082. doi:10.1093/schbul/sbn117 PMID:18791076

Vogel, M., Meier, J., Grönke, S., Waage, M., Schneider, W., Freyberger, H. J., & Klauer, T. (2011). Differential effects of childhood abuse and neglect: Mediation by posttraumatic distress in neurotic disorder and negative symptoms in schizophrenia? *Psychiatry Research*, *189*(1), 121–127. doi:10.1016/j.psychres.2011.01.008 PMID:21295351

Zanelli, J., Mollon, J., Sandin, S., Morgan, C., Dazzan, P., Pilecka, I., Reis Marques, T., David, A. S., Morgan, K., Fearon, P., Doody, G. A., Jones, P. B., Murray, R. M., & Reichenberg, A. (2019). Cognitive Change in Schizophrenia and Other Psychoses in the Decade Following the First Episode. *The American Journal of Psychiatry*, *176*(10), 811–819. doi:10.1176/appi.ajp.2019.18091088 PMID:31256609

Chapter 7
Attention-Deficit/Hyperactivity Disorder in Older Adults:
A Narrative Review and Clinical Insights

Felippe Mendonca
D'Or Institute for Research and Education, Brazil

Paulo Mattos
D'Or Institute for Research and Education, Brazil

Felipe Kenji Sudo
https://orcid.org/0000-0003-4738-1368
D'Or Institute for Research and Education, Brazil

ABSTRACT

Although it has now been accepted that most individuals with attention-deficit/hyperactivity disorder (ADHD) remain symptomatic in adulthood, the idea that the condition may persist in those aged 60+ has only recently garnered consideration from clinicians and researchers. Consequently, little is known about how the condition manifests in late life, and diagnosis is challenging in those in this age group. This review provides a comprehensive examination of the current state of knowledge on ADHD in old age, shedding light on its epidemiology, diagnostic challenges, clinical manifestations, comorbidities, and treatment options.

DOI: 10.4018/979-8-3693-0851-6.ch007

SUMMARY

ADHD is a chronic disorder with onset in childhood. Current epidemiological data suggested that symptoms may persist throughout the lifespan until old age. However, awareness of the condition affecting this age group is low among both the general population and the health professionals. In this chapter, we aimed to address the main challenges to the diagnosis and management of the disorder in older patients, including some of the particularities regarding its clinical features and treatment strategies in late life. In addition, we sought to summarize the evidence on these topics to provide recommendations for clinicians when caring for older patients with ADHD.

1. Attention-Deficit/Hyperactivity Disorder (ADHD) in Older Age Is a Burdensome and Underdiagnosed Condition

While Attention-Deficit/Hyperactivity Disorder (ADHD) may remit in childhood or adolescence, many adults may present either the full-blown disorder or subthreshold symptoms throughout their lifespan (J. J. S. Kooij, Bijlenga, Salerno, et al., 2019). If unrecognized and untreated until old age, it may impose a life-long struggle on individuals due to impairment in several areas, including low academic and career attainment, reduced income, relationship problems, and comorbid psychiatric disorders (J. J. S. Kooij, Bijlenga, & Michielsen, 2019).

Alternatively, ADHD individuals with previous partial remission may reexperience functional impairment in late adulthood. Some changing conditions in old age, such as widowing or becoming a caregiver, demand an array of organizational skills to care for oneself or for others, which heavily rely on executive functioning. These shifts in the dynamics of the environment may unveil subthreshold ADHD symptoms, especially if the subjects' difficulties had been securely managed by others throughout their adult lives (Eakin et al., 2004). As a result, those individuals may be susceptible of feeling overburdened or having troubles for managing everyday life tasks (Michielsen et al., 2018). Concurrently, aging-related cognitive changes may defy the maintenance of long-established adjustment and coping strategies, leading to disability. Unawareness of the diagnosis may deceitfully suggest that these issues are attributed to neurodegenerative disorders, instead of an exacerbation of previous ADHD cognitive features relative to normal aging (Mendonca et al., 2021).

Despite sharing many common features with the condition in childhood and adolescence, ADHD in elderly patients present unique characteristics that may pose challenges to healthcare providers. Given this scenario, the present chapter aims to summarize the main clinical, diagnostic, prognostic, and management aspects of ADHD in old age.

2. Attention-Deficit/Hyperactivity Disorder (ADHD) Is Prevalent in Younger Older Adults

Few studies have investigated the prevalence of the condition in older population (subjects aged 60 or older (United Nations & Human Rights Office of the high commisioner, 2021)). The Longitudinal Aging Study Amsterdam (LASA) estimated that 2.8% (95% CI 0.86–4.64) of individuals over 60 years old attended full DSM-IV-TR diagnostic criteria for ADHD, with higher rates observed in younger elderly adults (60-70 years, 4.0%) than in older groups (71-94 years, 1.1%) (Michielsen et al., 2012). In fact, limited evidence suggested that ADHD symptoms persist but decrease with age (Das et al., 2014), while other data indicated that the severity of symptoms remain stable throughout life (Müller et al., 2023). When the occurrence of subsyndromal ADHD was assessed, prevalence was estimated at 4.2% (95% CI 2.05–6.39) of overall older population, 5.8% of younger elderly subjects, and 2.1% of older elderly adults (Michielsen et al., 2012). The effect of sex on the estimates was not tested in the study due to small sample size (Michielsen et al., 2012). In addition, a Swedish study indicated that 3.3% of individuals aged 65-80 years reported current and childhood ADHD symptoms (Guldberg-Kjär et al., 2013), whereas 2.2% of an Australian sample between 68-74 years referred clinically significative ADHD features (Das et al., 2014). A recent meta-analysis showed that prevalence of ADHD in elderly populations differed largely according to classification criteria: 2.18% (95% CI = 1.51, 3.16) based on validated scales used in research, 0.23% (95% CI = 0.12, 0.43) based on full ADHD diagnostic criteria, and 0.09% (95% CI = 0.06, 0.15) relying on related ADHD treatment (Dobrosavljevic et al., 2020). Given these conflicting results, defining the appropriate diagnostic method is crucial to allow further advances in the knowledge of the disorder in this age group. This issue will be discussed in the following section.

3. Diagnosing ADHD in Older Age: A Challenging but Feasible Task

People with ADHD may be diagnosed in different times of life, or not at all. Regardless of the age of diagnosis, some of the benefits of acknowledging the condition include understanding of own difficulties, improving self-knowledge, reducing self-blame, and engaging in assertive strategies to cope with functional problems (Fleischmann & Fleischmann, 2012; Hansson Halleröd et al., 2015; Nyström et al., 2020).

Diagnosis of ADHD is based on a systematic assessment of lifetime symptoms and impairments in different domains (J. J. S. Kooij, Bijlenga, Salerno, et al., 2019). For this purpose, both self-report and collateral information from family members are relevant for the identification of core clinical features of the condition (Table

1) (American Psychiatric Association, 2013; J. J. S. Kooij, Bijlenga, Salerno, et al., 2019; World Health Organization, 2018). Remarkably, screening for ADHD should be performed in all adults presenting chronic complaints of inattentive or impulsive behaviors, as well as those with emotional instability (J. J. S. Kooij, Bijlenga, Salerno, et al., 2019).

Table 1. Official diagnostic criteria for ADHD in adults

Domains	DSM-5	ICD-11
Inattention	*At least 5 or more of the following:* (1) Often fails to give close attention to details or makes careless mistakes; (2) often has trouble holding attention on tasks; (3) often does not seem to listen when spoken to directly; (4) often does not follow through on instructions and fails to finish duties in the workplace; (5) often has trouble organizing tasks and activities; (6) often avoids, dislikes, or is reluctant to do tasks that require mental effort over a long period of time; (7) often loses things necessary for tasks and activities; (8) is often easily distracted; (9) is often forgetful in daily activities.	*Several symptoms of following 3 clusters*: (1) Difficulty sustaining attention to tasks that do not provide a high level of stimulation or reward or require sustained mental effort; lacking attention to detail; making careless mistakes in school or work assignments; not completing tasks; (2) easily distracted by extraneous stimuli or thoughts not related to the task at hand; often does not seem to listen when spoken to directly; frequently appears to be daydreaming or to have mind elsewhere; (3) loses things; is forgetful in daily activities; has difficulty remembering to complete upcoming daily tasks or activities; difficulty planning, managing and organizing schoolwork, tasks and other activities.
Hyperactivity / impulsivity	*At least 5 or more of the following*: (1) Often fidgets with or taps hands or feet, or squirms in seat; (2) often leaves seat in situations when remaining seated is expected; (3) often feels restless; (4) often unable to play or take part in leisure activities quietly; (5) Is often "on the go" acting as if "driven by a motor"; (6) often talks excessively; (7) often blurts out an answer before a question has been completed; (8) often has trouble waiting their turn; (9) often interrupts or intrudes on others.	*Several symptoms from the following 4 clusters:* (1) Excessive motor activity; (2) difficulty engaging in activities quietly; talks too much; (3) blurts out comments at work; difficulty waiting turn in conversation, games, or activities; interrupts or intrudes on others conversations or games; (3) a tendency to act in response to immediate stimuli without deliberation or consideration of risks and consequences.
Age of onset	Prior to age 12.	Prior to age 12.
Functional status	Clear evidence that the symptoms interfere with, or reduce the quality of, social, school, or work functioning.	Symptoms are persistent, and sufficiently severe that they have a direct negative impact on academic, occupational, or social functioning.
Exclusion criteria	Symptoms are not better explained by another mental disorder (such as a mood disorder, anxiety disorder, dissociative disorder, or a personality disorder). The symptoms do not happen only during the course of schizophrenia or another psychotic disorder.	Symptoms are not better accounted for by another mental disorder (e.g., an Anxiety or Fear-Related Disorder, a Neurocognitive Disorder such as Delirium). Symptoms are not because of a substance (e.g., cocaine) or medication (e.g., bronchodilators, thyroid replacement medication) on the central nervous system, including and withdrawal effects, and are not due to a Disease of the Nervous System.

Although it has now been accepted that most individuals with ADHD remain symptomatic (whether below diagnostic threshold or not) in adulthood, the idea that the condition may persist in those aged 60+ has only recently gardened consideration from clinicians and researchers (J. J. S. Kooij, Bijlenga, Salerno, et al., 2019; Sharma et al., 2021). Consequently, little is known about how it manifests in late-life, and current notion about the theme is mostly extrapolation from evidence obtained in young adults (J. J. S. Kooij, Bijlenga, Salerno, et al., 2019). In fact, several shortcomings exist when it comes to identifying the disorder in elderly population. Firstly, there are no validated scales or diagnostic interviews for the assessment of people within this demographics (J. J. S. Kooij, Bijlenga, & Michielsen, 2019). Specialists have recommended the use of instruments formulated for general adults, including the DIVA 2.0 and the Conners' Adult ADHD Diagnostic Interview for DSM-IV (CAADID) (J. J. S. Kooij, Bijlenga, & Michielsen, 2019), but further studies are needed to investigate the psychometric properties and the limitations of these tools for the characterization of ADHD in late-life.

Furthermore, since ADHD is a relatively new category, initially described in the DSM-II (1968) as "hyperkinetic reaction of childhood", most affected older subjects have not been diagnosed during youth (Sharma et al., 2021). Although severity of typical ADHD symptoms may remain stable throughout life (Müller et al., 2023), many older individuals may seek for mental health care due to concurrent psychiatric conditions, including mood disorders, substance-related disorders, sleep disorders or relationship problems (J. J. S. Kooij et al., 2016). Alternatively, general practitioners may be approached by these patients because of comorbid general medical conditions, especially those related to unhealthy lifestyles, such as obesity and uncontrolled type 2 diabetes mellitus (J. J. S. Kooij et al., 2016). In addition, impaired self-knowledge and deficits in autobiographical memory may contribute to poor awareness of the disorder in this age group (Sharma et al., 2021). As a result, typical complaints of inattention or hyperactivity-impulsivity behavior may not be spontaneously communicated by elderly patients during the clinical interview (Sharma et al., 2021). On the other hand, given that ADHD runs in families, it is possible that some older individuals may recognize their own symptoms after their children or grandchildren are diagnosed and treated for the disorder (J. J. S. Kooij, Bijlenga, & Michielsen, 2019).

Conversely, retrospectively establishing the onset of the condition before age of 12 may be difficult for several reasons. Differently from younger patients, corroborating evidence about the clinical manifestations obtained from family members or school records may not be accessible with older patients (Sharma et al., 2021). Spouses and children of these individuals may provide information about the enduring aspect of the symptoms, but details about childhood-specific issues are rarely available (Sharma et al., 2021). As mentioned, relying solely on the patients' self-report of early ADHD features could lead to biased conclusions (Sharma et al., 2021). To solve this conundrum, authors have suggested that documenting the chronicity of the symptoms, without necessarily focusing on the onset prior to 12 years old, could be sufficient to distinguish ADHD from other prevalent condition in older patients, especially from neurodegenerative disorders (Sharma et al., 2021).

However, it is possible that individuals perceive ADHD symptoms only later in life. It has been well established that the disorder is underdiagnosed in highly intelligent subjects because they use more compensatory strategies (J. J. S. Kooij, Bijlenga, Salerno, et al., 2019). Coping abilities may mask functional impairments and behavioral problems throughout the person's life (J. J. S. Kooij, Bijlenga, Salerno, et al., 2019). Moreover, some cases may find jobs that are suited for their symptom profile, whereas others may count with spouses or partners to assist them with everyday life tasks. Some may excel in certain aspects of life, while still showing impairments in others, such as paying bills, maintaining stable relationships or driving vehicles (J. J. S. Kooij, Bijlenga, Salerno, et al., 2019). Awareness of deficits may arise only when demands exceed capacity, which may only occur when circumstances change in old age - for instance, widowing, getting divorced, or getting retired (Michielsen et al., 2018). With these situations, subjects may unexpectedly be required to conduct activities that are highly dependent on executive function and organizational skills, which may prompt the emergence of functional disturbances. Correlation of onset of symptoms with shifts in the life of the patient should be investigated to allow inferences about environmental issues affecting compensatory strategies.

Also, normal aging is related to decline in an array of cognitive abilities, including executive function and working memory (Ferguson et al., 2021). In ADHD, these cognitive changes may overlap with preexisting difficulties, exacerbating functional deficits (Callahan et al., 2017). To both the subject and their families, acknowledging those seemingly "novel" cognitive and functional abnormalities may erroneously lead to the conclusion that these symptoms are associated with neurodegenerative disorders, such as Alzheimer's Disease (AD). Actually, studies indicated that late-life ADHD may mimic neuropsychological profiles of Mild Cognitive Impairment (MCI) due to AD (Callahan et al., 2017; Mendonca et al., 2021). These findings highlight the relevance of a comprehensive clinical interview for the differentiation between previous cognitive problems and pathological cognitive decline in old age.

As a matter of fact, suffering from ADHD might increase the risk of neurocognitive disorders in older persons. Recognized risk factors for dementia are remarkably coincidental with clinical outcomes associated with ADHD, such as lower educational level, high prevalence of traumatic brain injury, obesity, smoking, alcohol-related disorders, depression, social isolation, and physical inactivity(Livingston et al., 2020; Mendonca et al., 2021). Notably, longitudinal data found 1.62 to 3.41-fold increased odds of dementia, and 2.54 higher risk of MCI, associated with lifetime ADHD after adjustments for potential confounders (Dobrosavljevic et al., 2022; Levine et al., 2023; Tzeng et al., 2017). Clinicians ought to pay attention to steep cognitive, behavioral, and functional changes in those patients, to implement timely diagnostic and management measures.

Table 2. Summary of recommendations for the diagnosis of ADHD in older adults

1. Screening for ADHD should be performed in all adults presenting chronic complaints of inattentive or impulsive behaviors, as well as those with emotional instability.
2. Given that the condition runs in families, diagnosis should be suspected in relatives of patients with ADHD.
3. Diagnostic assessment is based on self and ancillary report of chronic ADHD symptoms and functional impairments.
4. Documenting the chronicity of symptoms, without focusing specifically on the onset before 12 years old, is sufficient for the diagnosis of ADHD in late life.
5. Adults with ADHD may only perceive symptoms late in life, when demands exceed capacity. Correlation of onset of symptoms with changes in life requiring executive and organizational skills should be investigated.
6. Age-related cognitive changes may overlap with preexisting cognitive difficulties in ADHD, leading to functional impairment. These seemingly "new" cognitive and functional could be confounded with neurodegenerative disorders. A comprehensive clinical interview may allow the differentiation between previous cognitive problems and pathological patterns of cognitive decline.
7. Neuropsychological profile of ADHD might mimic typical cognitive findings of Mild Cognitive Impairment due to Alzheimer's Disease.
8. Established risk factors for dementia coincide with clinical outcomes associated with ADHD. Clinicians ought to pay attention to steep cognitive, behavioral, and functional changes in those patients, so as to implement timely diagnostic and management measures.

4. Social impact and Quality of Life

Limited data suggested that cumulative consequences of ADHD in older individuals encompass enduring problems in different areas of life including work, relationships, personal finances, and health (Brod, Pohlman, et al., 2012; Brod, Schmitt, et al., 2012; Guldberg-Kjär & Johansson, 2009). The burden of the condition could be comparable to that of younger adults according to some authors(Philipp-Wiegmann et al., 2016), while qualitative data implied that it may get worse with age (Ginsberg et al., 2014).

Older subjects with ADHD were more likely to be divorced or never married, having fewer family members in their network, and reporting emotional loneliness than elderly individuals without ADHD (Michielsen et al., 2015). Quality of life was significantly lower in older ADHD participants compared to controls (Lensing et al., 2015; Thorell et al., 2019), but severity of this finding was similar to that of younger ADHD subjects (Thorell et al., 2019). Moreover, mounting negative impacts of the condition in elderly ADHD persons might be associated with perceived low financial stability, educational underachievement, and poor job performance (Brod, Schmitt, et al., 2012).

5. Before Starting Treatment: Assessing Comorbidities

Comorbidity with psychiatric disorders is the rule rather than the exception for ADHD patients, with lifetime rates between 60-80% (Fayyad et al., 2017). In fact, having three or more mental disorders might be associated with 10-fold higher odds of suffering from ADHD (Fayyad et al., 2017). In addition, common general medical comorbidities include metabolic diseases, asthma, migraine and autoimmune disorders (J. J. S. Kooij, Bijlenga, & Michielsen, 2019). Establishing all the comorbidities presented by the patient is important to correctly decide which interventions should be adopted and prioritized (J. J. S. Kooij, Bijlenga, Salerno, et al., 2019).

When managing older ADHD patients with multimorbidity, it is mandatory to carefully verify the risk of drug-drug and drug-disease interactions. One study indicated that polypharmacy, defined as using 5 or more medication classes, was nearly two times more common in older ADHD subjects than in general elderly population (Zhang et al., 2021). Furthermore, psychostimulants, such as methylphenidate, have been depicted as the psychotropic drug category most frequently associated with cardiocirculatory deaths over the last two decades (0.029–0.332 cases per 100,000 circulatory deaths) (Kelly & Vuolo, 2023). In the next section, relevant issues related to the management of ADHD in elderly cases will be addressed.

6. Treatment of ADHD in Older Subjects: Particularities and Recommendations

Treatment of ADHD in older patients should follow a multimodal approach, encompassing psychoeducation, pharmacotherapy, and psychotherapy (J. J. S. Kooij, Bijlenga, & Michielsen, 2019; J. J. S. Kooij, Bijlenga, Salerno, et al., 2019). As previously mentioned, all comorbid disorders should be established and the most impairing conditions should be prioritized prior to initiating interventions (J. J. S. Kooij, Bijlenga, Salerno, et al., 2019). As a rule, typical ADHD symptoms are chronic, and interventions can be implemented only when major risks are controlled. Specifically, eating disorders, substance-related disorders, severe mood disorders, and psychotic disorders ought to be treated first, while inattention and hyperactive/impulsive behavior could be managed solely after those conditions are remitted (J. J. S. Kooij, Bijlenga, & Michielsen, 2019; J. J. S. Kooij, Bijlenga, Salerno, et al., 2019).

Somatic comorbidities, namely cardiovascular diseases, should be considered before initiating pharmacologic strategies, given that they might worsen blood pressure and arrythmia (J. J. S. Kooij, Bijlenga, & Michielsen, 2019; J. J. S. Kooij,

Bijlenga, Salerno, et al., 2019). Particularly, psychostimulants have been associated with higher risk of death due to vascular complications (Kelly & Vuolo, 2023). Hence, all cardiovascular comorbidities should be treated before the start of specific drug treatment for ADHD, and stimulant medications should be avoided until risks are controlled (J. J. S. Kooij et al., 2016). Moreover, food refusal and odds of malnutrition ought to be assessed in elderly patients because of the potential negative effects of psychostimulants on body weight and appetite (J. J. S. Kooij, Bijlenga, Salerno, et al., 2019).

In addition, sleep problems, especially insomnia and drowsiness, are common in late life (Miner & Kryger, 2017). ADHD people suffer more frequently from delayed sleep onset and chronic insomnia than those without the condition (J. S. Kooij & Bijlenga, 2013). Finally, longer sleep latency and poorer sleep efficiency have been recognized as adverse effects of stimulant treatment (Snitselaar et al., 2017). Therefore, addressing sleep issues might be inevitable when treating older patients with ADHD. Psychoeducational approaches may improve sleep in those cases due to effects on daytime structure and maintenance of regular physical activity (J. J. S. Kooij, Bijlenga, Salerno, et al., 2019).

Psychoeducation. Improving the subjects' insight about the disorder and its outcomes might help with the treatment acceptance, but also with the quality of their relationships and psychological well-being (J. J. S. Kooij, Bijlenga, & Michielsen, 2019; J. J. S. Kooij, Bijlenga, Salerno, et al., 2019). Other recommendations may contribute to reducing functional impairments. For example, establishing routines and structured daily schedules through the application of calendars, reminders, and task lists could be useful for patients to stay organized and meet deadlines, compensating for executive function deficits (Solanto, 2011). Lifestyle changes, such as regular exercise, adequate sleep, and a balanced diet, can complement pharmacological and psychosocial interventions (Den Heijer et al., 2017).

Pharmacotherapy. Although no randomized controlled study has focused on the effects of psychostimulants (methylphenidate and lisdexamfetamine, for example) in older ADHD population, it is unlikely that these medications are less effective in those subjects compared to younger subjects. Therefore, this class of medications is the first choice for ADHD within this demographics, while atomoxetine and bupropion are second and third choices, respectively (J. J. S. Kooij, Bijlenga, & Michielsen, 2019). The main side effects of psychostimulants are dose-related elevations in blood pressure and heart rate, and reduced sleep and appetite (J. J. S. Kooij, Bijlenga, & Michielsen, 2019; J. J. S. Kooij, Bijlenga, Salerno, et al., 2019). Low dose betablockers may be introduced to diminish persistent elevated heart rates (J. J. S. Kooij, Bijlenga, & Michielsen, 2019). Relative contraindications of the use of these drugs are epilepsy, arrythmia, glaucoma, hyperthyroidism, severe hypertension, and anxiety – these conditions ought to be treated before considering stimulant treatment for ADHD (J. J. S. Kooij, Bijlenga, & Michielsen, 2019). In cases with risk of psychosis, atomoxetine or bupropion should be preferred (J. J. S. Kooij, Bijlenga, & Michielsen, 2019). While complications are rarely severe, closely monitoring adverse events in this age group, before and during treatment, is advised (J. J. S. Kooij, Bijlenga, & Michielsen, 2019; J. J. S. Kooij, Bijlenga, Salerno, et al., 2019). Additionally, lower dosages might be sufficient to reduce ADHD symptoms in patients with 60+ years. Starting low and slowly increasing dosages based on clinical evaluation of effectivity and adverse effects is recommended (J. J. S. Kooij, Bijlenga, & Michielsen, 2019).

Psychotherapy and support. Although pharmacological treatment is effective, some patients with ADHD might continue to suffer from significant symptoms. Due to lack of systematic empirical evidence of non-pharmacological interventions to the condition in this age group, it is recommended to offer support in the same manner as in younger adults, and to personalize the approaches to specific concerns of the patients (Goodman et al., 2016). Cognitive-Behavioral Therapy (CBT) might reduce emotional dysregulation, anxiety and depression (J. J. S. Kooij, Bijlenga, Salerno, et al., 2019). Older patients might be responsive to CBT group interventions on inattention (J. J. S. Kooij, Bijlenga, & Michielsen, 2019). In addition, since depressive symptoms and loneliness are particularly correlated in older ADHD subjects (Michielsen et al., 2015), strategies aiming to facilitate social participation and to improve familial connections might effectively ameliorate mood abnormalities in this group (Stickley et al., 2017). Coaching or counselling, offered individually or in group, may allow advances in the treatment of clinical features commonly associated with the disorder, such as diagnosis denial, low self-esteem, relationship issues, and depressive symptoms (J. J. S. Kooij, Bijlenga, & Michielsen, 2019). Other interventions, such as aerobic exercise and hatha yoga, may reduce impulsivity in adults with ADHD (Dinu et al., 2023).

Table 3. Summary of recommendations for the treatment of ADHD in older adults

General recommendations	
1.	Treatment of ADHD in older patients should follow a multimodal approach, encompassing psychoeducation, pharmacotherapy, and psychotherapy.
2.	Severe comorbid conditions, such as eating disorders, psychotic disorders, and substance-related disorders, should be treated before managing attentional or hyperactive/impulsive features.
3.	Severe uncontrolled cardiovascular diseases and low body weight should be considered before initiating pharmacologic strategies. The use of psychostimulants may be precluded in some cases.
Psychoeducation	
4.	Improving the patients' insight about ADHD and its outcomes may improve treatment acceptance, quality of relationships and psychological well-being.
5.	Establishing routines, using calendars, reminders, and task lists may improve functional difficulties in those patients. Interventions targeting lifestyle changes may help with ADHD symptoms and comorbidities.
Pharmacotherapy	
6.	Psychostimulants are the first choice of treatment for ADHD in late life.
7.	Low dose betablockers may alleviate persistent elevated heart rates associated with psychostimulants.
8.	Lower dosages might be sufficient to reduce ADHD symptoms in older patients. Starting low, slowly increasing dosages, and closely monitoring adverse effects is recommended.
Psychotherapy and support	
9.	It is recommended to offer support in the same manner as in younger adults, and to personalize the approaches to specific concerns of the older patients.
10.	Individual or group Cognitive-Behavioral Therapy (CBT) might reduce mood symptoms and inattention.
11.	Interventions aiming to facilitate social participation might improve depressive symptoms in older ADHD people.

Accessibility of healthcare. Organization of care for a lifespan disorder requires cooperation between child, adult, and geriatric professionals and/or institutions. Given that expertise in diagnosing and treating ADHD is far more advanced in child than in adult or geriatric psychiatry, transitions between specialized healthcare services should be articulated to avoid discontinuation of support (J. J. S. Kooij, Bijlenga, & Michielsen, 2019). Additionally, the growing recognition of ADHD as a lifespan disorder has been accompanied by the need of training of geriatric practitioners, as well as primary care professionals, to identify and manage the condition (J. J. S.

Kooij et al., 2016). The development of specific guidelines is necessary to enhance awareness of the disorder among clinicians, reduce misdiagnosis, standardize practices, and provide direction to policies aimed at promoting health care to older subjects with ADHD.

7. CONCLUSIONS AND FUTURE DIRECTIONS

ADHD is a chronic disorder with onset in childhood. Current epidemiological data indicated that persistence of symptoms into older age is not rare, especially when subsyndromal cases are considered. Despite that, underdiagnosis, misdiagnosis and undertreatment is the rule rather than the exception in this age group. One of the contributing factors hindering access to adequate healthcare is the insufficient medical training to recognize and treat the condition among geriatric practitioners. Summing to that, the lack of diagnostic criteria addressing the particularities of the disorder in this age group to allow inclusion of those with undocumented chronic symptoms, and the absence of validated tools to assess the condition in older patients are significant barriers to the identification and monitoring of ADHD in this population. Finally, the scarcity of evidence on the effectiveness and safety of treatment strategies for older ADHD may impact on the clinicians' confidence in intervening in the disorder. Future studies ought to address these shortcomings through randomized controlled trials on the effects of pharmacological and non-pharmacological therapies to older patients and validation of instruments designed to assess the condition in those over 60 years old.

REFERENCES

American Psychiatric Association. (2013). Diagnostic and statistical manual of mental disorders: DSM-5 (5th ed). American Psychiatric Association.

Brod, M., Pohlman, B., Lasser, R., & Hodgkins, P. (2012). Comparison of the burden of illness for adults with ADHD across seven countries: A qualitative study. *Health and Quality of Life Outcomes*, *10*(1), 47. doi:10.1186/1477-7525-10-47 PMID:22583562

Brod, M., Schmitt, E., Goodwin, M., Hodgkins, P., & Niebler, G. (2012). ADHD burden of illness in older adults: A life course perspective. *Quality of Life Research: An International Journal of Quality of Life Aspects of Treatment, Care and Rehabilitation*, *21*(5), 795–799. doi:10.1007/s11136-011-9981-9 PMID:21805205

Callahan, B. L., Bierstone, D., Stuss, D. T., & Black, S. E. (2017). Adult ADHD: Risk Factor for Dementia or Phenotypic Mimic? *Frontiers in Aging Neuroscience*, *9*, 260. doi:10.3389/fnagi.2017.00260 PMID:28824421

Das, D., Cherbuin, N., Easteal, S., & Anstey, K. J. (2014). Attention Deficit/ Hyperactivity Disorder Symptoms and Cognitive Abilities in the Late-Life Cohort of the PATH through Life Study. *PLoS One*, *9*(1), e86552. doi:10.1371/journal.pone.0086552 PMID:24489743

Den Heijer, A. E., Groen, Y., Tucha, L., Fuermaier, A. B. M., Koerts, J., Lange, K. W., Thome, J., & Tucha, O. (2017). Sweat it out? The effects of physical exercise on cognition and behavior in children and adults with ADHD: A systematic literature review. *Journal of Neural Transmission (Vienna, Austria)*, *124*(S1), 3–26. doi:10.1007/s00702-016-1593-7 PMID:27400928

Dinu, L. M., Singh, S. N., Baker, N. S., Georgescu, A. L., Singer, B. F., Overton, P. G., & Dommett, E. J. (2023). The Effects of Different Exercise Approaches on Attention Deficit Hyperactivity Disorder in Adults: A Randomised Controlled Trial. *Behavioral Sciences (Basel, Switzerland)*, *13*(2), 129. doi:10.3390/bs13020129 PMID:36829357

Dobrosavljevic, M., Solares, C., Cortese, S., Andershed, H., & Larsson, H. (2020). Prevalence of attention-deficit/hyperactivity disorder in older adults: A systematic review and meta-analysis. *Neuroscience and Biobehavioral Reviews*, *118*, 282–289. doi:10.1016/j.neubiorev.2020.07.042 PMID:32798966

Dobrosavljevic, M., Zhang, L., Garcia-Argibay, M., Du Rietz, E., Andershed, H., Chang, Z., Faraone, S., & Larsson, H. (2022). Attention-deficit/hyperactivity disorder as a risk factor for dementia and mild cognitive impairment: A population-based register study. *European Psychiatry*, *65*(1), e3. doi:10.1192/j.eurpsy.2021.2261 PMID:34924079

Eakin, L., Minde, K., Hechtman, L., Ochs, E., Krane, E., Bouffard, R., Greenfield, B., & Looper, K. (2004). The marital and family functioning of adults with ADHD and their spouses. *Journal of Attention Disorders*, *8*(1), 1–10. doi:10.1177/108705470400800101 PMID:15669597

Fayyad, J., Sampson, N. A., Hwang, I., Adamowski, T., Aguilar-Gaxiola, S., Al-Hamzawi, A., Andrade, L. H. S. G., Borges, G., De Girolamo, G., Florescu, S., Gureje, O., Haro, J. M., Hu, C., Karam, E. G., Lee, S., Navarro-Mateu, F., O'Neill, S., Pennell, B.-E., Piazza, M., ... Kessler, R. C. (2017). The descriptive epidemiology of DSM-IV Adult ADHD in the World Health Organization World Mental Health Surveys. *Attention Deficit and Hyperactivity Disorders*, *9*(1), 47–65. doi:10.1007/s12402-016-0208-3 PMID:27866355

Ferguson, H. J., Brunsdon, V. E. A., & Bradford, E. E. F. (2021). The developmental trajectories of executive function from adolescence to old age. *Scientific Reports*, *11*(1), 1382. doi:10.1038/s41598-020-80866-1 PMID:33446798

Fleischmann, A., & Fleischmann, R. H. (2012). Advantages of an ADHD Diagnosis in Adulthood: Evidence From Online Narratives. *Qualitative Health Research*, *22*(11), 1486–1496. doi:10.1177/1049732312457468 PMID:22910591

Ginsberg, Y., Beusterien, K. M., Amos, K., Jousselin, C., & Asherson, P. (2014). The unmet needs of all adults with ADHD are not the same: A focus on Europe. *Expert Review of Neurotherapeutics*, *14*(7), 799–812. doi:10.1586/14737175.2014.926220 PMID:24894408

Goodman, D. W., Mitchell, S., Rhodewalt, L., & Surman, C. B. H. (2016). Clinical Presentation, Diagnosis and Treatment of Attention-Deficit Hyperactivity Disorder (ADHD) in Older Adults: A Review of the Evidence and its Implications for Clinical Care. *Drugs & Aging*, *33*(1), 27–36. doi:10.1007/s40266-015-0327-0 PMID:26659731

Guldberg-Kjär, T., & Johansson, B. (2009). Old people reporting childhood AD/HD symptoms: Retrospectively self-rated AD/HD symptoms in a population-based Swedish sample aged 65–80. *Nordic Journal of Psychiatry*, *63*(5), 375–382. doi:10.1080/08039480902818238 PMID:19308795

Guldberg-Kjär, T., Sehlin, S., & Johansson, B. (2013). ADHD symptoms across the lifespan in a population-based Swedish sample aged 65 to 80. *International Psychogeriatrics*, *25*(4), 667–675. doi:10.1017/S1041610212002050 PMID:23237190

Hansson Halleröd, S. L., Anckarsäter, H., Råstam, M., & Hansson Scherman, M. (2015). Experienced consequences of being diagnosed with ADHD as an adult – a qualitative study. *BMC Psychiatry*, *15*(1), 31. doi:10.1186/s12888-015-0410-4 PMID:25884685

Kelly, B. C., & Vuolo, M. (2023). Trends in Psychotropic Drug–Implicated Cardiovascular Mortality: Patterns in U.S. Mortality, 1999–2020. *American Journal of Preventive Medicine*, *65*(3), 377–384. doi:10.1016/j.amepre.2023.02.016 PMID:36894483

Kooij, J. J. S., Bijlenga, D., & Michielsen, M. (2019). Assessment and Treatment of ADHD in People Over 60. *The ADHD Report*, *27*(4), 1–7, 15–16. doi:10.1521/adhd.2019.27.4.1

Kooij, J. J. S., Bijlenga, D., Salerno, L., Jaeschke, R., Bitter, I., Balázs, J., Thome, J., Dom, G., Kasper, S., Nunes Filipe, C., Stes, S., Mohr, P., Leppämäki, S., Casas, M., Bobes, J., Mccarthy, J. M., Richarte, V., Kjems Philipsen, A., Pehlivanidis, A., ... Asherson, P. (2019). Updated European Consensus Statement on diagnosis and treatment of adult ADHD. *European Psychiatry*, *56*(1), 14–34. doi:10.1016/j.eurpsy.2018.11.001 PMID:30453134

Kooij, J. J. S., Michielsen, M., Kruithof, H., & Bijlenga, D. (2016). ADHD in old age: A review of the literature and proposal for assessment and treatment. *Expert Review of Neurotherapeutics*, *16*(12), 1371–1381. doi:10.1080/14737175.2016.1204914 PMID:27334252

Kooij, J. S., & Bijlenga, D. (2013). The circadian rhythm in adult attention-deficit/hyperactivity disorder: Current state of affairs. *Expert Review of Neurotherapeutics*, *13*(10), 1107–1116. doi:10.1586/14737175.2013.836301 PMID:24117273

Lensing, M. B., Zeiner, P., Sandvik, L., & Opjordsmoen, S. (2015). Quality of Life in Adults Aged 50+ With ADHD. *Journal of Attention Disorders*, *19*(5), 405–413. doi:10.1177/1087054713480035 PMID:23514676

Levine, S. Z., Rotstein, A., Kodesh, A., Sandin, S., Lee, B. K., Weinstein, G., Schnaider Beeri, M., & Reichenberg, A. (2023). Adult Attention-Deficit/Hyperactivity Disorder and the Risk of Dementia. *JAMA Network Open*, *6*(10), e2338088. doi:10.1001/jamanetworkopen.2023.38088 PMID:37847497

Livingston, G., Huntley, J., Sommerlad, A., Ames, D., Ballard, C., Banerjee, S., Brayne, C., Burns, A., Cohen-Mansfield, J., Cooper, C., Costafreda, S. G., Dias, A., Fox, N., Gitlin, L. N., Howard, R., Kales, H. C., Kivimäki, M., Larson, E. B., Ogunniyi, A., ... Mukadam, N. (2020). Dementia prevention, intervention, and care: 2020 report of the Lancet Commission. *Lancet*, *396*(10248), 413–446. doi:10.1016/S0140-6736(20)30367-6 PMID:32738937

Mendonca, F., Sudo, F. K., Santiago-Bravo, G., Oliveira, N., Assuncao, N., Rodrigues, F., Soares, R., Calil, V., Bernardes, G., Erthal, P., Drummond, C., Tovar-Moll, F., & Mattos, P. (2021). Mild Cognitive Impairment or Attention-Deficit/Hyperactivity Disorder in Older Adults? A Cross Sectional Study. *Frontiers in Psychiatry*, *12*, 737357. doi:10.3389/fpsyt.2021.737357 PMID:34616321

Michielsen, M., Comijs, H. C., Aartsen, M. J., Semeijn, E. J., Beekman, A. T. F., Deeg, D. J. H., & Kooij, J. J. S. (2015). The Relationships Between ADHD and Social Functioning and Participation in Older Adults in a Population-Based Study. *Journal of Attention Disorders*, *19*(5), 368–379. doi:10.1177/1087054713515748 PMID:24378286

Michielsen, M., De Kruif, J. Th. C. M., Comijs, H. C., Van Mierlo, S., Semeijn, E. J., Beekman, A. T. F., Deeg, D. J. H., & Kooij, J. J. S. (2018). The Burden of ADHD in Older Adults: A Qualitative Study. *Journal of Attention Disorders*, *22*(6), 591–600. doi:10.1177/1087054715610001 PMID:26515893

Michielsen, M., Semeijn, E., Comijs, H. C., Van De Ven, P., Beekman, A. T. F., Deeg, D. J. H., & Kooij, J. J. S. (2012). Prevalence of attention-deficit hyperactivity disorder in older adults in the Netherlands. *The British Journal of Psychiatry*, *201*(4), 298–305. doi:10.1192/bjp.bp.111.101196 PMID:22878132

Miner, B., & Kryger, M. H. (2017). Sleep in the Aging Population. *Sleep Medicine Clinics*, *12*(1), 31–38. doi:10.1016/j.jsmc.2016.10.008 PMID:28159095

Müller, M., Turner, D., Barra, S., Rösler, M., & Retz, W. (2023). ADHD and associated psychopathology in older adults in a German community sample. *Journal of Neural Transmission (Vienna, Austria)*, *130*(3), 313–323. doi:10.1007/s00702-022-02584-4 PMID:36617378

Nyström, A., Petersson, K., & Janlöv, A.-C. (2020). Being Different but Striving to Seem Normal: The Lived Experiences of People Aged 50+ with ADHD. *Issues in Mental Health Nursing*, *41*(6), 476–485. doi:10.1080/01612840.2019.1695029 PMID:32267788

Philipp-Wiegmann, F., Retz-Junginger, P., Retz, W., & Rösler, M. (2016). The intraindividual impact of ADHD on the transition of adulthood to old age. *European Archives of Psychiatry and Clinical Neuroscience*, *266*(4), 367–371. doi:10.1007/s00406-015-0644-7 PMID:26438010

Sharma, M. J., Lavoie, S., & Callahan, B. L. (2021). A Call for Research on the Validity of the Age-of-Onset Criterion Application in Older Adults Being Evaluated for ADHD: A Review of the Literature in Clinical and Cognitive Psychology. *The American Journal of Geriatric Psychiatry*, *29*(7), 669–678. doi:10.1016/j.jagp.2020.10.016 PMID:33191098

Snitselaar, M. A., Smits, M. G., Van Der Heijden, K. B., & Spijker, J. (2017). Sleep and Circadian Rhythmicity in Adult ADHD and the Effect of Stimulants: A Review of the Current Literature. *Journal of Attention Disorders*, *21*(1), 14–26. doi:10.1177/1087054713479663 PMID:23509113

Solanto, M. V. (2011). *Cognitive-behavioral therapy for adult ADHD: Targeting executive dysfunction*. The Guilford Press.

Stickley, A., Koyanagi, A., Takahashi, H., Ruchkin, V., & Kamio, Y. (2017). Attention-deficit/hyperactivity disorder symptoms and loneliness among adults in the general population. *Research in Developmental Disabilities*, *62*, 115–123. doi:10.1016/j.ridd.2017.01.007 PMID:28131008

Thorell, L. B., Holst, Y., & Sjöwall, D. (2019). Quality of life in older adults with ADHD: Links to ADHD symptom levels and executive functioning deficits. *Nordic Journal of Psychiatry*, *73*(7), 409–416. doi:10.1080/08039488.2019.1646804 PMID:31380715

Tzeng, N.-S., Chung, C.-H., Lin, F.-H., Yeh, C.-B., Huang, S.-Y., Lu, R.-B., Chang, H.-A., Kao, Y.-C., Yeh, H.-W., Chiang, W.-S., Chou, Y.-C., Tsao, C.-H., Wu, Y.-F., & Chien, W.-C. (2017). Risk of Dementia in Adults With ADHD: A Nationwide, Population-Based Cohort Study in Taiwan. *Journal of Attention Disorders*. doi:10.1177/1087054717714057 PMID:28629260

United Nations & Human Rights Office of the High Commisioner. (2021). *Update to the 2012 Analytical Outcome Study on the normative standards in international human rights law in relation to older persons. Working paper prepared by the Office of the High Commissioner for Human Rights*. United Nations. https://www.ohchr.org/en/documents/outcome-documents/ohchr-working-paper-update-2012-analytical-outcome-study-normative

World Health Organization. (2018). *International statistical classification of diseases and related health problems (11th Revision)*. https://icd.who.int/browse11/l-m/en

Zhang, L., Reif, A., Du Rietz, E., Lagerberg, T., Butwicka, A., D'Onofrio, B. M., Johnell, K., Pedersen, N. L., Larsson, H., & Chang, Z. (2021). Comedication and Polypharmacy With ADHD Medications in Adults: A Swedish Nationwide Study. *Journal of Attention Disorders*, *25*(11), 1519–1528. doi:10.1177/1087054720923725 PMID:32478603

Chapter 8
Parkinson's Disease:
Pathophysiology, Diagnosis, and Perspectives

Pedro Renato de Paula Brandão
https://orcid.org/0000-0002-1191-2078
University Hospital of Brasília, Brazil

ABSTRACT

This chapter offers a comprehensive review of Parkinson's disease (PD), spanning from epidemiology to advanced treatment options. It delves into the neuropathology and the evolution of etiological theories, elucidating the genetic mechanisms and clinical subtypes of PD. The text outlines criteria for clinical diagnosis, including essential markers and exclusion factors, while introducing innovative diagnostic frameworks. The chapter is also an integrated guide discussing future research avenues for optimizing patient treatment and outcomes.

INTRODUCTION

First described by James Parkinson in 1817 in his seminal work, "An Essay on the Shaking Palsy," Parkinson's Disease (PD) stands as the most prevalent cause of parkinsonism (Obeso et al., 2017; Parkinson, 2002). PD is a chronic degenerative condition that is increasingly seen as a looming public health concern, especially in light of the aging demographics in emerging nations (Kalia & Lang, 2015; Tysnes & Storstein, 2017). This chapter aims to provide a comprehensive understanding of PD, encompassing its historical background, epidemiological scope, pathological

DOI: 10.4018/979-8-3693-0851-6.ch008

mechanisms, etiological factors, genetic underpinnings, clinical manifestations, and diagnostic criteria.

Historical Background

Although partial accounts can be found in ancient Indian and Chinese texts, James Parkinson's 1817 monograph is considered the first full medical description of the disorder (Parkinson, 2002). The work succinctly delineates key features, as illustrated in the following translated excerpt: *"[...] Involuntary tremulous motion, with lessened muscular power, in parts not in action and even when supported; with a propensity to bend the trunk forwards, and to pass from a walking to a running pace: the senses and intellect being uninjured."* Remarkably, only six cases were described in Parkinson's initial work, three of which were observed on the streets of London and one solely from a distance.

James Parkinson displayed remarkable accuracy and attention to detail. His observations covered a multitude of clinical facets that are still relevant today. Firstly, Parkinson's description of resting tremors, distinguishing them from those occurring during voluntary motion, was noteworthy. He recognized the unilateral onset of rest tremor, a key element in modern clinical diagnostic criteria. Parkinson also noted that rest tremors did not necessarily hinder fine motor acts, illustrating the complex nature of this symptom (Obeso et al., 2017). Moreover, he accurately documented gait abnormalities in PD, including shuffling, reduced step length, festination, and the forward-flexed posture now known as camptocormia.

However, Parkinson's essay did not explicitly mention rigidity or muscle stiffness, a cardinal feature of PD recognized later by Charcot (Li & Le, 2017). Additionally, he misinterpreted bradykinesia as weakness, failing to identify it as a distinct disease characteristic. Furthermore, Parkinson's essay offered insights into nonmotor symptoms, particularly in advanced disease stages, such as sleep disturbances, bowel issues, and urinary problems. While he couldn't examine these symptoms comprehensively due to limited patient interaction, his observations hinted at their presence. Lastly, Parkinson's account highlighted the slow and nearly imperceptible progression of the disease, emphasizing that patients often struggled to pinpoint its onset, a phenomenon still relevant in modern clinical practice.

Epidemiology

The epidemiological landscape of PD exhibits variability across different geographies and demographic subgroups (Tysnes & Storstein, 2017). Current data suggest that the global annual incidence of PD fluctuates between 4.5 and 16 cases per 100,000 population (Ascherio & Schwarzschild, 2016). This variance may be attributed to methodological differences in data collection, as well as disparities in healthcare access and reporting across regions. In terms of prevalence, the numbers are likewise disparate, ranging from 17 to 227 cases per 100,000 individuals. Variations in study design, sample size, and population demographics, including age distribution, can partially explain this broad range. The prevalence of PD often shows an upward trend with increasing age, reinforcing its characterization as a neurodegenerative disorder commonly associated with older populations (Fearnley & Lees, 1991).

Mortality rates in individuals diagnosed with PD present another critical aspect of its epidemiology. Compared to age-matched controls without PD, those with the disease generally experience mortality rates of approximately twice as high (Miller & Cronin-Golomb, 2010). This elevated mortality can be attributed to various factors, including increased susceptibility to pneumonia, complications arising from immobility, and other comorbidities often associated with PD, such as cognitive decline and cardiovascular disease.

The gender distribution in PD also offers noteworthy insights. Multiple studies have observed a modest male predominance, with an odds ratio of approximately 1.5 (Miller & Cronin-Golomb, 2010). The mechanisms underlying this gender disparity are still a topic of debate but may involve both hormonal and environmental factors. For instance, estrogen is hypothesized to confer some neuroprotective effects, which could partly explain the lower incidence rates among females.

Furthermore, familial clustering has been observed in the epidemiology of PD. Individuals with a family history of the disease are at a higher risk of developing PD themselves, with odds ratios ranging from 2.7 to 3.7 (Thacker & Ascherio, 2008). While the genetic contributions are complex and not fully understood, this observation underscores the importance of genetic factors in the etiology of the disease.

Neuropathology

PD is characterized by two consistent cellular pathologies: the loss of dopaminergic neurons in the ventrolateral region of the substantia nigra (SN) and the presence of Lewy bodies primarily in the brain stem (Kalia & Lang, 2015). While the loss of dopaminergic cells in the SN remains a critical diagnostic feature, the distribution and quantity of Lewy bodies display notable variability.

In 1893, Bloq and Marinesco were the first to propose the involvement of the SN, a concept subsequently supported by ongoing research (Obeso et al., 2017). The significance of cellular inclusion bodies in PD was underscored by Friedrich Heinrich Lewy in 1912, and in 1919, Constantin Trétiakoff connected these inclusion bodies with SN pathology, highlighting their prevalence in most PD cases.

The central role of SN degeneration in PD was firmly established through two pivotal discoveries: Arvid Carlsson's initial elucidation of dopamine's role in the brain (Carlsson et al., 1958) and Oleh Hornykiewicz's demonstration of abundant dopaminergic neurons within the SN (Hornykiewicz, 2002; Lees et al., 2015). The degeneration of these pigmented dopaminergic neurons in the SN consistently appears as a neuropathological hallmark in all PD patients, forming one of the two essential pathological criteria for a definitive diagnosis. It is crucial to note that while this specific neuronal loss is a necessary feature, it is not exclusive to PD, as it occurs in other neurodegenerative conditions presenting with parkinsonism. Additionally, Lewy bodies, primarily composed of alpha-synuclein, serve as another critical diagnostic marker in PD, predominantly localized within the brain stem. Although Lewy bodies are not unique to PD, they exhibit distinct patterns of distribution among different patient groups, influenced by factors such as the presence of Alzheimer's pathology.

In summary, anatomically, PD is characterized by the loss of neurons in the A9 region of the substantia nigra *pars compacta*, with more severe involvement observed in the ventrolateral region projecting to the putamen. Macroscopically, this results in the depletion of neuromelanin in the substantia nigra and locus coeruleus, accompanied by astrocytic and microglial proliferation, which manifests as a characteristic pallor in axial sections of the mesencephalon and pons. On microscopic examination, there is evidence of dopaminergic nerve terminal loss in the striatum, particularly in the posterior putamen, which may precede the neuronal loss in the substantia nigra, suggesting a "dying-back" mechanism of degeneration. Relevant histopathological findings include the Lewy bodies, which are hyaline cytoplasmic inclusions, and intraneuritic Lewy neurites. Immunohistochemical studies confirm alpha-synuclein as the main component in Lewy bodies and neurites. The neuropathology of PD extends beyond the substantia nigra, affecting various regions in the nervous system, including monoaminergic nuclei, basal forebrain cholinergic nuclei, hypothalamus, spinal cord, sympathetic ganglia, plexuses, and cortical areas (G. M. Halliday et al., 1990, 2014).

Historical Evolution of Etiological Theories

The quest for understanding the PD's etiology has evolved since its initial description by James Parkinson in 1817, who posited that the symptomatology could be ascribed to a spinal cord lesion. Nearly a century later, the Von Economo encephalitis epidemic in 1916 led to the hypothesis that viral infections could trigger PD (Dickman, 2001). However, histological examinations of the substantia nigra revealed neurofibrillary tangles rather than Lewy bodies, complicating this supposition. Subsequent hypotheses posited heavy metal intoxication as a causal factor (Bjorklund et al., 2018), inspired by increased incidence rates among metallurgists in the United States.

Between 1979 and 1982, a seminal discovery was made in California where multiple cases of synthetic intravenous drug users developed parkinsonism highly analogous to idiopathic PD (Langston, 2017). Investigations revealed the presence of the neurotoxin MPTP (1-methyl-4-phenyl-1,2,3,6-tetrahydropyridine), inadvertently synthesized during an attempt to produce the synthetic opioid MPPP. This neurotoxin inhibited Complex I of the mitochondrial chain, leading to oxidative stress damage. These cases were known as the "frozen addicts."

Genetic Insights and Molecular Mechanisms

In 1997, a pivotal shift in PD research occurred when Polymeropoulos et al. identified alpha-synuclein gene mutations in a hereditary PD case (PARK1), marking the beginning of genetic insights into PD (Polymeropoulos et al., 1997). Concurrently, Spillantini et al. provided histopathological evidence of alpha-synuclein in Lewy bodies, solidifying the link between alpha-synuclein and PD (Spillantini et al., 1997). These discoveries laid the foundation for the multifactorial etiological theory of PD.

This theory highlights key factors contributing to PD pathogenesis (Li & Le, 2017; Ye et al., 2023), including oxidative stress causing cumulative damage, mitochondrial dysfunction impairing cellular energy production and increasing ROS, disrupted protein clearance systems compromising protein degradation, and the accumulation of toxic proteins, particularly alpha-synuclein, which forms Lewy bodies. Additionally, the prion-like behavior of alpha-synuclein, where aggregates can propagate misfolding, has emerged as a transformative concept, driving extensive research to elucidate the underlying mechanisms (Dunning et al., 2012). These landmark findings have significantly advanced our understanding of PD's complex molecular and genetic underpinnings, opening avenues for potential therapeutic interventions.

Braak´s Theory Explained

In 2002, a pivotal moment in PD research occurred when Braak and colleagues introduced a hypothesis that reshaped understanding about this condition (Braak et al., 2002, 2004; Braak & Del Tredici, 2008). Braak's hypothesis introduced a new perspective, suggesting that an elusive pathogen or toxin, potentially residing in the gut could trigger sporadic PD. This departure from previous theories marked a significant shift, linking PD's origins to events in the gastrointestinal tract.

Additionally, Braak and his research team introduced a comprehensive staging system for PD, anchored in a distinct pattern of alpha-synuclein (αSyn) propagation within the CNS (Burke et al., 2008; G. Halliday et al., 2012). This staging system has provided valuable insights into the disease's progression and offers a structured framework for comprehending its various stages. Subsequently, the dual-hit hypothesis expanded upon Braak's initial concept, proposing that sporadic PD may originate in two distinct locations within the body: the neurons of the nasal cavity and those in the gut (Hawkes et al., 2007, 2009). These sites are believed to play critical roles in initiating the disease.

From these initial focal points, the pathology is thought to follow a predetermined course, with αSyn aggregates spreading along specific routes: the olfactory tract and the vagus nerve. These pathways guide the pathological progression toward and within the CNS. Braak's staging system for PD outlines six stages of pathological advancement. In Stage 1, the disease commences in lower brainstem structures and the olfactory system, predominantly affecting the dorsal motor nucleus of the vagus nerve and the anterior olfactory nucleus. Lewy neurites, filamentous alpha-synuclein aggregates, prevail at this stage. Stage 2 involves additional lesions in the medulla oblongata, in addition to the raphe nuclei and gigantocellular reticular nucleus, with the disease ascending to the locus coeruleus in the pontine tegmentum. Lewy neurites remain more abundant than Lewy bodies. In Stage 3, the SN is affected, and Lewy body lesions begin forming in the *pars compacta*, while acetylcholine-rich neurons in the basal nucleus of Meynert are also impacted. Stage 4 is characterized by severe dopaminergic cell loss in the SN *pars compacta*, involvement of mesocortex and allocortex, and damage to the amygdala, thalamic subnuclei, and anterior olfactory nucleus. Stage 5 sees the disease invading the neocortex and spreading to temporal, parietal, and frontal lobes, causing cell loss in several brainstem structures. Finally, in Stage 6, the neocortex is massively affected, impacting both motor and sensory areas, representing the most severe stage of the disease.

The Braak staging system has been the subject of ongoing research and discussion. Some studies have suggested that the staging system may not fully capture the heterogeneity of PD pathology, as it may not apply to all cases (Outeiro et al., 2023). Additionally, other factors, such as the presence of comorbid Alzheimer's disease pathology, can influence the clinical presentation and progression of PD (Compta et al., 2011).

Genetic Polymorphisms and Clinical Subtypes

The influence of genetic polymorphisms on idiopathic forms of PD has been extensively investigated, revealing multiple candidate genes implicated in the its pathogenesis. This suggests that a polygenic architecture underlies the phenotypic expression of PD.

The classification of putative Mendelian inheritance cases has been denoted as PARK, a numbering system initiated with the first SNCA mutation. Although diverse chromosomal regions have been designated with the prefix PARK, accumulating evidence indicates that fewer genes exhibit indisputable monogenic mechanisms sufficient for phenotypic manifestation. The original system of using locus symbols to identify chromosomal regions linked to familial disorders had significant limitations, such as ambiguity in distinguishing disease-causing genes from genetic risk factors. To address this, a International Parkinson and Movement Disorder Society (MDS) Task Force initiated a new naming system in 2016, which offers criterion-based lists for confirmed monogenic movement disorders, along with stringent evidence requirements for gene-disease associations (Marras et al., 2012, 2016).

Classical PD

PARK-SNCA (Alpha-Synuclein): PARK 1/4

Both PARK1 and PARK4 subtypes are associated with mutations in the α-synuclein gene (SNCA) located at chromosome 4q21-q23 (Polymeropoulos et al., 1997). These mutations encompass missense variants (such as Ala53Thr, Ala40Pro, and E46K) (Krüger et al., 1998; Zarranz et al., 2004) or involve gene multiplication (PARK4) (Chartier-Harlin et al., 2004; Singleton et al., 2003). These genetic aberrations were initially described in the Contursi kindred and are thought to induce a deleterious gain of function. Studies spanning over two decades have revealed an elevated incidence of cognitive impairments and dementia in affected individuals, underlining the substantial impact of SNCA mutations on cognitive function (Fuchs et al., 2007; Nishioka et al., 2006). Notably, mutations in SNCA can engender classical parkinsonism, yet what adds intrigue to this gene is its capacity to produce a spectrum of PD manifestations. This spectrum ranges from the more conventional presentation observed in the majority of PD cases to a more severe form characterized by early-onset parkinsonism and prominent dementia. Alpha-synuclein's pivotal role extends beyond genetic forms, encompassing risk variants and its integral involvement in the neuropathological processes that underpin PD, rendering it a subject of extensive investigation in the field.

PARK-LRRK2 (Leucine-Rich Repeat Kinase 2), PARK8

PARK8 stands out as the most prevalent monogenic form of PD, primarily arising from mutations within the Leucine-Rich Repeat Kinase 2 (LRRK2) gene (Paisán-Ruiz et al., 2013). This genetic variation leads to the production of the dardarin protein, which is believed to function as a cytoplasmic kinase, playing a role in the phosphorylation of membrane proteins, vesicular structures, and microtubules. LRRK2 mutations have been detected in individuals with late-onset PD, particularly those with a familial history of the disease, constituting approximately 1.5-2.5% of this population (Santos-Lobato et al., 2021).

Clinically, LRRK2-associated PD closely resembles the idiopathic form, with typical motor symptoms. However, some distinctions are noted, including a potentially higher prevalence of tremors, particularly in the lower extremities. Of particular significance is the p.G2019S mutation within LRRK2, distinguished by its prevalence among affected individuals.

PARK-VPS35 (Vacuolar Protein Sorting 35), PARK17

Mutations within the VPS35 gene are associated with clinically typical PD and have shed light on the intricate molecular pathways involved in PD pathogenesis (Rahman & Morrison, 2019). VPS35, located on chromosome 16, has emerged as a causative factor in late-onset familial PD, designated as PARK17. This gene encodes a 796-residue protein found throughout human tissues, with a particular presence on endosomes, where it forms the retromer complex alongside other peripheral membrane proteins (Sassone et al., 2021; Williams et al., 2017). The precise mechanisms by which VPS35 mutations lead to dopaminergic neuron degeneration remain under investigation. One hypothesis suggests that these mutations disrupt lysosomal and autophagy function by interfering with the recycling of receptors involved in hydrolase transport to the lysosome. Furthermore, recent research indicates that VPS35 and the retromer complex also impact mitochondrial homeostasis, implying a potential link between VPS35 mutations and mitochondrial dysfunction.

Early-Onset PD

PARK-PRKN (Parkin RBR E3 Ubiquitin Protein Ligase), PARK2

Parkin-associated PD is particularly prevalent in the Japanese population, and often characterized by symmetric parkinsonism, early onset before the age of 50, slow disease progression, early-onset dystonia, and postural instability (von Coelln et al., 2004). The normal Parkin protein appears to have a crucial role in protein ubiquitination (Tanaka et al., 2001), marking proteins for degradation. PARK2, representing approximately 15% of early-onset PD cases and over 60% of cases with onset before age 20, is typically inherited in an autosomal recessive manner (Mata et al., 2004). Dystonia, characterized by sustained muscle contractions leading to abnormal postures, is a distinctive feature in some cases. Mutations in Parkin can vary from single base pair substitutions to larger deletions, with the prevailing hypothesis suggesting that these mutations result in a loss of Parkin function. This loss of function may occur through various mechanisms, including nonsense-mediated decay, which destabilizes truncated transcripts, as well as missense mutations that reduce catalytic activity, disrupt ubiquitination, and impair proteasomal degradation or destabilize the Parkin protein itself.

PARK-PINK1 (PTEN-Induced Putative Kinase 1), PARK6

PARK-PINK1 (PTEN-Induced Putative Kinase 1), designated as PARK6, is associated with early-onset PD characterized by a slower disease progression and a remarkable responsiveness to levodopa treatment (Valente et al., 2004). The underlying pathophysiology is attributed to loss-of-function mutations leading to mitochondrial dysfunction (Vizziello et al., 2021; Wang et al., 2020). Notably, this genetic subtype is often linked to psychiatric features, with patients presenting mood disorders, depression, or other psychiatric symptoms in addition to the classic motor symptoms of PD. The relationship between PINK1 and psychiatric manifestations is an active area of research, shedding light on the complex interplay between genetics and neuropsychiatric aspects of PD. Furthermore, PINK1-related PD may manifest with unusual clinical characteristics, including hyperreflexia, dystonia at onset, early dyskinesias induced by L-dopa, and various behavioral disturbances. Importantly, PINK1-related PD exhibits a higher rate of cognitive impairment, primarily affecting attention and executive functions, compared to other monogenic PD.

PARK-DJ1 (Parkinsonism Associated Deglycase), PARK7

Similar to PARK6, DJ-1 pathogenic variants are associated with early-onset PD characterized by a slow disease progression and a favorable response to levodopa treatment (Lesage et al., 2020). The pathophysiological mechanism likely involves an increase in oxidative stress. DJ-1, encoded by the PARK7 gene, is a small, highly conserved protein expressed ubiquitously and typically exists in a dimeric state under normal conditions (Repici & Giorgini, 2019). Clinically, individuals with DJ-1 mutations present with early-onset motor symptoms such as dyskinesia, rigidity, and tremors, followed by the later emergence of psychiatric symptoms, including psychotic disturbances, anxiety, and cognitive decline. The neuropathological aspects of DJ-1-linked PD remain relatively unexplored, with limited brain analyses, although one case showed Lewy body pathology, indicating a potential connection between DJ-1 and α-synucleinopathy. Moreover, DJ-1 mutation carriers exhibit a high percentage (57%) of non-motor symptoms, suggesting that DJ-1 function influences not only motor symptoms but also other aspects of PD symptomatology, such as anxiety, cognitive decline, and psychotic symptoms.

Genetic Susceptibility Factors

Glucocerebrosidase Gene (GBA1)

Homozygous mutations in the glucocerebrosidase gene (GBA1) are the etiological basis of Gaucher's disease, a lysosomal storage disorder. These mutations result in diminished enzymatic efficacy of glucocerebrosidase (GCase), leading to intracellular accumulation of glycosphingolipids, glucosylceramide, and glucosylsphingosine. Intriguingly, heterozygous mutations in GBA1 have emerged as a salient genetic risk factor for PD, contributing to 5–15% of cases (Gan-Or et al., 2018). The mutations in GBA1 appear to compromise lysosomal function, subsequently affecting the autophagic-lysosomal pathway, resulting in intraneuronal accumulation of α-synuclein.

Over the past two decades, diverse population studies have explored this association, revealing a higher prevalence of GBA mutations among PD patients compared to non-affected individuals (Aharon-Peretz et al., 2004; Riboldi & Di Fonzo, 2019; Sidransky et al., 2009; Smith & Schapira, 2022). Particular GBA mutations, such as c.1226A > G (N370S) and c.1448T > C (L444P), have been identified as the commonest variants associated with PD. Interestingly, the frequency of these mutations varies with ancestry; for example, the c.1226A > G (N370S) mutation is most prevalent among PD patients of Eastern European Ashkenazi Jewish ancestry.

From a clinical perspective, GBA-associated PD (GBA-PD) manifests similarly to sporadic PD but certain mutations were shown to correlate with a more aggressive disease phenotype, including an earlier age of onset and rapid progression. Non-motor symptoms such as depression, sleep disturbances, and anosmia also appear to be more pronounced in GBA-PD patients. Neuropathologically, GBA-PD and sporadic PD are comparable, both characterized by nigrostriatal dopamine loss and the presence of Lewy bodies.

However, GBA mutations act as risk factors rather than deterministic agents for PD (Smith & Schapira, 2022). Among GBA mutation carriers, approximately 9.1% develop PD, though this figure may vary based on the specific mutation and ancestry. The incomplete penetrance of these mutations remains an area of active research. In terms of pathogenesis, GBA mutations are believed to impact endolysosomal and autophagic pathways, both crucial for the degradation of alpha-synuclein. Mutations affect different functional domains of the glucocerebrosidase enzyme, influencing its stability, interaction with other proteins, and overall functionality. Although significant strides have been made in understanding how GBA mutations influence the risk and course of PD, the exact mechanism linking specific mutations to the expression and progression of the disease is not yet fully elucidated.

Clinical PD Diagnosis

The Diagnosis of PD has been an area of sustained focus and refinement. MDS has formulated comprehensive guidelines to enable a reliable and standardized approach to diagnosis (Postuma et al., 2015). Below we transcribe these diagnostic criteria, including the recommended essential and supportive criteria, the various red flags and absolute exclusionary elements that currently guide clinicians in diagnosing PD.

Essential Criterion: Parkinsonism

The *sine qua non* for PD diagnosis is the presence of parkinsonism, characterized by bradykinesia in combination with either rest tremor or rigidity. A thorough examination should be conducted using the MDS-Unified Parkinson's Disease Rating Scale (MDS-UPDRS) (Goetz et al., 2007) to document these cardinal manifestations.

Categories of PD Diagnosis

Upon confirmation of parkinsonism, the diagnosis can be categorized into two distinct groups (Postuma et al., 2015). Firstly, there is Clinically Established PD, characterized by the absence of absolute exclusion criteria, the presence of a minimum of two supportive criteria, and the absence of any red flags. On the other hand, Clinically Probable PD is identified by the absence of absolute exclusion criteria, and the presence of red flags, which can be offset by the presence of supportive criteria. Expressly, one red flag necessitates one supportive criterion, while two red flags necessitate two supportive criteria. It's important to note that a maximum of two red flags is permissible in this category.

Supportive Criteria

1. Dramatic and Clear Beneficial Response to Dopaminergic Therapy: The patient returns to normal or near-normal function during initial treatment. This can be quantified as a greater than 30% improvement in UPDRS III with medication adjustment or a well-documented history of significant changes from a reliable caregiver.
2. Presence of Levodopa-Induced Dyskinesia
3. Documented Rest Tremor of a Limb
4. Olfactory Loss or Cardiac Sympathetic Denervation on MIBG Scintigraphy

Summarized Absolute Exclusion Criteria

The presence of any of these features precludes a PD diagnosis:

1. Unequivocal cerebellar abnormalities
2. Downward supranuclear gaze palsy or selective slowing of downward vertical saccades
3. Diagnosis of probable frontotemporal dementia variant or primary progressive aphasia within the first five years
4. Parkinsonian features confined to the lower limbs for over three years
5. Medication-induced parkinsonism
6. Lack of observable response to high-dose levodopa
7. Unequivocal cortical sensory loss, ideomotor limb apraxia, or progressive aphasia
8. Normal presynaptic dopaminergic functional neuroimaging
9. Alternative condition explaining parkinsonism

Summarized Red Flags

1. Rapid progression requiring wheelchair use within 5 years
2. Absence of symptom progression over 5 years
3. Early bulbar dysfunction
4. Inspiratory respiratory dysfunction
5. Severe autonomic dysfunction within 5 years
6. Frequent falls within 3 years
7. Distonic anterocollis or hand/foot contractions within 10 years
8. Absence of common non-motor symptoms despite 5 years of disease duration
9. Unexplained pyramidal tract signs
10. Symmetrical bilateral parkinsonism

Are We Going Toward a Biological-Only Diagnosis and Classification?

The evolution of PD research has ushered in an era of exploration focused on identifying and utilizing biomarkers, giving rise to a paradigm shift towards biomarker-driven diagnostics, focusing on substances, structures, or biological processes that reflect the disease's underlying pathology. The imperative to develop disease-modifying treatments that target the molecular substrates of neurodegeneration underscores the exigency of biomarker discovery in PD. Since clinically defined PD harbors heterogeneous etiological underpinnings, burgeoning attention is directed toward biomarkers delineating pathologies, primarily those associated with α-synuclein (α-syn). This line of inquiry has emerged as a focal point in the scientific discourse surrounding PD.

The Neuronal Synuclein Disease Integrated Staging System

A paradigmatic proposal emanates from The Michael J. Fox Foundation for Parkinson's Research, introducing the Neuronal Synuclein Disease Integrated Staging System (NSD-ISS) (Chahine et al., 2023). This framework articulates a continuum of disorders, including PD, Dementia with Lewy Bodies (DLB), and their concomitant conditions, collectively termed Neuronal α-synuclein Disease (NSD). Within NSD, stages are biologically demarcated irrespective of clinical phenotypes:

- Stage 0: Fully penetrant pathogenic SNCA gene variants encoding α-syn.
- Stage 1A: Isolated presence of n-α-syn biomarker.
- Stage 1B: Conjoint presence of n-α-syn biomarker and dopaminergic neuronal dysfunction.
- Stage 2: Incipient clinical manifestations without functional incapacitation, further classified into 2A (n-α-syn only) and 2B (n-α-syn plus dopaminergic dysfunction).
- Stages 3-6: Concomitant biomarkers with graded escalations in functional disability.

The NSD-ISS aspires to catalyze targeted therapeutic advancements by amalgamating biological anchors—Synuclein presence/absence and dopaminergic loss—with gradations in functional impairment. A draft manuscript, which was previously made accessible online to solicit public input, was open for contributions until recently (https://www.michaeljfox.org/StagingManuscript). The finalized version of the document is anticipated to be imminently released.

The SyNerGe Framework: A Biological Codification

Höglinger et al. have conceptualized the SyNerGe system, joining information from genetic variants, α-syn pathology, neurodegenerative markers, and clinical syndromes (Höglinger et al., 2023). This proposition aligns with the emerging paradigm that advocates classifying neurodegenerative diseases, including PD, based on biological or pathophysiological criteria. According to the authors, the preclinical onset of PD is already well-substantiated, as is the significant clinical variability. Additionally, α-syn Lewy pathology has been linked to a gamut of both motor and non-motor symptoms.

This biologically oriented classification, coined 'G-S-N-C' or 'SyNerGe', encompasses:

- G (Ge, Gene): Pathogenic or predisposing gene variants.
- S (Sy, Synuclein): Tissue deposition or seeding assay-based α-syn pathology.
- N (Ner, Neurodegeneration): Neuroimaging-ascertained neurodegenerative evidence.
- C (Clinical): Clinically associated syndromes with varying specificity levels.

The framework proposes biological designations that encapsulate both sporadic and genetic forms of PD, thereby broadening the diagnostic purview. Such designations would serve to refine our understanding of PD's biological complexities, aligning with the shift towards pathophysiology-based definitions aimed at enhancing therapeutic precision and addressing extant diagnostic limitations.

Research on Disease-Modifying Drugs

Aggregated α-syn is implicated in synaptic vesicle dysfunction, lysosomal and mitochondrial anomalies, and intracellular protein transport impairment (Wong & Krainc, 2017). Consequently, disease-modifying therapies (DMTs) aimed at mitigating α-syn aggregation have garnered significant attention. Immunotherapeutic interventions targeting α-syn, initially introduced in 2005, have shown beneficial histopathological outcomes in mouse models (Masliah et al., 2005). Numerous subsequent investigations substantiate the efficacy of active and passive immunization against α-syn in both in vitro and in vivo models of PD (Höllerhage et al., 2022; Knecht et al., 2022). Several vaccine candidates, including PD01A and PD03A, have completed initial Phase 1 trials, while a new formulation, ACI-7104, is poised for Phase 2 trials (NCT06015841).

Passive immunotherapy, however, has not yet demonstrated consistent clinical efficacy, with certain trials falling short of primary endpoints. BIIB054/cinpanemab from Biogen was examined in the SPARK study, enrolling 357 early-stage PD patients, but failed to meet its primary and secondary endpoints concerning changes in the MDS-UPDRS and DaT-SPECT imaging (Lang et al., 2022). Notably, PRX002/prasinezumab, In a Phase 2 trial involving early-stage PD patients, also did not demonstrate a significant impact on the primary endpoint of MDS-UPDRS scores or secondary imaging measures via SPECT compared to placebo (Pagano et al., 2022). The study also reported that prasinezumab did not substantially differ from the placebo group in most clinical secondary endpoints. AbbVie's ABBV-0805, another candidate, demonstrated good tolerance and safety in Phase I but was withdrawn during a Phase Ib trial for strategic reasons.

Antisense oligonucleotides specific to α-syn have proven effective in murine PD models (Alarcón-Arís et al., 2018) and are now being tested in Phase 1 trials for Multiple System Atrophy (MSA) (NCT04165486). Small molecules such as anle138b and NPT200-11 have also shown some promise in interfering with α-syn aggregation and are progressing through clinical trials (NCT04685265, NCT04875962, NCT04658186).

Neuroinflammatory processes, hallmarked by innate and adaptive immune responses and blood-brain barrier dysfunction, are also considered critical in PD pathogenesis (Tansey et al., 2022). While conventional anti-inflammatory agents and statins have yielded unsatisfactory outcomes (Stevens et al., 2022), other strategies targeting microglial activation appear promising and are undergoing clinical evaluation (EudraCT: 2021-003879-34; EudraCT: 2018-003089-14).

Mitochondrial dysfunction, particularly in the context of Parkin and PINK1 gene mutations, contributes to neuronal loss. Despite the limited success of some agents targeting this pathway (NET-PD Investigators et al., 2015), ongoing trials focus on alternative mitochondrial modulators, such as Ursodeoxycholic Acid (UDCA) (NCT03840005).

LRRK2 is also studies as a therapeutic target. LRRK2 inhibitors such as DNL151 have entered clinical trials, examining their potential for disease modification (NCT05418673; NCT05348785). In parallel, GBA gene mutations are being targeted by gene therapy approaches and enzyme activity modulators, such as ambroxol (NCT05778617).

CONCLUSION

In conclusion, this chapter has endeavored to provide a comprehensive and nuanced understanding of PD, a neurodegenerative disorder of escalating global importance. Beginning with an analysis of its epidemiology, we made an overview of the complex landscape of its neuropathology and genetics, mentioning especially dominant and recessive forms and susceptibility factors. The chapter further explored the essential criteria for clinical diagnosis and addressed the limitations therein, highlighting the emerging frameworks that aim to bring us closer to a biological diagnosis of PD.

It is evident that the scientific understanding of PD is in a state of dynamic evolution, influenced by advancing technologies and a growing body of molecular and genetic research. Future directions include validating biological definitions for PD and the development of more effective disease-modifying drugs. As both clinical and basic research communities continue to unravel PD-related complexities, interdisciplinary collaboration will remain pivotal in translating these scientific insights into optimized patient outcomes.

REFERENCES

A New Drug Development Tool for Parkinson's Disease and Dementia with Lewy Bodies: The Neuronal Synuclein Disease Integrated Staging System. (n.d.). Retrieved September 20, 2023, from https://www.michaeljfox.org/StagingManuscript

Aharon-Peretz, J., Rosenbaum, H., & Gershoni-Baruch, R. (2004). Mutations in the glucocerebrosidase gene and Parkinson's disease in Ashkenazi Jews. *The New England Journal of Medicine, 351*(19), 1972–1977. doi:10.1056/NEJMoa033277 PMID:15525722

Alarcón-Arís, D., Recasens, A., Galofré, M., Carballo-Carbajal, I., Zacchi, N., Ruiz-Bronchal, E., Pavia-Collado, R., Chica, R., Ferrés-Coy, A., Santos, M., Revilla, R., Montefeltro, A., Fariñas, I., Artigas, F., Vila, M., & Bortolozzi, A. (2018). Selective α-Synuclein Knockdown in Monoamine Neurons by Intranasal Oligonucleotide Delivery: Potential Therapy for Parkinson's Disease. *Molecular Therapy, 26*(2), 550–567. doi:10.1016/j.ymthe.2017.11.015 PMID:29273501

Ascherio, A., & Schwarzschild, M. A. (2016). The epidemiology of Parkinson's disease: Risk factors and prevention. *Lancet Neurology, 15*(12), 1257–1272. doi:10.1016/S1474-4422(16)30230-7 PMID:27751556

Bjorklund, G., Stejskal, V., Urbina, M. A., Dadar, M., Chirumbolo, S., & Mutter, J. (2018). Metals and Parkinson's Disease: Mechanisms and Biochemical Processes. *Current Medicinal Chemistry, 25*(19), 2198–2214. doi:10.2174/0929867325666171129124616 PMID:29189118

Braak, H., & Del Tredici, K. (2008). Invited Article: Nervous system pathology in sporadic Parkinson disease. *Neurology, 70*(20), 1916–1925. doi:10.1212/01.wnl.0000312279.49272.9f PMID:18474848

Braak, H., Del Tredici, K., Bratzke, H., Hamm-Clement, J., Sandmann-Keil, D., & Rüb, U. (2002). Staging of the intracerebral inclusion body pathology associated with idiopathic Parkinson's disease (preclinical and clinical stages). *Journal of Neurology, 249*(3), 1-5.

Braak, H., Ghebremedhin, E., Rüb, U., Bratzke, H., & Del Tredici, K. (2004). Stages in the development of Parkinson's disease-related pathology. *Cell and Tissue Research, 318*(1), 121–134. doi:10.1007/s00441-004-0956-9 PMID:15338272

Burke, R. E., Dauer, W. T., & Vonsattel, J. P. G. (2008). A critical evaluation of the Braak staging scheme for Parkinson's disease. *Annals of Neurology*, *64*(5), 485–491. doi:10.1002/ana.21541 PMID:19067353

Carlsson, A., Lindqvist, M., Magnusson, T., & Waldeck, B. (1958). On the presence of 3-hydroxytyramine in brain. *Science*, *127*(3296), 471. doi:10.1126/science.127.3296.471.a PMID:13529006

Chahine, L. M., Merchant, K., Siderowf, A., Sherer, T., Tanner, C., Marek, K., & Simuni, T. (2023). Proposal for a Biologic Staging System of Parkinson's Disease. *Journal of Parkinson's Disease*, *13*(3), 297–309. doi:10.3233/JPD-225111 PMID:37066922

Chartier-Harlin, M.-C., Kachergus, J., Roumier, C., Mouroux, V., Douay, X., Lincoln, S., Levecque, C., Larvor, L., Andrieux, J., Hulihan, M., Waucquier, N., Defebvre, L., Amouyel, P., Farrer, M., & Destée, A. (2004). Alpha-synuclein locus duplication as a cause of familial Parkinson's disease. *Lancet*, *364*(9440), 1167–1169. doi:10.1016/S0140-6736(04)17103-1 PMID:15451224

Compta, Y., Parkkinen, L., O'Sullivan, S. S., Vandrovcova, J., Holton, J. L., Collins, C., Lashley, T., Kallis, C., Williams, D. R., de Silva, R., Lees, A. J., & Revesz, T. (2011). Lewy- and Alzheimer-type pathologies in Parkinson's disease dementia: Which is more important? *Brain*, *134*(5), 1493–1505. doi:10.1093/brain/awr031 PMID:21596773

Dickman, M. S. (2001). Von Economo encephalitis. *Archives of Neurology*, *58*(10), 1696–1698. doi:10.1001/archneur.58.10.1696 PMID:11594935

Dunning, C. J. R., Reyes, J. F., Steiner, J. A., & Brundin, P. (2012). Can Parkinson's disease pathology be propagated from one neuron to another? *Progress in Neurobiology*, *97*(2), 205–219. doi:10.1016/j.pneurobio.2011.11.003 PMID:22115849

Fearnley, J. M., & Lees, A. J. (1991). Ageing and Parkinson's disease: Substantia nigra regional selectivity. *Brain*, *114*(5), 2283–2301. doi:10.1093/brain/114.5.2283 PMID:1933245

Fuchs, J., Nilsson, C., Kachergus, J., Munz, M., Larsson, E.-M., Schüle, B., Langston, J. W., Middleton, F. A., Ross, O. A., Hulihan, M., Gasser, T., & Farrer, M. J. (2007). Phenotypic variation in a large Swedish pedigree due to SNCA duplication and triplication. *Neurology*, *68*(12), 916–922. doi:10.1212/01.wnl.0000254458.17630.c5 PMID:17251522

Gan-Or, Z., Liong, C., & Alcalay, R. N. (2018). GBA-Associated Parkinson's Disease and Other Synucleinopathies. *Current Neurology and Neuroscience Reports, 18*(8), 44. doi:10.1007/s11910-018-0860-4 PMID:29884970

Goetz, C. G., Fahn, S., Martinez-Martin, P., Poewe, W., Sampaio, C., Stebbins, G. T., Stern, M. B., Tilley, B. C., Dodel, R., Dubois, B., Holloway, R., Jankovic, J., Kulisevsky, J., Lang, A. E., Lees, A., Leurgans, S., LeWitt, P. A., Nyenhuis, D., Olanow, C. W., ... LaPelle, N. (2007). Movement Disorder Society-sponsored revision of the Unified Parkinson's Disease Rating Scale (MDS-UPDRS): Process, format, and clinimetric testing plan. *Movement Disorders, 22*(1), 41–47. doi:10.1002/mds.21198 PMID:17115387

Halliday, G., McCann, H., & Shepherd, C. (2012). Evaluation of the Braak hypothesis: How far can it explain the pathogenesis of Parkinson's disease? *Expert Review of Neurotherapeutics, 12*(6), 673–686. doi:10.1586/ern.12.47 PMID:22650170

Halliday, G. M., Leverenz, J. B., Schneider, J. S., & Adler, C. H. (2014). The neurobiological basis of cognitive impairment in Parkinson's disease. *Movement Disorders, 29*(5), 634–650. doi:10.1002/mds.25857 PMID:24757112

Halliday, G. M., Li, Y. W., Blumbergs, P. C., Joh, T. H., Cotton, R. G. H., Howe, P. R. C., Blessing, W. W., & Geffen, L. B. (1990). Neuropathology of immunohistochemically identified brainstem neurons in Parkinson's disease. *Annals of Neurology, 27*(4), 373–385. doi:10.1002/ana.410270405 PMID:1972319

Hawkes, C. H., Del Tredici, K., & Braak, H. (2007). Parkinson's disease: A dual-hit hypothesis. *Neuropathology and Applied Neurobiology, 33*(6), 599–614. doi:10.1111/j.1365-2990.2007.00874.x PMID:17961138

Hawkes, C. H., Del Tredici, K., & Braak, H. (2009). Parkinson's disease: The dual hit theory revisited. *Annals of the New York Academy of Sciences, 1170*(1), 615–622. doi:10.1111/j.1749-6632.2009.04365.x PMID:19686202

HöglingerG. U.AdlerC. H.BergD.KleinC.OuteiroT. F.PoeweW.PostumaR.StoesslJ. LangA. E. (2023). *Towards a Biological Definition of Parkinson's Disease* (No. 2023040108). Preprints. doi:10.20944/preprints202304.0108.v1

Höllerhage, M., Klietz, M., & Höglinger, G. U. (2022). Disease modification in Parkinsonism: Obstacles and ways forward. *Journal of Neural Transmission (Vienna, Austria), 129*(9), 1133–1153. doi:10.1007/s00702-022-02520-6 PMID:35695938

Hornykiewicz, O. (2002). Dopamine miracle: From brain homogenate to dopamine replacement. *Movement Disorders, 17*(3), 501–508. doi:10.1002/mds.10115 PMID:12112197

Kalia, L. V., & Lang, A. E. (2015). Parkinson's disease. *Lancet*, *386*(9996), 896–912. doi:10.1016/S0140-6736(14)61393-3 PMID:25904081

Kieburtz, K., Tilley, B. C., Elm, J. J., Babcock, D., Hauser, R., Ross, G. W., Augustine, A. H., Augustine, E. U., Aminoff, M. J., Bodis-Wollner, I. G., Boyd, J., Cambi, F., Chou, K., Christine, C. W., Cines, M., Dahodwala, N., Derwent, L., Dewey, R. B. Jr, Hawthorne, K., ... Wills, A.-M. (2015). Effect of creatine monohydrate on clinical progression in patients with Parkinson disease: A randomized clinical trial. *Journal of the American Medical Association*, *313*(6), 584–593. doi:10.1001/jama.2015.120 PMID:25668262

Knecht, L., Folke, J., Dodel, R., Ross, J. A., & Albus, A. (2022). Alpha-synuclein Immunization Strategies for Synucleinopathies in Clinical Studies: A Biological Perspective. *Neurotherapeutics; the Journal of the American Society for Experimental NeuroTherapeutics*, *19*(5), 1489–1502. doi:10.1007/s13311-022-01288-7 PMID:36083395

Krüger, R., Kuhn, W., Müller, T., Woitalla, D., Graeber, M., Kösel, S., Przuntek, H., Epplen, J. T., Schöls, L., & Riess, O. (1998). Ala30Pro mutation in the gene encoding alpha-synuclein in Parkinson's disease. *Nature Genetics*, *18*(2), 106–108. doi:10.1038/ng0298-106 PMID:9462735

Lang, A. E., Siderowf, A. D., Macklin, E. A., Poewe, W., Brooks, D. J., Fernandez, H. H., Rascol, O., Giladi, N., Stocchi, F., Tanner, C. M., Postuma, R. B., Simon, D. K., Tolosa, E., Mollenhauer, B., Cedarbaum, J. M., Fraser, K., Xiao, J., Evans, K. C., Graham, D. L., ... Dam, T. (2022). Trial of Cinpanemab in Early Parkinson's Disease. *The New England Journal of Medicine*, *387*(5), 408–420. doi:10.1056/NEJMoa2203395 PMID:35921450

Langston, J. W. (2017). The MPTP Story. *Journal of Parkinson's Disease*, *7*(s1), S11–S19. doi:10.3233/JPD-179006 PMID:28282815

Lees, A. J., Tolosa, E., & Olanow, C. W. (2015). Four pioneers of L-dopa treatment: Arvid Carlsson, Oleh Hornykiewicz, George Cotzias, and Melvin Yahr. *Movement Disorders*, *30*(1), 19–36. doi:10.1002/mds.26120 PMID:25488030

Lesage, S., Lunati, A., Houot, M., Romdhan, S. B., Clot, F., Tesson, C., Mangone, G., Toullec, B. L., Courtin, T., Larcher, K., Benmahdjoub, M., Arezki, M., Bouhouche, A., Anheim, M., Roze, E., Viallet, F., Tison, F., Broussolle, E., Emre, M., ... Brice, A. (2020). Characterization of Recessive Parkinson Disease in a Large Multicenter Study. *Annals of Neurology*, *88*(4), 843–850. doi:10.1002/ana.25787 PMID:33045815

Li, S., & Le, W. (2017). Milestones of Parkinson's Disease Research: 200 Years of History and Beyond. *Neuroscience Bulletin*, *33*(5), 598–602. doi:10.1007/s12264-017-0178-2 PMID:28895075

Marras, C., Lang, A., van de Warrenburg, B. P., Sue, C. M., Tabrizi, S. J., Bertram, L., Mercimek-Mahmutoglu, S., Ebrahimi-Fakhari, D., Warner, T. T., Durr, A., Assmann, B., Lohmann, K., Kostic, V., & Klein, C. (2016). Nomenclature of genetic movement disorders: Recommendations of the international Parkinson and movement disorder society task force. *Movement Disorders*, *31*(4), 436–457. doi:10.1002/mds.26527 PMID:27079681

Marras, C., Lohmann, K., Lang, A., & Klein, C. (2012). Fixing the broken system of genetic locus symbols. *Neurology*, *78*(13), 1016–1024. doi:10.1212/WNL.0b013e31824d58ab PMID:22454269

Masliah, E., Rockenstein, E., Adame, A., Alford, M., Crews, L., Hashimoto, M., Seubert, P., Lee, M., Goldstein, J., Chilcote, T., Games, D., & Schenk, D. (2005). Effects of alpha-synuclein immunization in a mouse model of Parkinson's disease. *Neuron*, *46*(6), 857–868. doi:10.1016/j.neuron.2005.05.010 PMID:15953415

Mata, I. F., Lockhart, P. J., & Farrer, M. J. (2004). Parkin genetics: One model for Parkinson's disease. *Human Molecular Genetics*, *13*(Spec No 1), R127–R133. doi:10.1093/hmg/ddh089 PMID:14976155

Miller, I. N., & Cronin-Golomb, A. (2010). Gender differences in Parkinson's disease: Clinical characteristics and cognition. *Movement Disorders*, *25*(16), 2695–2703. doi:10.1002/mds.23388 PMID:20925068

Nishioka, K., Hayashi, S., Farrer, M. J., Singleton, A. B., Yoshino, H., Imai, H., Kitami, T., Sato, K., Kuroda, R., Tomiyama, H., Mizoguchi, K., Murata, M., Toda, T., Imoto, I., Inazawa, J., Mizuno, Y., & Hattori, N. (2006). Clinical heterogeneity of alpha-synuclein gene duplication in Parkinson's disease. *Annals of Neurology*, *59*(2), 298–309. doi:10.1002/ana.20753 PMID:16358335

Obeso, J. A., Stamelou, M., Goetz, C. G., Poewe, W., Lang, A. E., Weintraub, D., Burn, D., Halliday, G. M., Bezard, E., Przedborski, S., Lehericy, S., Brooks, D. J., Rothwell, J. C., Hallett, M., DeLong, M. R., Marras, C., Tanner, C. M., Ross, G. W., Langston, J. W., ... Stoessl, A. J. (2017). Past, present, and future of Parkinson's disease: A special essay on the 200th Anniversary of the Shaking Palsy. *Movement Disorders*, *32*(9), 1264–1310. doi:10.1002/mds.27115 PMID:28887905

Outeiro, T. F., Alcalay, R. N., Antonini, A., Attems, J., Bonifati, V., Cardoso, F., Chesselet, M.-F., Hardy, J., Madeo, G., McKeith, I., Mollenhauer, B., Moore, D. J., Rascol, O., Schlossmacher, M. G., Soreq, H., Stefanis, L., & Ferreira, J. J. (2023). Defining the Riddle in Order to Solve It: There Is More Than One "Parkinson's Disease.". *Movement Disorders*, *38*(7), 1127–1142. doi:10.1002/mds.29419 PMID:37156737

Pagano, G., Taylor, K. I., Anzures-Cabrera, J., Marchesi, M., Simuni, T., Marek, K., Postuma, R. B., Pavese, N., Stocchi, F., Azulay, J.-P., Mollenhauer, B., López-Manzanares, L., Russell, D. S., Boyd, J. T., Nicholas, A. P., Luquin, M. R., Hauser, R. A., Gasser, T., Poewe, W., ... Bonni, A. (2022). Trial of Prasinezumab in Early-Stage Parkinson's Disease. *The New England Journal of Medicine*, *387*(5), 421–432. doi:10.1056/NEJMoa2202867 PMID:35921451

Paisán-Ruiz, C., Lewis, P. A., & Singleton, A. B. (2013). LRRK2: Cause, Risk, and Mechanism. *Journal of Parkinson's Disease*, *3*(2), 85–103. doi:10.3233/JPD-130192 PMID:23938341

Parkinson, J. (2002). An essay on the shaking palsy (Reprinted). *The Journal of Neuropsychiatry and Clinical Neurosciences*, *14*(2), 223–236. doi:10.1176/jnp.14.2.223 PMID:11983801

Polymeropoulos, M. H., Lavedan, C., Leroy, E., Ide, S. E., Dehejia, A., Dutra, A., Pike, B., Root, H., Rubenstein, J., Boyer, R., Stenroos, E. S., Chandrasekharappa, S., Athanassiadou, A., Papapetropoulos, T., Johnson, W. G., Lazzarini, A. M., Duvoisin, R. C., Di Iorio, G., Golbe, L. I., & Nussbaum, R. L. (1997). Mutation in the alpha-synuclein gene identified in families with Parkinson's disease. *Science*, *276*(5321), 2045–2047. doi:10.1126/science.276.5321.2045 PMID:9197268

Postuma, R. B., Berg, D., Stern, M., Poewe, W., Olanow, C. W., Oertel, W., Obeso, J., Marek, K., Litvan, I., Lang, A. E., Halliday, G., Goetz, C. G., Gasser, T., Dubois, B., Chan, P., Bloem, B. R., Adler, C. H., & Deuschl, G. (2015). MDS clinical diagnostic criteria for Parkinson's disease. *Movement Disorders*, *30*(12), 1591–1601. doi:10.1002/mds.26424 PMID:26474316

Rahman, A. A., & Morrison, B. E. (2019). Contributions of VPS35 Mutations to Parkinson's Disease. *Neuroscience*, *401*, 1–10. doi:10.1016/j.neuroscience.2019.01.006 PMID:30660673

Repici, M., & Giorgini, F. (2019). DJ-1 in Parkinson's Disease: Clinical Insights and Therapeutic Perspectives. *Journal of Clinical Medicine*, *8*(9), 1377. doi:10.3390/jcm8091377 PMID:31484320

Riboldi, G. M., & Di Fonzo, A. B. (2019). GBA, Gaucher Disease, and Parkinson's Disease: From Genetic to Clinic to New Therapeutic Approaches. *Cells*, *8*(4), 364. doi:10.3390/cells8040364 PMID:31010158

Santos-Lobato, B. L., Schumacher-Schuh, A., Mata, I. F., Letro, G. H., Braga-Neto, P., Brandão, P. R. P., Godeiro-Junior, C. O., Coletta, M. V. D., Camargos, S. T., Borges, V., Rieder, C. R. M., & Tumas, V. (2021). Genetics of Parkinson's disease in Brazil: A systematic review of monogenic forms. *Arquivos de Neuro-Psiquiatria*, *79*(7), 612–623. doi:10.1590/0004-282x-anp-2020-0409 PMID:34468500

Sassone, J., Reale, C., Dati, G., Regoni, M., Pellecchia, M. T., & Garavaglia, B. (2021). The Role of VPS35 in the Pathobiology of Parkinson's Disease. *Cellular and Molecular Neurobiology*, *41*(2), 199–227. doi:10.1007/s10571-020-00849-8 PMID:32323152

Sidransky, E., Nalls, M. A., Aasly, J. O., Aharon-Peretz, J., Annesi, G., Barbosa, E. R., Bar-Shira, A., Berg, D., Bras, J., Brice, A., Chen, C.-M., Clark, L. N., Condroyer, C., De Marco, E. V., Dürr, A., Eblan, M. J., Fahn, S., Farrer, M., Fung, H.-C., ... Ziegler, S. G. (2009). Multi-center analysis of glucocerebrosidase mutations in Parkinson disease. *The New England Journal of Medicine*, *361*(17), 1651–1661. doi:10.1056/NEJMoa0901281 PMID:19846850

Singleton, A. B., Farrer, M., Johnson, J., Singleton, A., Hague, S., Kachergus, J., Hulihan, M., Peuralinna, T., Dutra, A., Nussbaum, R., Lincoln, S., Crawley, A., Hanson, M., Maraganore, D., Adler, C., Cookson, M. R., Muenter, M., Baptista, M., Miller, D., ... Gwinn-Hardy, K. (2003). Alpha-Synuclein locus triplication causes Parkinson's disease. *Science*, *302*(5646), 841. doi:10.1126/science.1090278 PMID:14593171

Smith, L., & Schapira, A. H. V. (2022). GBA Variants and Parkinson Disease: Mechanisms and Treatments. *Cells*, *11*(8), 1261. doi:10.3390/cells11081261 PMID:35455941

Spillantini, M. G., Schmidt, M. L., Lee, V. M., Trojanowski, J. Q., Jakes, R., & Goedert, M. (1997). Alpha-synuclein in Lewy bodies. *Nature*, *388*(6645), 839–840. doi:10.1038/42166 PMID:9278044

Stevens, K. N., Creanor, S., Jeffery, A., Whone, A., Zajicek, J., Foggo, A., Jones, B., Chapman, R., Cocking, L., Wilks, J., Webb, D., Carroll, C., Inches, J., Underwood, D., Frost, J., James, A., Schofield, C., James, R., O'Reilly, C., ... Rose, J. (2022). Evaluation of Simvastatin as a Disease-Modifying Treatment for Patients With Parkinson Disease. *JAMA Neurology*, *79*(12), 1232–1241. doi:10.1001/jamaneurol.2022.3718 PMID:36315128

Tanaka, K., Suzuki, T., Chiba, T., Shimura, H., Hattori, N., & Mizuno, Y. (2001). Parkin is linked to the ubiquitin pathway. *Journal of Molecular Medicine (Berlin, Germany), 79*(9), 482–494. doi:10.1007/s001090100242 PMID:11692161

Tansey, M. G., Wallings, R. L., Houser, M. C., Herrick, M. K., Keating, C. E., & Joers, V. (2022). Inflammation and immune dysfunction in Parkinson disease. *Nature Reviews. Immunology, 22*(11), 11. Advance online publication. doi:10.1038/s41577-022-00684-6 PMID:35246670

Thacker, E. L., & Ascherio, A. (2008). Familial aggregation of Parkinson's disease: A meta-analysis. *Movement Disorders, 23*(8), 1174–1183. doi:10.1002/mds.22067 PMID:18442112

Tysnes, O.-B., & Storstein, A. (2017). Epidemiology of Parkinson's disease. *Journal of Neural Transmission, 124*(8), 901–905. doi:10.1007/s00702-017-1686-y

Valente, E. M., Abou-Sleiman, P. M., Caputo, V., Muqit, M. M. K., Harvey, K., Gispert, S., Ali, Z., Del Turco, D., Bentivoglio, A. R., Healy, D. G., Albanese, A., Nussbaum, R., González-Maldonado, R., Deller, T., Salvi, S., Cortelli, P., Gilks, W. P., Latchman, D. S., Harvey, R. J., ... Wood, N. W. (2004). Hereditary early-onset Parkinson's disease caused by mutations in PINK1. *Science, 304*(5674), 1158–1160. doi:10.1126/science.1096284 PMID:15087508

Vizziello, M., Borellini, L., Franco, G., & Ardolino, G. (2021). Disruption of Mitochondrial Homeostasis: The Role of PINK1 in Parkinson's Disease. *Cells, 10*(11), 3022. doi:10.3390/cells10113022 PMID:34831247

von Coelln, R., Dawson, V. L., & Dawson, T. M. (2004). Parkin-associated Parkinson's disease. *Cell and Tissue Research, 318*(1), 175–184. doi:10.1007/s00441-004-0924-4 PMID:15503153

Wang, N., Zhu, P., Huang, R., Wang, C., Sun, L., Lan, B., He, Y., Zhao, H., & Gao, Y. (2020). PINK1: The guard of mitochondria. *Life Sciences, 259*, 118247. doi:10.1016/j.lfs.2020.118247 PMID:32805222

Williams, E. T., Chen, X., & Moore, D. J. (2017). VPS35, the Retromer Complex and Parkinson's Disease. *Journal of Parkinson's Disease, 7*(2), 219–233. doi:10.3233/JPD-161020 PMID:28222538

Wong, Y. C., & Krainc, D. (2017). α-synuclein toxicity in neurodegeneration: Mechanism and therapeutic strategies. *Nature Medicine, 23*(2), 1–13. doi:10.1038/nm.4269 PMID:28170377

Ye, H., Robak, L. A., Yu, M., Cykowski, M., & Shulman, J. M. (2023). Genetics and Pathogenesis of Parkinson's Syndrome. *Annual Review of Pathology*, *18*(1), 95–121. doi:10.1146/annurev-pathmechdis-031521-034145 PMID:36100231

Zarranz, J. J., Alegre, J., Gómez-Esteban, J. C., Lezcano, E., Ros, R., Ampuero, I., Vidal, L., Hoenicka, J., Rodriguez, O., Atarés, B., Llorens, V., Gomez Tortosa, E., del Ser, T., Muñoz, D. G., & de Yebenes, J. G. (2004). The new mutation, E46K, of alpha-synuclein causes Parkinson and Lewy body dementia. *Annals of Neurology*, *55*(2), 164–173. doi:10.1002/ana.10795 PMID:14755719

Chapter 9
Primary Progressive Aphasia in Northeast Brazil

Paulo Roberto de Brito Marques
University of Pernambuco, Brazil

ABSTRACT

All 29 patients underwent structural neuroimaging examination. With the advent of analysis cerebrospinal fluid biomarkers, the authors were able to perform this in 38% of cases, with 45% having Alzheimer's pathology, 36% tau protein pathology, 9% having normal biomarkers, and 9% having increased β-amyloid protein alone. With technological developments, 55% of cases underwent brain single photon perfusion scintigraphy (SPECT) studies (81%) and brain PET/SCAN (19%). The clinical results were 43% compatible with the logopenic variant, 36% with the semantic variant, and 21% with the non-fluent variant. Around 3% of cases occurred due to a mutation in exon 5 of the VCP gene. Around 10% of cases underwent post-mortem studies, with Alzheimer's disease (logopenic variant) confirmed in 33%, Pick's disease (non-fluent variant) in 33%, and cortico-basal degeneration (variant not fluent).

Primary progressive aphasia (PPA) is a disorder mainly of language, neurodegenerative in nature, with a progressive course and a lethal evolution, differing widely in evolution both in each type of variant and in the time of its course (Ulugut et al. 2022). Until now, we believe that PPA, as a syndrome, has an open definition, especially regarding its variants and etiological correlation. In some cases, complaints of memory lapses occur before the appearance of language symptoms, or perhaps some patients do not know how to differentiate memory deficit from anomia; memory deficit is forgetfulness and anomie is the inability to say the name of an object (Brito-Marques, 2018). Language symptoms generally take around 2 years before

DOI: 10.4018/979-8-3693-0851-6.ch009

the syndrome begins to express symptoms other than those of language, usually detected by objective assessment (Mesulam, 1982).

Acute and chronic aphasic syndromes differ in terms of their topography (Binder, 2015). When comparing the topography between both aphasias, it was observed that PPA presents more involvement with the language neural networks in the left temporal cortex, especially in the temporal pole, while in aphasias due to acute lesions of vascular etiology, they generally spare the temporal poles (Binder, 2015). The mechanism between both aphasias may be different but with similar results. Although the semiology of PPA is mainly characterized by language alterations, early deficits in social cognition, emotion recognition and recent memory also stand out. There are also changes in the theory of mind and empathy, mainly in the semantic and agrammatic variants (Fittipaldi et al. 2019).

Depending on the variant, PPA can evolve into a more generalized state of dementia after 7 years of diagnosis, with the logopenic variant being the most aggressive form of PPA (Grossman & Irwin, 2018), and functional neuroimaging is more sensitive in pointing out early hypoperfusion in the same region (Brito-Marques-Brito-Marques, 2023). However, the literature does not provide a more comprehensive explanation as to whether PPA is an independent pathological condition or more of an atypical syndrome found in neurodegenerative diseases (Rahul & Ponniah, 2019). We observed that PPA presents itself in Northeast Brazil with the same clinical characteristics as anywhere else, including cases with pathological confirmation. The biggest difficulties are little support for research even within universities and the delay for patients to reach university services. Cultural comparative studies with other centers must be carried out.

BRIEF HISTORY

In 1892, Arnold Pick first described a progressive language disorder. In this disorder, there was atrophy in the frontal and temporal regions of the left cerebral hemisphere which he called progressive non-fluent or agrammatic aphasia (Hopper et al. 2023). A century later, Mesulam carefully studied six right-handed patients with slowly progressive aphasia, without intellectual and behavioral disorders. In most of these cases, symptoms began in the pre-senile phase due to anomic aphasia in 5 patients, and speech deafness in 1 of them. In the 90s, Neary et al. published the criteria for frontotemporal lobar degeneration. Based on these criteria, it was observed that lesions affecting the right side of the brain had different clinical expressions than those in the left hemisphere. On the right, behavioral variant frontotemporal dementia (bv-FTD) stood out, while on the left, PPA. An impairment in the left frontotemporal region can give rise to two distinct forms of PPA: the non-fluent variant, and the semantic variant (Kertesz et al. 1999). In 2001, in Northeast Brazil, Brito-Marques & Vieira-Mello described a case of non-fluent variant PPA in a patient with a pathological diagnosis of Pick's disease (Figure 1). This was the first case of PPA published in Brazil with pathological confirmation.

Figure 1. Shows the upper left MRI image, showing significant left frontal, temporal, and parietal atrophy; on the right, upper functional neuroimaging of significant hypoperfusion in the frontal region, especially on the right; on the lower left, macroscopic image of the brain with moderate atrophy in the frontal region; and on the lower right the presence of Pick's corpuscles (Professor Paulo Brito Collection)

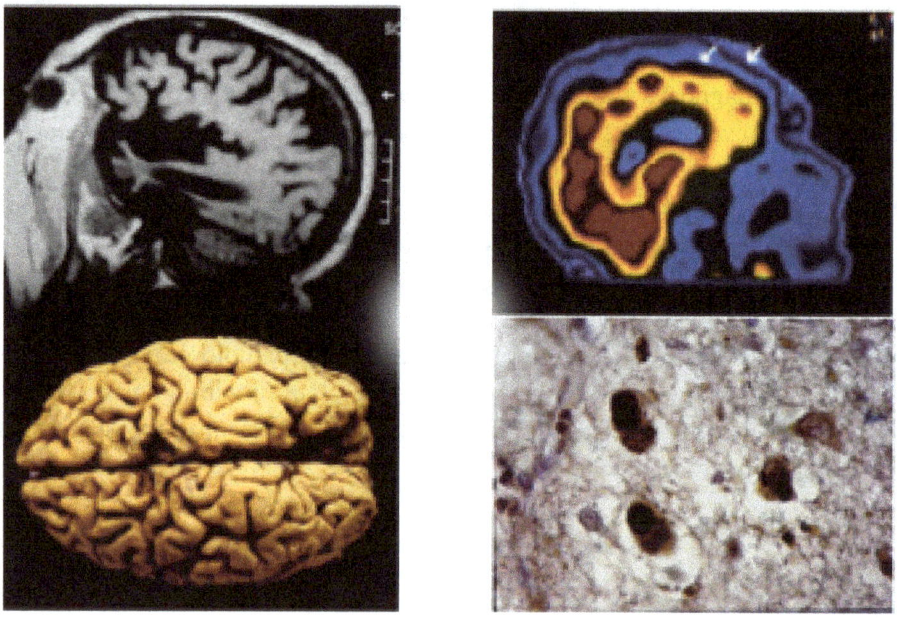

New studies were carried out in 2004 by Gorno-Tempini et al., describing the third variant of PPA, the logopenic variant of primary progressive aphasia (lv-PPA). After the three APP variants were classified, their frequency was described in 2006 by Amici et al. This group carried out a clinical, neuropsychological, and neuroimaging study showing that the most common variants of PPA in regressive order are: 1. Nonfluent variant primary progressive aphasia (nfv-PPA), 2. Primary progressive aphasia semantic variant (sv-PPA), and 3. Logopenic variant primary progressive aphasia (lv-PPA). With APP advances its complexity has also increased, to better study, in 2009, Rogalski et al. proposed the term mild cognitive impairment (MCI) of PPA, a specific designation for the border zone between normal aging and amnestic dementia and PPA. To organize the new APP classification, in 2011, Gorno-Tempini et al. published the clinical criteria for diagnosing PPA and its 3 main variants. In principle, they proposed that patients be diagnosed with PPA and then divided into subtypes according to the language characteristics of each subtype, in addition to supporting neuroimaging and genetic studies. In 2011, after updating the new criteria for PPA, in Northeast Brazil, Brito-Marques et al. published the second case of APP in Brazil with pathological confirmation. It was a case of nfv-PPA in a patient with corticobasal degeneration, whose clinical picture predominated with changes in executive functions, in praxis, and mainly in the language (Figure 2).

Figure 2. Shows upper left structural neuroimaging with significant frontoparietal atrophy, especially parietal; on the upper right, functional neuroimaging shows severe frontal hypoperfusion; the lower left shows a macroscopic image with moderate bilateral frontal atrophy; and on the lower right, an image of balloon cells is seen (Professor Paulo Brito Collection)

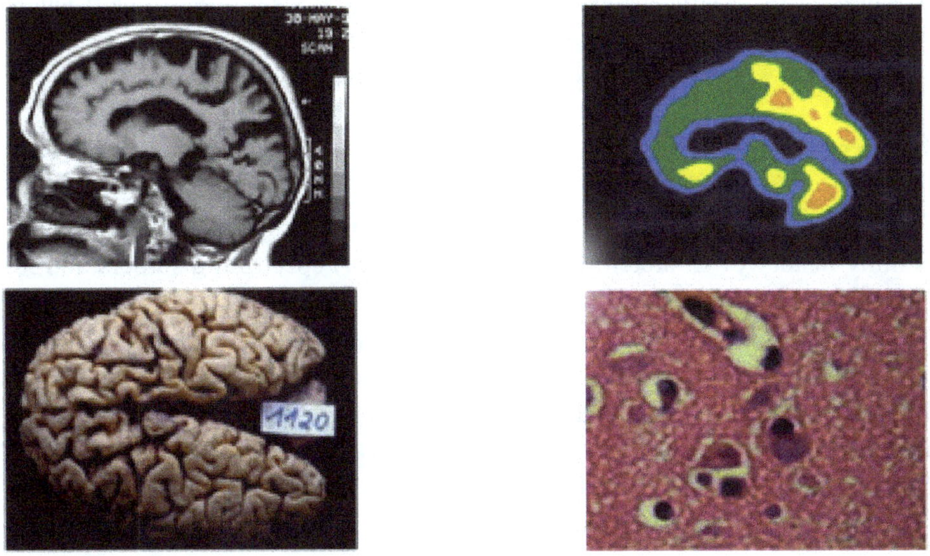

In 2013, another clinical and neuropsychological classification was carried out by Senaha et al. This group studied 100 patients with APP syndromes and obtained a percentage of 35% for APP-vs, 29% for nfv-PPA, and 16% for lv-PPA, confirming the higher frequency of sv-PPA. This result was different from the classification carried out in 2006 by Amici et al. In this same year, Leyton et al. published a clinical and neuropsychologic study. They concluded that patients with PPA-lv are on the verge of global dementia that typically develops very quickly, contrasting with the long period of circumscribed semantic impairment observed in patients with PPA-sv. In 2014, perhaps the largest neuropathological study on PPA was carried out by Mesulam et al. They studied 58 autopsies of patients with PPA and showed the distribution of biomarkers for each APP subtype: 1. Nfv-PPA, 65% showed the presence of tau protein, 18% TDP-43 (TDP-43 DNA binding protein), 12% Alzheimer's disease (AD), and 6% AD + Lewy body dementia; 2. Sv-PPA, 67% had the TDP-43 biomarker, and 33% AD; 3. Lv-PPA, 56% had DA, 25% TDP-43, 16% tau protein, 3% TDP-43, and DA; and 4. Mixed variant PPA (mv), 67% for AD, 17% tau-protein, and 17% TDP-43. In 2016, Rogalski et al. studied patients with lv-PPA with probable AD. The clinical subtypes of PPA were logopenic (n=13) and agrammatic (n=6). In spontaneous speech, word searching was the universal symptom. Language components such as fluency, repetition, naming, and agrammatism ranged from preserved to severely impaired. In 2018, Grossman et al. made an interesting comparison between acute and PPA showing an association between clinical and topographic characteristics in both etiologies. In 2023, Brito-Marques & Brito-Marques reported a case of nfv-PPA in a 60-year-old left-handed patient, showing that brain single photon emission computed tomography (SPECT) detects changes earlier than magnetic resonance imaging (MRI).

Understanding Genetics

Generally, less than a third of cases of primary progressive aphasia (PPA) have a family history of frontotemporal dementia (FTD) spectrum disorders and less than 10% of autosomal dominant inheritance. When subdividing PPA, was found a family history for nfv-PPA in 20-30% and for sv-PPA and lv-PPA in 10 to 20% in FTLD cases (Rohrer et al. 2009). Nfv-PPA is associated with the MAPT and GRN (progranulin) genes, mutations in the C9ORF72, TARDBP, SQSTM1, TBK1, and CHCHD10 genes; sv-PPA presents mutations in the MAPT (tau), C9ORF72, TARDBP, TBK1, TREM 2 and CHCHD10 genes (Delon & Miller, 2018) and lv-PPA is associated with the apolipoprotein-E ε4 gene (Gorno-Tempini et al. 2004).

Basic Classification of Frontotemporal Dementia

Frontotemporal dementia (FTD) is a collective term that encompasses a group of clinical syndromes caused by neurodegenerative diseases characterized by progressive changes in behavior, executive function, and language (Bang et al 2015). Thus, frontotemporal lobar dementia (FTLD) opens up space for symptoms on the right side of the brain of behavioral changes and the left side of the brain, executive dysfunction, and language symptoms as PPA and its variants: logopenic cases,

86% to 90% are due to AD, compared to only 11% to 16% of semantic cases and, 15% to 20% of non-fluent cases (Ossenkoppele et al. 2015). It is believed that the percentage of unclassified PPA cases varies from 30 to 40%, being considered as mv-PPA, due to deficits in grammar and comprehension. The nfv-PPA is clinically overdiagnosed, and there may be false-positive cases, thereby reducing the number of mv-PPA (Teichmann, 2021).

The inclusion criteria for the diagnosis of PPA at the beginning of the syndrome, result in difficulties with language and some interference in activities of daily living, especially in logopenic and semantic variants, while cases of bv-FTD are characterized by behavioral changes, loss of episodic memory and recent visual memory, in addition to visuospatial changes (Vandenberghe, 2011). Each of the disorders of both FTD and aphasic variants presents several specific epidemiological, clinical, laboratory, neuroimaging, neuropathological, clinical management, and evolutionary characteristics. Therefore, the neurologist can diagnose and differentiate its variants and consequent etiologies (Erkkinen et al. 2018).

In addition to the classic classification of the 3 aphasic variants of PPA, two subtypes of PPA can be found with characteristics of the same properties as in the official classification: the anomic ones, which are similar to the logopenic variety, but with an intact repetition, and the mixed ones, which present structural impairment. grammar and word comprehension, even at early stages (Mesulam et al. 2013). Regarding mixed cases of PPA, we can say that they are not differentiated from cases of primary progressive apraxia of speech (PPPA) and other cases of PPA, thus justifying the overdiagnosis of nfv-PPA (Teichmann, 2021).

Semiological Approach to Primary Progressive Aphasia

The speech of patients with primary progressive aphasia (PPA) has often been described simply as 'fluent' or 'non-fluent'. The primary symptoms of patients with PPA are word-finding difficulties, circumlocutions, and incorrect words as phonemic paraphasias (Mioto et al. 2012). Many patients also report problems with memory (Mioto, et al. 2012; Bertola, 2019). Detailed multidimensional quantification of the components of speech production is necessary to adequately characterize the differences between the speech patterns of each PPA variant. Motor speech and linguistic resources derived from a simple functional neuroimaging description are clinically useful in differentiating between PPA variants, localizing the topography, and assisting in the etiological diagnosis of neurodegenerative aphasias (Wilson et al. (2010). Non-verbal activities may point to the patient's level of insight (Figure 3).

Figure 3. Shows a copy of Rey's figure made by a patient, 68 years old, engineer with mixed variant primary progressive aphasia and 7 years of evolution since the first symptoms; the patient believed that his drawing of him looked a little like the model (Professor Paulo Brito Collection)

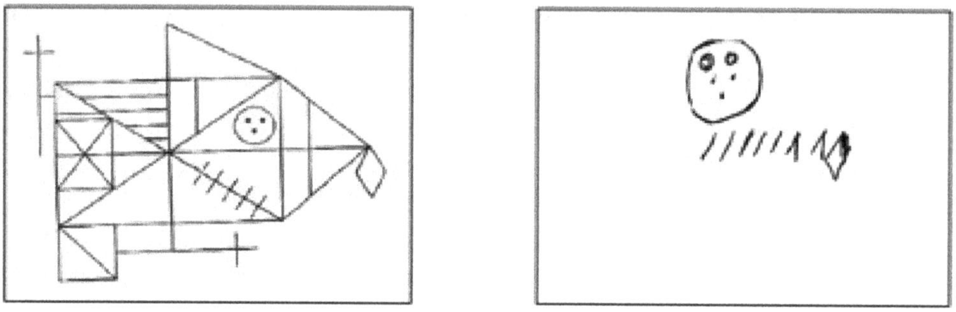

Neurophysiopathology

There is general agreement that, after initial processing in the unimodal sensory cortex, the processing pathways for speech and written language converge to access verbal meaning. However, the existing literature provides conflicting reports regarding the cortical location of this convergence. Most studies of stroke in aphasic topography the domain of verbal comprehension in the temporal-posterior and inferior parietal cortex, while there is evidence of focal cortical neurodegenerative syndromes implicated in the temporal-anterior cortex (Spitsyna et al. 2006).

Functional neuroimaging studies in normal subjects have failed to reconcile these opposing positions. A functional neuroimaging paradigm with spoken and written narratives demonstrated common activation during spoken and written language comprehension. The regions involved were the inferior and lateral portions of the temporal-anterior cortex and the junction of the temporal-occipital-parietal cortex, all on the left. These results indicate that verbal comprehension utilizes unimodal processing streams that converge on anterior and posterior hetero-modal cortical regions in the left temporal lobe (Spitsyna et al. 2006).

Studies out carried at CNADC – Northwestern University of Chicago on language and APP (2011) evaluated the essential and auxiliary characteristics of APP. The deficit depends on the most compromised pathway, whether the ventral (naming, understanding words, knowing people and objects, spelling the whole word) or the dorsal (fluency in searching for words, grammatical construction of sentences, understanding syntax, phonological decoding, word and sentence repetition). If the information goes from the grapheme to the phoneme, the stimulus follows the dorsal route, but if it seeks the entire orthographic form through the stock of acquired knowledge of the word as a whole, it follows the ventral semantic route. However, more details are needed to understand the subsystems and their countless variations suffered in neural networks during the degenerative process imposed by the pathology of PPA syndromes.

A more detailed study using functional neuroimaging showed the distribution of changes across APP subtypes (Mesulam, 2012). This investigation focused on the diagnosis, classification, and clinical-anatomical characteristics of the early and mild impairment stages in a group of 25 consecutively enrolled patients with PPA (Figs. 4a and 4b). In 2014, Mesulam et al. concluded that the anterior portion of the left temporal lobe, a region ignored by classical aphasiology, needs to be inserted into the language network with a critical role in the multisynaptic hierarchy underlying word comprehension and object naming. This statement, as a language neural network, strengthens the ambiguity between nfv-PPA and sv-PPA, at least at the onset of symptoms.

Figure 4. a. Atrophy patterns in PPA-G and PPA-S. Red and yellow areas designate peak atrophy sites at False Discovery Rate = 0.05, full-width at half-maximum = 20 when patient groups were compared to a normal control group of 27 subjects. FG = fusiform gyrus; IFG = inferior frontal gyrus; ITG = inferior temporal gyrus; MFG = middle frontal gyrus; MTG = middle temporal gyrus; OFC = orbitofrontal cortex; PHG = parahippocampal gyrus; STG = superior temporal gyrus; TPC = temporopolar cortex; TPJ = temporoparietal junction (Mesulam et al., 2012) and Figure 4b. Atrophy patterns in PPA-L and PPA-M. Red and yellow areas designate peak atrophy sites at False Discovery Rate = 0.05, full-width at half-maximum = 20 when patient groups were compared to a normal control group of 27 subjects. BA37 = Brodmann area 37; IFG = inferior frontal gyrus; ITG = inferior temporal gyrus; MTG = middle temporal gyrus; OFC = orbitofrontal cortex; PHG = parahippocampal gyrus; STG = superior temporal gyrus; TPC = temporopolar cortex; TOC = temporooccipital cortex; TPJ = temporoparietal junction (Mesulam et al., 2012)

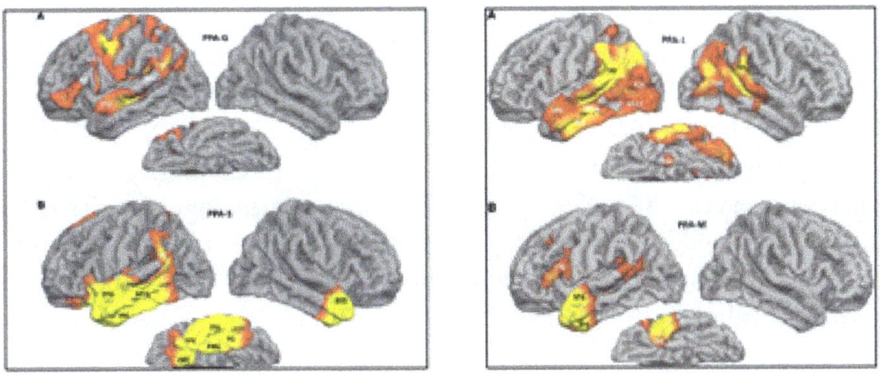

Neuropathology

Impairments in cholinergic cortical neurotransmission may contribute to the formation and deposition of β-amyloid plaques and increased phosphorylation of tau protein in Alzheimer's disease (AD). In both AD and PPA, there is moderate or severe loss of cholinergic neurons in the basal forebrain of the left hemisphere (nucleus basalis of Meynert), especially involving cortical areas related to language, attention, memory, etc. (Schliebs & Arent, 2006). The clinical and neuropathological categorization of patients with progressive aphasia is an area of controversy; its diagnosis must be made on a case-by-case basis (Knibb et al. 2006). These authors studied the post-mortem nfv-PPA (n 23) and sv-PPA (n 15) syndromes and their analysis showed that in the majority of cases, in both groups of pathologies, it was in the spectrum of FTLD: the most common were non-Alzheimer's tauopathy in cases of nfv-PPA (43%) and FTLD with ubiquitin-positive inclusions, tau-negative in cases of APP-vs (53%). Despite the strict exclusion of cases with clinically significant memory deficits or other cognitive impairments, AD pathology is present in approximately one-third of both PPA cases, albeit often with an atypical neuroanatomical distribution. On the other hand, in 2001, Brito-Marques et al. (2001) published a case of Pick's disease with positive ubiquitin and negative tau in a patient with nfv-PPA, different from the majority described by Knibb et al. (2006). According to Mesulam et al. (2014), the association of lv-PPA with AD pathology is much more modest than is suggested by the results of biomarker studies. Individual features of aphasia, such as agrammatism and impaired comprehension, were as informative of the underlying pathology as diagnoses of more laborious subtypes. At the single-patient level, no clinical pattern was pathognomonic of a specific type of neuropathology, highlighting the critical role of biomarkers in diagnosing the underlying disease.

Therapeutic and Non-Pharmacological Approach

In general, there are still no studies that can guide pharmacological treatment in PPAs. There is little indication for a cholinesterase inhibitor or memantine in patients with lv-PPA and nfv-PPA (where Alzheimer's pathology is considered), although any benefit may generally be modest, and there may still be an exacerbation of behavioral symptoms, and newer generation neuroleptics may be indicated to manage severe agitation or psychotic symptoms later in the illness (Marshall et al. 2018). Memantine appears to be well tolerated in sv-PPA and nfv-PPA, but there is no clear evidence of benefit (Boxer et al. 2009). Selective serotonin reuptake inhibitors and Trazodone should be considered in patients with comorbid depression or anxiety and may help resolve behavioral symptoms such as impulsivity, agitation, and aggressive sleep disorder, particularly in sv-PPA (Prodan et al. 2009). New-generation neuroleptics may be indicated to manage severe agitation or psychotic symptoms later in the course of the illness (Marshall et al. 2018). At the onset of symptoms, speech therapy and occupational therapy may have some relevance, depending on the clinical form of PPA and the personality profile of each patient.

NON-FLUENT PROGRESSIVE APHASIA

Non-fluent variant primary progressive aphasia (nfv-PPA) is a syndrome characterized by apraxia of speech, in 82% of cases, mild dysarthria, in 36%, and varying degrees of deficits in the complex processing of syntax, in addition to power is associated with atrophy of the insula and inferior frontal gyrus in right-handers (Gorno-Tempini et al. 2004). Symptoms of echolalia were also found in 63% of cases, with preserved repetition and impairment in comprehension caused by reduced inhibition of the medial frontal cortex and release of the anterior perisylvian area; generally, the other variants of echolalia do not present (Ota et al. 2021). Nfv-PPA may occur in cases of crossed aphasia, as reported in a right-handed patient with a lesion predominantly in the right cerebral hemisphere (Bessi et al. 2018; Canu et al. 2019). The definition of crossed aphasia occurs when brain damage results in the cerebral hemisphere ipsilateral to that of the dominant right hand (Spinelli et al. 2015). Interestingly, a 60-year-old left-handed patient with nfv-PPA, with 8 years of evolution, showed brain SPECT with hypoperfusion of moderate to severe intensity in the left cerebral hemisphere and little evidence of asymmetry in the temporal lobes on the MRI exam (Brito-Marques & Brito-Marques, 2023). On the other hand, it is believed that nfv-PPA is overdiagnosed due to the existence of cases of primary progressive apraxia of speech (PPPA), at least initially, classified as nfv-PPA (Teichmann, 2021).

Epidemiology

There are not many population studies on frontotemporal lobar degeneration, showing a prevalence of 15 to 22 cases per 100,000 inhabitants/year (Galvin et al. 2017). Regarding non-fluent variant primary progressive aphasia (nfv-PPA), the prevalence is 15% of FTLD cases (Erkkinen et al. 2018). Pathologies that fall within this spectrum include cases of Pick's disease (Brito-Marques et al. 1999), corticobasal degeneration (Brito-Marques et al. 2011), and progressive supranuclear palsy (PSP). Nfv-PPA is more common in females, starting between the sixth and seventh decade of life with median survival between 8 and 12 years (Erkkinen et al. 2018).

Simplified Clinical Criteria for Nonfluent Variant Primary Progressive Aphasia. At least one of the following key features must be present (Gorno-Tempini et al. 2011; Rascovsky et al. 2011).

1. Agrammatism in speech production
2. Laborious and interrupted speech with errors and inconsistent distortions of speech sound (apraxia of speech).

At least two of the following three other features must be present:

1. Impaired comprehension of syntactically complex sentences
2. Spared single-word comprehension
3. Spared object knowledge

Diagnostic Features

According to Gorno-Tempini et al. (2011), non-fluent variant primary progressive aphasia (nfv-PPA) is a neurodegenerative syndrome with an insidious clinical course and progressive and lethal evolution. According to these same authors, the clinical diagnosis of nfv-PPA must contain at least one of the main items: I. 1. Agrammatism in language production; and 2. Elaborate and hesitant speech with fluency up to 30% of normal with errors and sound distortions (apraxia of speech). At least 2 of the following 3 items must be present: 1. Impairment in understanding syntactically complex sentences; 2. Understanding preserved for isolated words; and 3. Knowledge of the object. II. The diagnosis of nfv-PPA with neuroimaging support. Both of the following criteria must be present: 1. Clinical diagnosis of nfv-PPA; and 2. Neuroimaging must show one or more of the following results: A. Predominant left frontoinsular atrophy on MRI or B. Predominant hypoperfusion (SPECT-cerebral) / hypometabolism (PET/SCAN-cerebral) in the frontoinsular region left rear. III. The nfv-PPA with pathology definition. The clinical diagnosis and criteria of the following 2 or 3 items must be present: 1. The clinical diagnosis of nfv-PPA, 2. Specific histopathological evidence of neurodegenerative disease (FTLD-tau, FTLD-TDP, AD, and others), and 3. The presence of a known pathogenic mutation (Figure 5).

Figure 5. Shows an image of a copy of Rey's figure made by a 57-year-old patient, PhD in biology with non-fluent variant primary progressive aphasia; the patient was aware of her difficulty and asked to repeat the copy of the model 7 times (Professor Paulo Brito Collection)

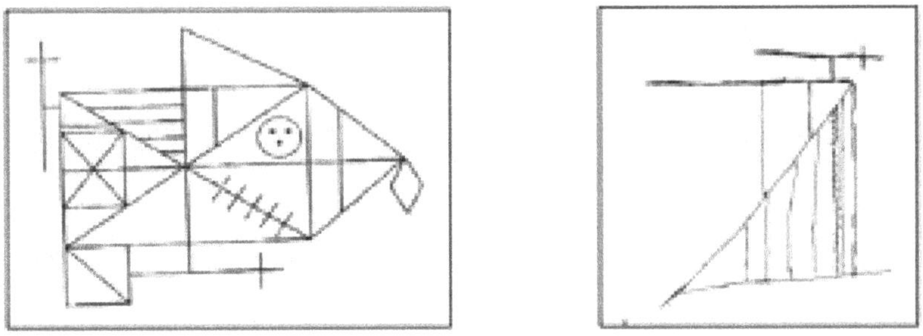

Differential Diagnosis

There are some difficulties in differentiating nfv-PPA in Northeast Brazil. Some phenotypes are not well evaluated at the beginning of the syndrome because more attention is paid to the etiology than the syndrome, such as in cases of progressive supranuclear palsy, Pick's disease (Brito-Marques et al. 2001) or corticobasal degeneration (Brito-Marques et al. 2011) or in cases of yet-to-be-defined amyotrophic lateral sclerosis (Brito-Marques et al. 1999). Second, on some occasions, the patient takes a long time to be taken to the Public Health Service and the initial semiology is based on family history (Brito-Marques et al. 2023). Another difficulty in differentiating it is with cases of progressive apraxia of speech, which may be present at the beginning in 88% of nfv-PPA cases. Finally, 30 to 40% of PPA cases are not classified, leaving them as mixed PPA or apraxia of speech or underdiagnosed with the other variants (Teichmann, 2021).

Complementary Exams

The apolipoprotein-E allele-ε4 is present in 20% of nfv-PPA cases (Gorno-Tempini et al. 2004), and may be associated with biomarkers for Alzheimer's pathology. Structural and functional neuroimaging examination is crucial to help identify the topography of the syndrome, which is sometimes surprising (Figure 6). Brain MRI shows signs of diffuse anterior cortical atrophy, especially in the left inferior frontal gyrus, insula, and anterior superior temporal region (at the beginning of the syndrome, structural MRI may be normal). The fMRI shows signs of atrophy when mapping the same region as the structural MRI. The single photon emission brain scintigraphy (SPECT) exam shows signs of hypoperfusion in the frontotemporal region, predominantly on the left, as well as the brain PET/SCAN exam. Cerebrospinal fluid shows biomarkers compatible with a tauopathy (FTLD-tau) in 52% of cases, AD in 25%, TDP-43 in 19%, and in the remaining cases with miscellany, in 4% (Marshal et al. 2018).

Figure 6. Shows on the left MRI, in the coronal section, of a sinister manic patient with nfv-PPA, shows important signs of frontotemporal cerebral atrophy of mild to moderate intensity with predominance on the left side; on the right shows a brain SPECT image with significant hypoperfusion in the left cerebral hemisphere (Professor Paulo Brito Collection)

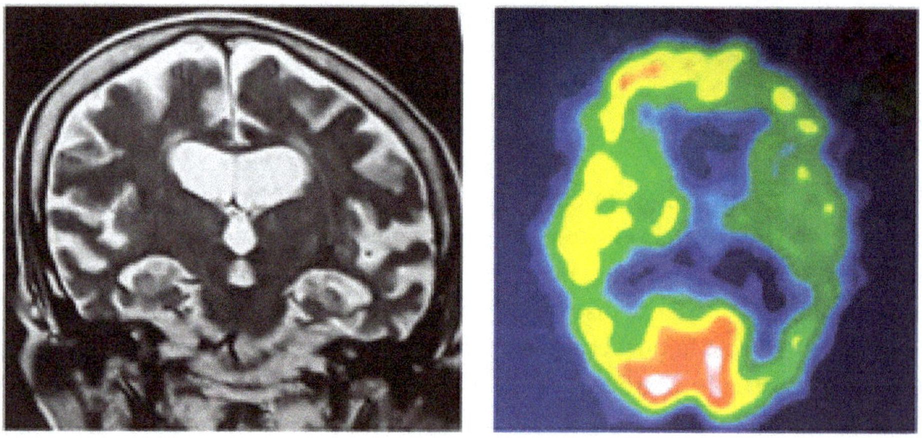

Neuropathology

The majority of cases submitted to autopsy revealed cerebral atrophy of moderate to severe intensity, predominantly in the left frontotemporal region (Mesulam et al. 2013). Under microscopy, neuronal inclusions are seen with the presence of the tau protein associated with microtubules, as in FTLD-tau (Mesulam et al. 2013). Under the macroscopic aspect, AD almost always led to peak atrophy that included the temporoparietal junction; TDP-C almost always led to severe anterior temporal atrophy (Mesulam et al. 2021); Pick's disease routinely caused combined atrophy of anterior temporal and prefrontal cortex (Brito-Marques et al. 2001); PSP and CBD tended to be associated with surprisingly modest cortical atrophy, usually in dorsal premotor or inferior frontal cortex (Mesulam et al. 2021).

Treatment

Plasticity is highly dependent on experience and as in any learning system, language reorganization involves a synergistic interaction between the organism-intrinsic, cognition, and brain, in addition to the organism's extrinsic variables and environmental stimuli (Thompson, 2019). A recent study showed that transcranial stimulation can improve, after 2 months of stimulation, word naming in nfv-APP carriers (Zhao et al. 2021). However, nfv-APP does not yet have an approved etiological treatment. At the beginning of symptoms, anxiolytics and antidepressants can be administered to help with symptoms of low mood, irritation, aggression, and sleep disorders. In 2004, Moretti et al. used acetylcholinesterase inhibitors in 20 patients with FTD, achieving a reduction in behavioral symptoms. In 2008, Kertesz et al. showed that galantamine was not behaviorally effective in patients with VC-FTD, but a trend of efficacy was shown in the aphasic subgroup, which may be clinically significant and safe. Memantine appears to be well tolerated in nfv-PPA with Alzheimer's pathology, but there is no clear evidence of benefit (Marshall et al. 2018).

PRIMARY PROGRESSIVE APRAXIA OF SPEECH

The topic of primary progressive apraxia (PPA) was introduced in this part to help the reader differentiate it from the nfv-PPA. Progressive Primary Apraxia of Speech impairs the planning of programming of sensorimotor commands necessary to direct the movements that result in speech and prosody. Thus, it limits communication due to the reduction in speech fluency, substitution of sounds, articulatory attempts, false speech starts and restarts, and syllable segmentation, in addition to an increase in difficulty depending on the increase in the utterance (Beber et al. 2018).

This clinical entity occurs in 20% of all progressive speech and language disorders, in addition to sharing nfv-PPA with agrammatism. The predominance of phonetic impairment (sound level errors with distorted substitutions or additions) or prosodic impairment (such as slow speed or segmented speech) are the main sources of heterogeneity in the disorder. This has led to the recognition of APPF subtypes such as phonetic (formally Type 1), prosodic (formally Type 2), and mixed (formally Type 3) (Josephs et al. 2014; Botha & Josephs, 2019). The classic complaint of the PPAS patient is: I know what I want to say, but I can't get the words out. During clinical assessment, there are changes in verbal and semantic fluency and articulation with dysarthria and episodes of stuttering. Around 75% of cases present buccolabial apraxia when clicking the tongue, releasing a kiss, blowing, eating a pine cone, and separating the seeds in the mouth (Josephs et al. 2014; Botha & Josephs, 2019).

In 2019, Botha & Josephs revised the previous criteria (Josephs et al. 2012) and formulated new criteria for diagnosing primary progressive apraxia of speech.

Inclusion

1. Insidious and slowly progressive onset of speech disorder.
2. Apraxia of speech is the only symptom or dominates the speech disorder.
3. Dysarthria may be present, but less intense than apraxia.
4. Any evidence of aphasia is considered erroneous.

Exclusion

1. Deficits occur more frequently in other non-neurodegenerative diseases.
2. Cognitive disorders are best seen in psychiatric disorders.
3. Evidence of detailed aphasia on neuropsychological assessment (patient may meet criteria for primary progressive aphasia).
4. Dysarthria is considered more serious than apraxia of speech.
5. Initial predominance of impairment with episodic memory, and visual and visuospatial memory (the patient may present criteria for dementia in typical or atypical Alzheimer's disease).
6. Initial predominance of behavioral disorder (patient may meet criteria for behavioral variant frontotemporal dementia).
7. Initial predominance of parkinsonism or eye movement disorder (patient may meet criteria for progressive supranuclear palsy).
8. Initial predominance of ideomotor apraxia, parkinsonism, dystonia, or other asymmetric cognitive or motor features excluding speech/language (patient may meet criteria for corticobasal syndrome).
9. Predominance of impairment in upper or lower motor neurons (the patient may meet criteria for motor neuron disease.

SEMANTIC VARIANT OF FLUENT PRIMARY PROGRESSIVE APHASIA

Semantic variant primary progressive aphasia (sv-PPA) is a disorder of fluent language, loss of the concept of words and objects, presence of anomia, and difficulty understanding a word in the semantic language, in addition to an impoverishment of vocabulary and facial recognition (Snowden et al. 1989). Sv-PPA was first described by Snowden et al. (1989) as a form of circumscribed cerebral atrophy.

However, it is important to recognize that some patients with semantic dementia, despite sharing the same pathological associations, may not meet the criteria for PPA (Botha & Josephs, 2019).

In a study carried out by Caramazza et al. (1982), there is a difference between the semantic structure of words, the visual-perceptive process, and the categorical process. Many aphasic people have difficulty speaking or recognizing the names of objects, and some aphasics, have perceptual similarity of pairs of objects but are unable to classify the objects into their lexical categories. According to these same actors, classification disorder results from impairment at the level of semantic perceptual analysis which, guided by the characteristics of categories and objects, has a verbal code of information represented in basic motor and perceptual modalities. Each feature provides flexible memory values for words and meanings. Caramazza et al. (1982), different people can classify objects differently and the same people can change this classification over time. If objects are classified by their characteristics, they can change when new characteristics are learned, needing to be within the limits of categories to flexibly guide the use of concepts. According to the topography and the association with other symptoms, APP-vs may present with different clinical characteristics. Its combination with other symptoms that are not part of it indicates, in terms of evolution, the emergence of different etiologies (Caramazza et al. 1982).

A study carried out by Warrington et al. (1984) showed that patients with herpetic encephalitis described a new classification of semantic deficits. There was selective deficiency and preservation of certain categories of visual stimuli for living things and foods, which are widely spoken, identified, and understood verbally. Living things and food appeared to suffer selective harm from confrontation while inanimate objects were selectively preserved. According to these authors, the real objects were within categories and these contained the characteristics common to a group. There was the possibility of category specificity in the organization of meaning systems that are also semantic and modality-specific systems.

Specific category deficiencies require explicit category organization in semantic memory. The organization of specific categories can emerge from the perceptual, functional, and associative characteristics with information about objects that are maintained in the language process. The emergence of semantic categories in neural networks may emerge from the inherent nature of semantics and present itself, but not by the special internal categorical organization of semantic memory (Small et al. 1995). Categories such as fruits, vegetables, and other types of food depend more on their colors than any other characteristic. For example, the patient doesn't know when asked to describe rice, but if asked what color it is, he will say white. Vehicles depend more on their composition and type of movement than on other details. Many kitchen tools and items depend greatly on the number of their limbs or practical movements. Animal identification depends on a variety of characteristics without a single predominant contribution (Small et al. 1995).

Cognitive deficits in sv-PP-vs have been attributed to damage to the gray matter of the anterior temporal lobe; The key aspects of the syndrome may be due to altered anatomical connectivity between the language networks that involve this lobe, but not solely due to cortical impairment (Agosta et al. 2010). In sv-PPA, significant changes are more transmitted by the inferior longitudinal fasciculus and less by the arcuate and uncinate fasciculi. Therefore, this results from greater anatomical damage to the white matter of the superior and inferior temporal gyrus, left hemisphere connections involved in semantic and lexical processes, although there is relative preservation of the superior longitudinal frontoparietal fasciculus. According to Agosta et al. (2010), these data contribute to the understanding of the anatomical changes that occur in sv-PPA and may help explain the dissociation between a single word and some, although it spares phonology and fluency in cases of sv-PPA.

Epidemiology

There are still no population studies on the epidemiological aspects of APP and variants, although the incidence of FTLD is between 1 and 17 cases per 100 thousand inhabitants/year in patients under 70 years of age (Erkkinen et al. 2018). Among cases of FTLD, APP-vs is between 2 and 7% of cases, and DFT-vc is three times more common. The onset of symptoms is between the sixth and eighth decade of life and the median survival is between 10 and 12 years. The gender ratio is similar (Erkkinen et al. 2018).

Simplified clinical criteria for semantic primary progressive aphasia

Both items below must be present (Gorno-Tempini et al. 2011; Rascovsky et al. 2011).

1. Appointment due to impaired confrontation and
2. Impaired single-word comprehension

At least three of the following other diagnostic features below must be present:

1. Impaired object concept, especially for items of low frequency or familiarity
2. Dyslexia or surface dysgraphia
3. Spared repetition
4. Spared speech production (both grammar and speech motor skills)

Clinical Features

Semantic variant - Primary Progressive Aphasia is a stereotypical syndrome characterized by fluent aphasia and empty speech, severe anomie, reduced vocabulary, and impairment in understanding isolated words. It progresses to virtual and complete dissolution of the semantic components of language, progressing to dementia (Hodges et al. 1992). There is a marked reduction in the generation of exemplars of selective semantic categories (animals, fruits, vegetables, furniture, tools, etc.), which is a consistent and initial characteristic of this syndrome. Semantic memory demonstrates poverty over a variety of living and artificial items. In contrast, phonology, spoken language grammar, and repetition are preserved, as is the understanding of complex syntactic commands. In reading there is a pattern of surface dyslexia/dysgraphia. Autobiographical, episodic, and nonverbal memory are relatively maintained. When the right temporal lobe is affected, usually 5 years after the onset of symptoms, depression, difficulty recognizing facial emotions, prosopagnosia, loss of criticism, or appropriate social behavior may occur; other less frequent symptoms are topoagnosia, associative agnosia, and behavioral and personality changes with characteristics of Klüver-Bucy syndrome (Hodges et al. 1992). In the older literature, such cases would have been submitted under the rubric of Pick's disease (Hodges et al. 1992).

Differential Diagnosis

The symptoms that form sv-APP must be differentiated from other symptoms that are found in several distinct syndromes such as Alzheimer's syndrome, frontotemporal dementia, Gerstmann syndrome, progressive mixed aphasia, etc., as follows:

1. APP-vs must be differentiated from bv-FTD because, throughout the clinical picture, the former may present behavioral symptoms similar to those of bv-FTD. However, bv-FTD does not present early changes in semantic memory and behavioral changes occur from the beginning of the clinical picture.
2. Sv-PPA must be differentiated from Alzheimer's disease (AD) due to the complaint of forgetfulness reported by the family and/or patient, confusing lexical amnesia with the loss of the concept of words and objects. The topography of atrophy can also be confusing: in sv-PPA it predominates in the anterolateral temporal cortex while in AD, in the medial portion of the temporal lobe.
3. Gerstmann syndrome can mimic semantic aphasia. Different explanations for Gerstmann syndrome have been proposed: 1. The pathological process localized in the left parietal white matter disconnecting the cortical networks

and 2. Disturbance in the ability to verbally mediate some spatial knowledge. Both explanations seem coherent (Ardila, 2020).

Complementary Exams

Generally, the apolipoprotein-E allele-ε4 is not found in sv-PPA (Gorno-Tempini et al. 2004). The topographic diagnosis of the syndrome depends on structural and functional neuroimaging exams. Magnetic resonance imaging (MRI) of the skull shows asymmetric atrophy in the anterior-inferior and mesial temporal lobe, especially on the left (Marshal et al. 2018), mimicking the saber-cut appearance seen in cases of Pick's disease. The pattern of atrophy in sv-PPA predominates in the ventral-anterolateral portion of the temporal lobe increased by atrophy of the temporal cortex, but not by atrophy of the mesial portion of the temporal lobe, as seen in cases of AD. Simple positron emission tomography (SPECT) and positron emission tomography (PET/SCAN) indicate hypoperfusion or hypometabolism, respectively, predominating in the anterior portion of the dominant temporal lobe (Figure 7). Cerebrospinal fluid examination is recommended in the search for changes in biomarkers that, in sv-PPA, show trends towards FTLD-TDP-43 in 69%, Alzheimer's disease in 25% and Pick's disease with FTLD- tau (+) in 6% (Gorno-Tempini et al. 2004). When biomarkers are compatible with Alzheimer's pathology, there is a therapeutic indication. However, sometimes the etiology of APP-vs surprises us by appearing in FTD caused by a genetic mutation in a case of myopathy (Figure 8).

Figure 7. Shows on the left a structural MRI image with signs of atrophy in the anterior portion of the temporal lobe, predominantly on the left, and on the right, a cerebral SPECT image with significant hypoperfusion in the anterior portion of the temporal lobe, also predominantly on the left (Professor Paulo Brito Collection)

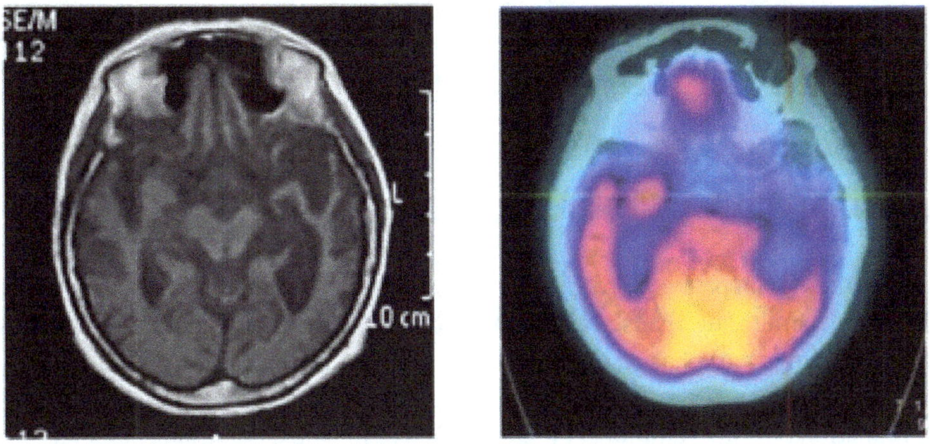

Figure 8. Biceps muscle biopsy: A) Haematoxylin and eosin staining showing inflammatory reaction and macrophagy; B) Modified Gomori Trichrome: rimmed vacuoles and wide range of fibers sizes; C-E) Immunohistochemistry by peroxidase method: Labeling of lymphocytes CD4, CD8 and macrophages CD68, respectively. Optical microscopy; Lens=40x (Professor Carolina Cunha Collection)

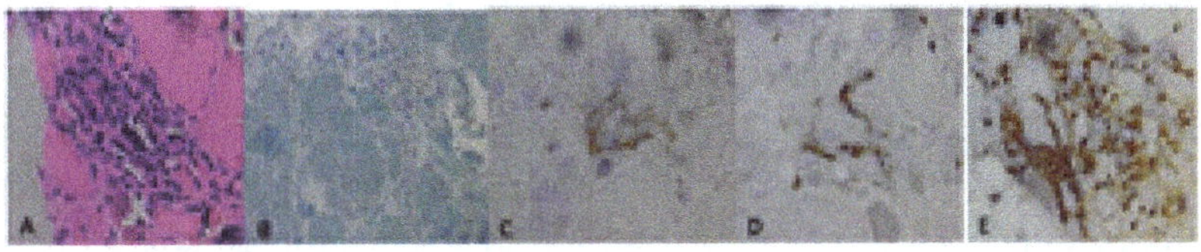

Neuropathology

When examining the brain, macroscopically, diffuse atrophy was observed, a little more than expected for the age group, and severe atrophy in the anterior and ventral portion of the temporal lobes, especially on the left. The majority of sv-PPA cases (67%) present intracellular inclusions of the TDP-43 protein and, to a lesser extent (33%), AD markers are found (Mesulam et al. 2014).

Treatment

There is no pharmacological treatment for sv-PPA, but speech therapy rehabilitation is necessary from the beginning of the diagnosis (Marschall et al. 2018a). When changes in mood and behavior appear, selective serotonin reuptake inhibitors and/ or Trazodone are used (Prodan et al. 2009). Newer generation neuroleptics may be indicated to manage severe agitation or psychotic symptoms later in the illness (Marshall et al. 2018).

LOGOPENIC VARIANT PRIMARY PROGRESSIVE APHASIA

Logopenic variant primary progressive aphasia (lv-PPA) is a syndrome that presents fluent speech, although paused and with many hesitations due to searching for words. Their phonemic paraphasias occur a lot during spontaneous language and selective naming. Dysnomia may be the only significant objective finding on cognitive examination (Gorno-Tempini et al. 2004). One of the most common characteristics is the difficulty of repeating both words and phrases. Grammar and comprehension are, at first, relatively preserved. Lv-PPA was first described by the group led by Gorno-Tempini, et al. (2004).

Lv-PPA remains less understood than the two other PPA variants. This may be due to the lack of sufficient studies to cover the cognitive, topographic, and biological aspects, leaving the current diagnosis with the label of "logopenic aphasia complex" (Teichmann et al. 2013). It is considered a distinct syndrome from primary progressive aphasia (Rogalsky et al. 2016); Some patients begin the syndrome with anxiety, lack of concentration, and loss of episodic memory followed by language changes.

Functional neuroimaging indicates deficits in the functions of the phonological loop in the left posterior temporal lobe, which may point to the essence of the underlying mechanism. Structural neuroimaging may be normal early in the syndrome. Studies suggest that AD biomarkers can indicate the underlying pathology in most cases, although their pathological distribution still awaits further studies. However, cases with Lv-PPA point more to a clinical presentation of atypical AD than to a presentation with clinical characteristics of typical AD (Ahmed et al. 2012).

A study (Petroi et al. 2020) investigated the cortical regions associated with phonological errors in patients with lv-PPA. He suggested that the most relevant phonological errors were related to temporoparietal atrophy, including the left supramarginal gyrus. On the other hand, Cabrera-Martin et al. (2016) described a case of lv-PPA in a right-handed patient with a right brain injury. For these authors, APP-vl should not be considered as a lesion that starts in the left hemisphere, but a syndrome characterized by a brain asymmetry with a preference for areas and neural networks involved in language. According to Mesulam et al. (2021), asymmetric neurodegeneration is a unique and consistent feature in all PPA variants found, but in lv-PPA the asymmetry appears to be less distinct.

Epidemiology

There is no longitudinal study with lv-PPA, so the experience is still individualized for each author. Its clinical diagnosis varies both in the appearance of symptoms and in its evolution. It is known that it is a rare syndrome that expresses itself before the age of 65, pre-senile, perhaps in most cases, as an atypical form of AD (Gorno-Tempini et al. 2008). About 86 to 90% of cases of lv-PPA are due to Alzheimer's pathology, compared with 11% of cases of sv-PPA and 15 to 20% with nfv-PPA (Ossenkoppele et al. 2015).

Classification

The characteristics of lv-PPA can be divided into two subvariants according to 1. The degree of severity of the disease, nature, and distribution of the lesions have important implications for the diagnosis and management of the patient, and 2. The potential future with the use of drugs. Lv-PPA, caused by Alzheimer's pathology, differs from its two variants by demonstrating greater hypoperfusion in the regions of the inferior parietal and posterosuperior and middle temporal cortex (Teichmann et al. 2011). In this case, the syndrome progresses to dementia within 12 months, while sv-PPA progresses within 3 years, regardless of education level and gender. The decline is most accelerated not only in tasks that rely heavily on verbal skills, such as language, fluency, and verbal episodic memory but also in attention/orientation and visuospatial domains (Leyton et al. 2013).

Within the same lv-PPA syndrome, there are variants according to the severity of the aphasia and the degree of cognitive impairment and temporoparietal atrophy. However, these variations do not change the duration of the disease. According to (Machulda et al. 2013) the proposed variants are 1. Patients with mild aphasia in short-term illness (lv-PPA typical mild form); 2. Patients with mild aphasia and long-lasting disease (lv-PPA atypical mild form); and 3. Patients with severe aphasia and moderate disease duration (lv-PP typical severe form). All three forms show left temporoparietal atrophy. The mild atypical group shows less atrophy, despite the longer duration of the disease, in addition to mild neuropsychological deficit, being a slowly progressive variant of lv-PP. To Henri & Gorno-Tempini (2010) the logopenic variant is a distinct subtype of progressive aphasia that may have value as a predictor of underlying tauopathy-like pathology and, especially, atypical AD.

Diagnostic criteria for APP-logopenic variant (Henry & Gorno-Tempini, 2010; Gorno-Tempini et al. 2011).

I. For the clinical diagnosis of APP-vl, both of the essential features below must be present:

1. Impairment in word search in both spontaneous speech and naming;
2. Impairment in repeating phrases and sentences.

At least 3 of the following characteristics must be present:

1. Phonological errors in both spontaneous speech and naming;
2. Understanding isolated words and knowledge of the object;
3. Normal speech expression;
4. Absence of clear agrammatism.

Clinical Features

The most common clinical features in APP-vl are the following: spontaneous speech is relatively fluent with a slow tempo, but filler words and circumlocutions are frequent; hesitant with pauses looking for words, impaired repetition (show the patient a clock and, after naming it, ask him to repeat the word clock) of words and phrases, even with the help of a clue, repetition is impaired; selective naming is moderately affected. Generally, grammar, articulation, and episodic memory are preserved. Although comprehension is reduced, semantic and single-word knowledge, grammatical structure, and speaking skills are all spared, which distinguishes lv-PPA from sv-PPA and nfv-PPA. Three or more altered language components are expected: phonological errors in speech, misunderstanding of more than 4 words, and failure to recognize objects (Gorno-Tempini et al. 2011).

Some individuals may not present any cognitive impairment other than language. Others demonstrate more pronounced cognitive impairment in verbally mediated memory tasks (e.g., as in Rey's Auditory Verbal Learning or Logical Memory task), and still others demonstrate more global cognitive impairment (Owens et al. 2018). As the disease progresses, the patient speaks in single words, non-words, letters, and digits until he acquires a condition of mutism, which is accompanied by grunts or unintelligible noises.

Differential Diagnosis

Generally, the differential diagnosis of lv-PPA is challenging, especially regarding deficiencies in long sentence repetition skills, semantic reversibility, and syntactic complexity, two psycholinguistic variables known to place additional demands on phonological working memory (Figure 9) (Macoir et al. al. 2021). The speech of patients with PPA has often been described simply as 'fluent' or 'non-fluent', but it must be remembered that fluency is a complex construct that encompasses characteristics such as speed and timing of speech, articulatory agility, and syntactic structure, which they do not always decline in parallel. Motor speech and linguistic features, derived from a simple image description, are clinically useful in differentiating between PPA variants and have clear anatomical localization power that can aid diagnosis (Mesulam, 2012). However, some patients may develop, throughout the clinical course, signs consistent with both nfv-PPA and sv-PPA, making the diagnosis of lv-PPA syndrome difficult (Wilson et al. 2010).

Figure 9. Shows a patient with logopenic variant primary progressive aphasia. The patient changed the position of Rey's figure, saying it was better his way. His language trying to say what he was doing was unintelligible (Professor Paulo Brito Collection)

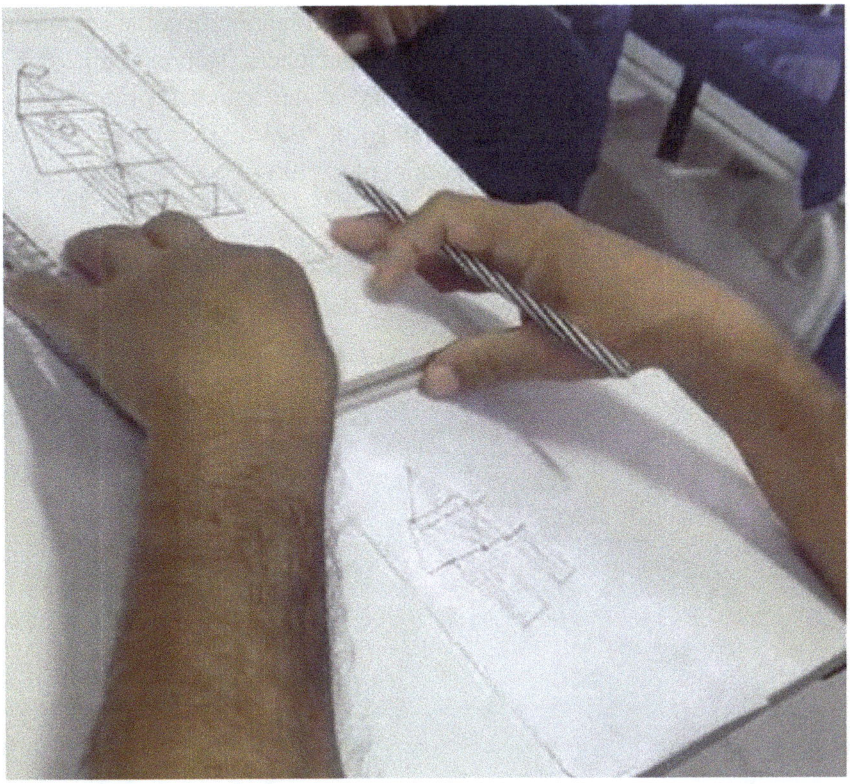

Complementary Exams

The apolipoprotein-E (APO-E) ε4 allele is present in 67% of APP-vl cases (Gorno-Tempini et al. 2004). MRI examination of the skull may show atrophy predominantly in the posterior third of the left temporal superomedial and parietal posteroinferior gyri (Marshal et al. 2018). However, this predominance of the left temporoparietal junction follows a relatively poor standard between sensitivity and specificity for diagnosis. Brain SPECT or brain PET/SCAN examination reveals hypoperfusion or hypometabolism, respectively, of the temporoparietal junction and superior temporal gyrus (Figures 10 and 11) with greater concentration in area 42 and, to a lesser extent, in frontal areas 8, 9 and 46 from Brodmann; the hippocampi are not affected. APP-vl, due to Alzheimer's pathology, is more aggressive and extensive, causing linguistic disorders of the lexicon, semantic and phonemic sequence, and cognitive processes in working memory, verbal and ideomotor apraxia (Teichmann et al. 2013). A longitudinal study with MRI volumetry of the skull showed that, at the beginning of the disease, there is involvement of both hemispheres, but as the disease progresses with worsening of anomia, repetition, and comprehension of sentences, the changes in the neuroimaging examination become asymmetrical with a clear predominance of the left cerebral hemisphere (Rohrer et al. 2013). The examination of cerebrospinal fluid is indicated in the search for biomarkers that point to the possible substrate of the pathology of AD, in 50%; DLFT-TDP-43, in 38%, and DLFT-tau, in 12% (Gorno-Tempini et al. 2004).

Figure 10. Shows cerebral PET/SCAN imaging with moderate glycolytic hypometabolism in the left temporal lobe and to a mild degree in the left frontal, parietal, and occipital lobes. Patient with logopenic variant resulting from Alzheimer's pathology (Professor Paulo Brito Collection)

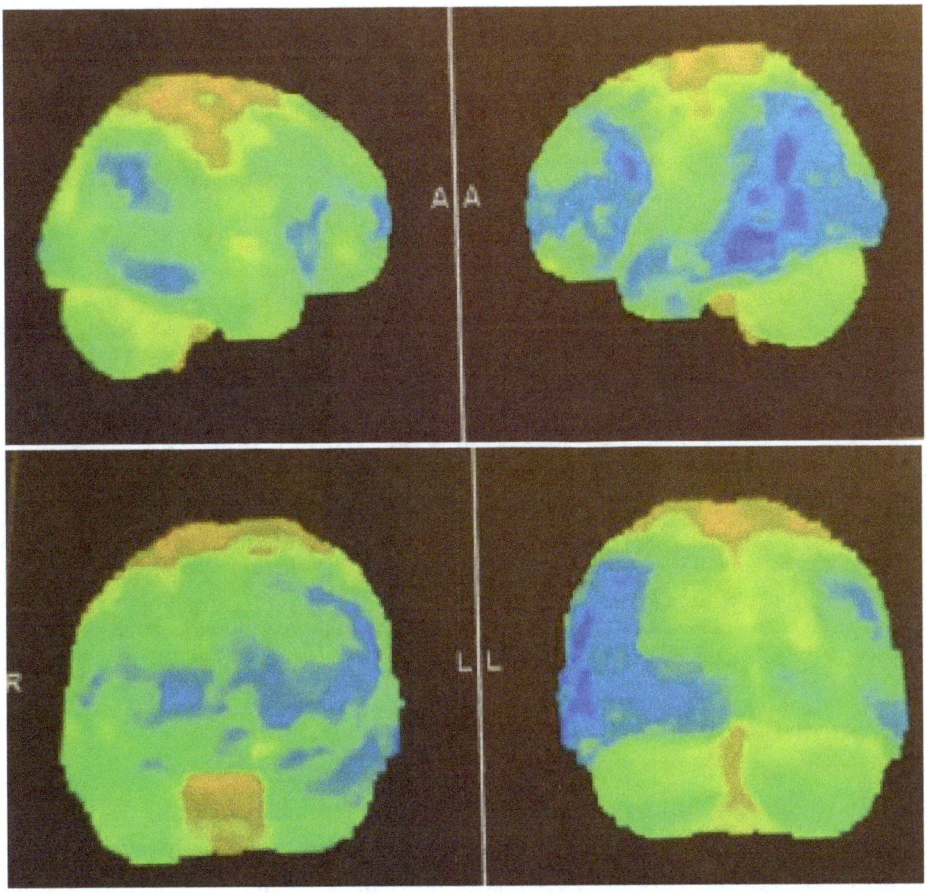

Figure 11. Brain PET/SCAN examination shows asymmetrical glycolytic hypometabolism in the temporoparietal region, mainly on the left, resulting from tauopathy suggestive of APP-vl, although with a clinical picture and evolution compatible with mv-PPA or logopenic aphasia complex (see Figure 13) (Professor Paulo Brito Collection)

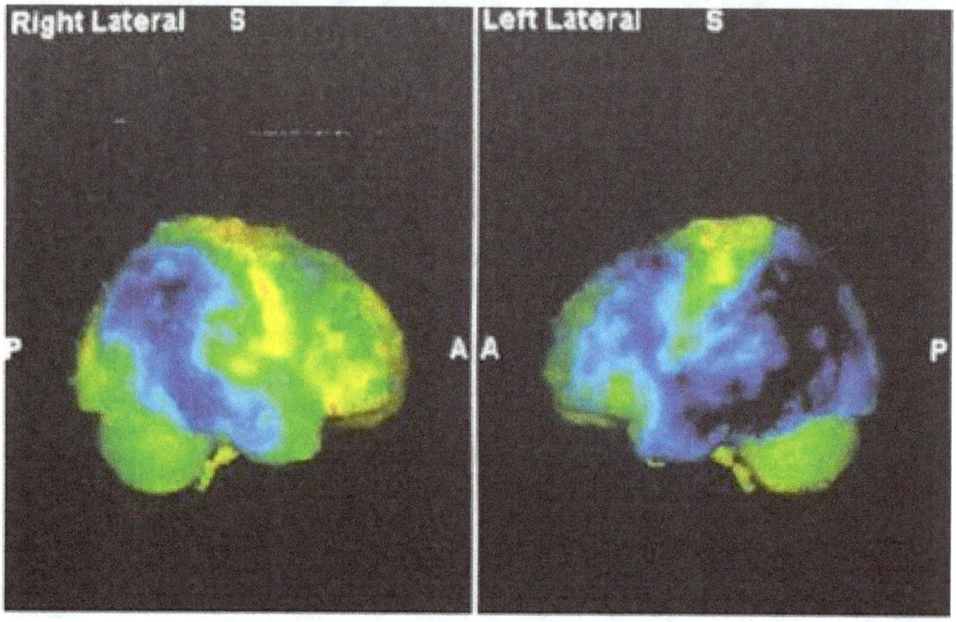

Neuropathology

Autopsy studies are still modest. In 2014, Mesulam et al. performed 58 autopsies on patients with PPA and their histopathology showed that 45% had Alzheimer's disease (AD) and 55% had FTLD. On macroscopy, there was temporoparietal atrophy, mainly on the left. AD pathology in PPA exhibited multiple atypical features. The male gender tended to predominate with more intense neurofibrillary degeneration in the left hemisphere, the hippocampus-entorhinal Braak pattern was inclined in favor of the neocortex. The association of APP-vl with AD pathology was much more modest than was suggested by the results of in vivo amyloid imaging studies and the APO-E ε4 allele would not be a risk factor. Individual features of aphasia such as agrammatism and impaired comprehension were as informative of the underlying pathology as more diagnostically burdensome subtypes. At the single-patient level, no clinical pattern was pathognomonic of a specific type of neuropathology, highlighting the critical role of biomarkers in diagnosing the underlying disease (Figure12).

Figure 12. Shows a microscopic image of a patient who had a logopenic variant of primary progressive aphasia with a very long evolution of symptoms lasting 15 years until clinical diagnosis. The patient started the syndrome by persevering common nouns. The autopsy revealed markers of Alzheimer's disease (Professor Paulo Brito Collection)

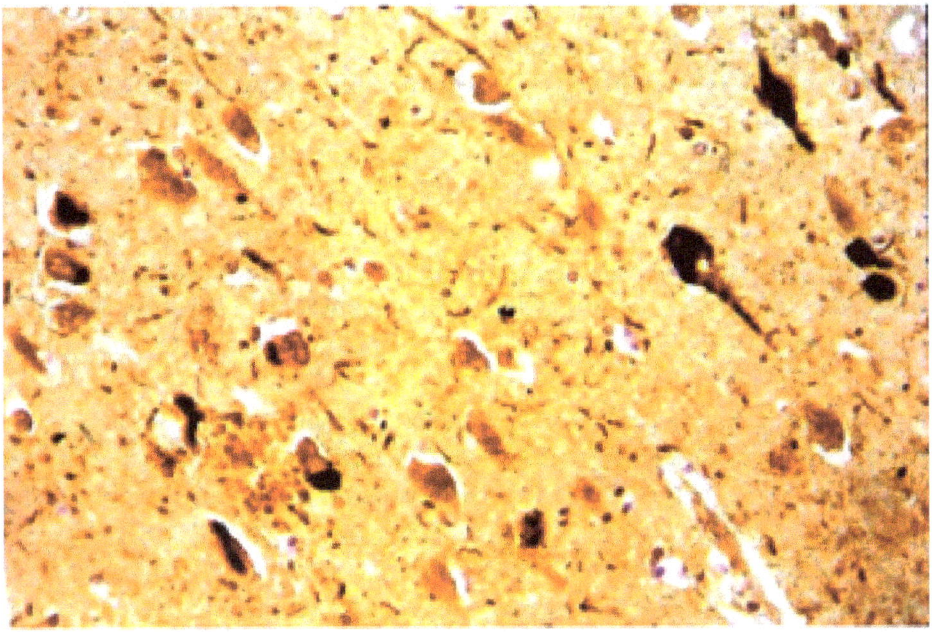

Treatment

In principle, there is no pharmacological treatment for cases of PPA that do not present significant changes in the cerebrospinal fluid such as those found in Alzheimer's pathology. There is a low threshold for trying a cholinesterase inhibitor or memantine in patients with lv-PPA and nfv-PPA (where Alzheimer's pathology is considered), although any benefit is generally modest, there may be an exacerbation of behavioral symptoms (Marshall et al. 2018). Selective serotonin reuptake inhibitors and trazodone should be considered in patients with depression or anxiety and may help resolve behavioral symptoms of impulsivity, agitation, aggressiveness, and sleep disorder (Prodan et al. 2009). Newer generation neuroleptics may be indicated for severe agitation or psychotic symptoms later in the course of the illness (Marshall et al. 2018).

MIXED VARIANT PRIMARY PROGRESSIVE APHASIA OR UNCLASSIFIED

Mixed variant primary progressive aphasia (mv-PPA) is a language disorder that presents clinical characteristics consisting of agrammatism with deficits in word comprehension, even in the early stages of the syndrome (Mesulam et al. 2012). Furthermore, it is usually accompanied by poor fluency and frequent paraphasias (Mesulam et al. 2009). Another more recent definition suggests that: 1. Mv-PPA is associated with the language profiles of lv-PPA, sv-PPA, or nfv-PPA; 2. Your involvement must be taken into account, especially with APP-vl; 3. Have an incomplete language profile; 4. Be considered in a very early stage of PPA; and 5. Genetic forms must also be considered (Teichmann, 2021). During the neuropsychological assessment, the patient's level of education must be taken into account to establish the level of testing (Mesulam et al. 2009). A study showed mild impairment with the mv-PPA phenotype that may emerge as a distinct clinical disease (Mesulam et al. 2012). According to this phenotype of this study, the greatest emphasis on atrophy was located in the most anterior portion of the left temporal lobe and the left inferior frontal gyrus (in this last topography there is a deficit in word comprehension and consequent anomia) (Figure 4b). Other patients in this study were not classified at baseline but exhibited relatively isolated but severe anomia. Two of these progressed to meet the criteria of APP-vs. Possibly, the third party can also do it in time. However, there is a possibility that some patients will maintain this profile, representing an anomic variant of nfv-PPA (Figure 13).

Figure 13. Shows a 67-year-old patient, an engineer, with an 8-year history since the first symptoms, significant anomia, unintelligible spoken language, except for exclamations, reasonable contextual understanding, normal repetition and reading, dysgraphia and acalculia, verbal and semantic fluency no response, MEM=6/30, Pfeffer=15/30. When trying to build the WAIS cubes, the patient believes that their construction is similar to the model. A patient who does not meet the criteria for the time of evolution, cognitive impairment, and develops normal repetition and reading, despite showing glycolytic hypometabolism in the left temporoparietal lobes would have mv-PPA or be part of the logopenic aphasia complex (Teichmann et al. 2013).

Patients with the mv-PPA label accumulate symptoms of semantic, lexical variants, and syntactic disorders. The crucial issue would be to apply such specific implicit testing measures in clinical practice to delineate genuine language impairment, individualizing the most damaged language component and assigning a clear language variant diagnosis of PPA cases (Teichmann, 2021). The mv-PPA call should demonstrate a predominant impairment of a language component that leads to a clear variant classification. Concomitant but less pronounced damage to other components of language may simply reflect that the degenerative process is extending over time, affecting supplementary components of language, or extralinguistic domains (Teichmann, 2021).

Summary

In conclusion, primary progressive aphasia is a syndrome that is characterized by a slow and progressive loss of language in the first years and gradually compromises other cognitive domains. Family history of PPA does not exceed 30% of cases, especially in the non-fluent and semantic variants; the APO-E allele ε4 gene may be associated with cases of logopenic variant Alzheimer's pathology. Depending on its clinical form, it may progress more quickly, like the logopenic variant, or not, like the non-fluent variant. Its topography is located in the left cerebral hemisphere and predominates in the inferior temporal lobe (nfv-PPA), in the anterior and inferior temporal lobe (sv-PPA), and in the temporoparietal junction (lv-PPA). The most common etiologies are Alzheimer's disease, FTLD-tau, PSP, CBD, Pick's disease, etc. There is no etiological treatment, however, the use of palliatives is necessary, especially for mood disorders; at the onset of symptoms, speech therapy rehabilitation should be indicated.

REFERENCES

Agosta, F., Henry, R.G., & Migliaccio, R. (2010). *Language networks in semantic dementia*. Academic Press.

Amici, S., Gorno-Tempini, M. L., Ogara, J. M., Dronkers, H. F., & Millera, B. L. (2006). An overview on Primary Progressive Aphasia and its variants. *Behavioural Neurology*, *17*(2), 77–87. doi:10.1155/2006/260734 PMID:16873918

Ardila, A. (2020). Gerstmann Syndrome. *Current Neurology and Neuroscience Reports*, *20*(11), 48. Advance online publication. doi:10.1007/s11910-020-01069-9 PMID:32852667

Bang, J., Spina, S., & Miller, B. (2015). Frontotemporal dementia. *Lancet*, *386*(10004), 1672–1682. doi:10.1016/S0140-6736(15)00461-4 PMID:26595641

Beber, B. C., Berbert, M. C. B., Grawer, R. S., & Cardoso, M. C. A. F. (2018, March). Rate and rhythm control strategies for apraxia of speech in nonfluent primary progressive aphasia. *Dementia & Neuropsychologia*, *12*(1), 80–84. doi:10.1590/1980-57642018dn12-010012 PMID:29682238

Bertola, L. (2019). *Psicometria e Estatística Aplicadas à Neuropsicologia Clínica. Editora*. Editora Pearson Clinical Brasil.

Bessi, V., Piaceri, I., & Padiglioni, S. (2018). Crossed aphasia in nonfluent variant of primary progressive aphasia carrying a GRN mutation. *J Neurological Sciences*. Doi: . 2018.06.026 doi:10.1016 / j.jns

Binder, J R. (2015). The Wernicke area. *Neurology, 85*(24), 2170–2175. doi:.00000000 00002219 doi:10.1212/wnl

Botha, H., & Josephs, K.A. (2019). Primary progressive aphasias and apraxia of speech. *Continuum, 25*(1), 101–127.

Boxer, A. L., Lipton, A. M., Womack, K., Merrilees, J., Neuhaus, J., Pavlic, D., Gandhi, A., Red, D., Martin-Cook, K., Svetlik, D., & Miller, B. L. (2009). An open-label study of memantine treatment in 3 subtypes of frontotemporal lobar degeneration. *Alzheimer Disease and Associated Disorders, 23*(3), 211–217. doi:10.1097/WAD.0b013e318197852f PMID:19812461

Brito-Marques, P. R. (2002). Frontotemporal cortical atrophy with gliosis in the Gray mather of cerebral cortex. Case report. *Arquivos de Neuro-Psiquiatria, 60*, 462–468. doi:10.1590/S0004-282X2002000300023 PMID:12131951

Brito-Marques, P. R. (2018). *Desordens Cognitivas e Demências: Diagnóstico Diferencial e Tratamento*. EDUPE.

Brito-Marques, P. R., & Brito-Marques, J. M. A. M. (2023). A left-handed woman with a non-fluent variant of primary progressive aphasia with a left asymmetric brain SPECT. *Dementia & Neuropsychologia, 17*, e2022009. doi:10.1590/1980-5764-dn-2022-0095

Brito-Marques, P. R., & Cabral-Filho, J. E. (2004). The role of education on in Mini-Mental State Examination: A study in Northeast Brazil. *Arquivos de Neuro-Psiquiatria, 62*(2a), 206–211. doi:10.1590/S0004-282X2004000200003 PMID:15235718

Brito-MarquesP. R.Vieira-MelloR. (1999). Amyotrophic lateral Sclerosis with dementia. *Arq. Neuro-Psiquiatr., 57*(2A). https://doi.org/ doi:10.1590/S0004-282X1999000 2000 18

Brito-Marques, P.R., Vieira-Mello, R., & Montenegro, L. (2001). *Classic Pick´s disease type with ubiquitine positive and tau negative inclusions.* Academic Press.

Brito-Marques, P. R., Vieira-Mello, R. J., Montenegro, M. F. V., & Aragão, M. F. V. (2011). Clinicopathologic analysis of progressive non-fluent aphasia and corticobasal degeneration, Case report and review. *Dementia & Neuropsychologia, 5*(2), 135–141. doi:10.1590/S1980-57642011DN05020013 PMID:29213736

Cabrera-Martin, M. N., Matías-Guiub, J. A., & Yus-Fuertes, M. (2016). PET-TC 18F-FDG y RM funcional en un caso de afasia progresiva primaria logopénica cruzada. *Revista Española de Medicina Nuclear e Imagen Molecular, 35*(6), 394–397. Advance online publication. doi:10.1016/j.remn.2016.05.002 PMID:27321169

Canu, E., Bessi, V., Leocadi, M., Padiglioni, S., Nacmias, B., Sorbi, S., Filippi, M., & Agosta, F. (2019). Crossed aphasia confrmed by fMRI in a case with nonfuent variant of primary progressive aphasia carrying a GRN mutation. *Journal of Neurology, 266*(5), 1274–1279. Advance online publication. doi:10.1007/s00415-019-09298-w PMID:30923936

Caramazza, A., Berndt, R.S., & Brownell, H.H. (1982). *The semantic deficit hypothesis: perceptual parsing and object classification by aphasic patients.* Academic Press.

Damasio, A., & Damasio, H. (1992). Grain and Language. Mind and Brain, 89-95.

Erkkinen, M.G., Kim, M.O., & Geschwind, M.D. (2018). Clinical Neurology and Epidemiology of the Major Neurodegenerative Diseases. *Cold Springer Harbor Perspectives in Biol, 10*(4).

Fittipaldi, S., Ibanez, A., Baez, S., Manesa, F., Sedeno, L., & Adolfo, A. M. (2019). More than words: Social cognition across variants of primary progressive aphasia. *Neuroscience and Biobehavioral Reviews, 100*, 263–284. doi:10.1016/j.neubiorev.2019.02.020 PMID:30876954

Galvin, J. E., Howard, D. H., Denny, S. S., Dickinson, S., & Tatton, N. (2017). The social and economic burden of frontotemporal degeneration. *Neurology, 89*(20), 2049–2056. doi:10.1212/WNL.0000000000004614 PMID:28978658

Gorno-Tempini, M.L., Brambati, S.M., & Ginex, V. (2008). *The logopenic – phonological variant of primary progressive aphasia.* Academic Press.

Gorno-Tempini, M. L., Dronkers, N. F., Rankin, K. P., Ogar, J. M., Phengrasamy, L., Rosen, H. J., Johnson, J. K., Weiner, M. W., & Miller, B. L. (2004, March). Cognition and Anatomy in Three Variants of Primary Progressive Aphasia. *Annals of Neurology, 55*(3), 335–346. doi:10.1002/ana.10825 PMID:14991811

Gorno-Tempini, M. L., Hills, A. E., & Weintraub, S. (2011). Classification of primary progressive aphasia and its variants. *Neurology, 76*(11), 1006–1014. doi:10.1212/WNL.0b013e31821103e6 PMID:21325651

Grossman, M., & Irwin, D. J. (2018). Primary progressive aphasia and stroke aphasia. *Continuum, 24*(3), 745–767. PMID:29851876

Henry, M.L., & Gorno-Tempini, M.L. (2010). *The logopenic variant of primary progressive aphasia.* Academic Press.

Hodges, J. R. (2001). Frontotemporal dementia (Pick's disease): Clinical features and assessment. *Neurology*, *56*(Suppl 4), S6–S10. doi:10.1212/WNL.56.suppl_4.S6 PMID:11402143

Hodges, J. R., Patterson, K., Oxbury, S., & Funnell, E. (1992). Semantic dementia. Progressive fluent aphasia with temporal lobe atrophy. *Brain*, *115*(6), 1783–1806. doi:10.1093/brain/115.6.1783 PMID:1486461

Jager, C.A., Haigh, A.M.F., & Garrad, P. (2012). *Logopenic aphasia in Alzheimer's disease: clinical variant or clinical feature?* Academic Press.

Josephs, K. A., Duffy, J. R., Strand, E. A., Machulda, M. M., Senjem, M. L., Gunter, J. L., Schwarz, C. G., Reid, R. I., Spychalla, A. J., Lowe, V. J., Jack, C. R. Jr, & Whitwell, J. L. (2014). The evolution of primary progressive apraxia of speech. *Brain*, *137*(10), 2783–2795. doi:10.1093/brain/awu223 PMID:25113789

Josephs, K. A., Duffy, J. R., Strand, E. A., Machulda, M. M., Senjem, M. L., Master, A. V., Lowe, V. J., Jack, C. R. Jr, & Whitwell, J. L. (2012). Characterizing a neurodegenerative syndrome: Primary progressive apraxia of speech. *Brain*, *135*(pt5), 1522–1536. doi:10.1093/brain/aws032 PMID:22382356

Kertesz, A., Davidson, W., & Munoz, D. G. (1999). Clinical and pathological overlap between frontotemporal dementia, primary progressive aphasia and corticobasal degeneration: The Pick complex. *Dementia and Geriatric Cognitive Disorders*, *10*(Suppl. 1), 46–49. doi:10.1159/000051212 PMID:10436340

Kertesz, A., Morlog, D., Light, M., Blair, M., Davidson, W., Jesso, S., & Brashear, R. (2008). Galantamine in Frontotemporal Dementia and Primary Progressive Aphasia. *Dementia and Geriatric Cognitive Disorders*, *25*(2), 178–185. doi:10.1159/000113034 PMID:18196898

Knibb, J. A., Xereb, J. H., Patterson, K., & Hodgges, J. R. (2006). Clinical and Pathological Characterization of Progressive Aphasia. *Annals of Neurology*, *59*(1), 156–165. doi:10.1002/ana.20700 PMID:16374817

Leyton, C. E., Hsieh, S., Mioshi, E., & Hodges, J. R. (2013). Cognitive decline in logopenic aphasia: More than losing words. *Neurology*, *80*(10), 897–93. doi:10.1212/WNL.0b013e318285c15b PMID:23390170

Machulda, M.M., Whitwell, J.L., & Duffy, J.R. (2013). *Identification of an atypical variant of logopenic progressive aphasia.* Academic Press.

Macoir, J., Martel-Sauvageau, V., Bouvier, L., Laforce, R., & Monetta, L. (2021). Heterogeneity of repetition abilities in logopenic variant primary progressive aphasia. *Dementia & Neuropsychologia*, *15*(3), 405–412. doi:10.1590/1980-57642021dn15-030014 PMID:34630930

Marshall, C. R., Hardy, C. J. D., Volkmer, A., Russell, L. L., Bond, R. L., Fletcher, P. D., Clark, C. N., Mummery, C. J., Schott, J. M., Rossor, M. N., Fox, N. C., Crutch, S. J., Rohrer, J. D., & Warren, J. D. (2018). Primary progressive aphasia: A clinical approach. *Journal of Neurology*, *265*(6), 1474–1490. doi:10.1007/s00415-018-8762-6 PMID:29392464

Mesulam, M., Coventry, C., Bigio, E. H., Geula, C., Thompson, C., Bonakdarpour, B., Gefen, T., Rogalski, E. J., & Weintraub, S. (2021). Nosology of Primary Progressive Aphasia and the Neuropathology of Language. *Advances in Experimental Medicine and Biology*, *1281*, 33–49. doi:10.1007/978-3-030-51140-1_3 PMID:33433867

Mesulam, M. M. (1982). Slowly progressive aphasia without generalized dementia. *Annals of Neurology*, *11*(6), 592–598. doi:10.1002/ana.410110607 PMID:7114808

Mesulam, M. M. (2001). Article. *Primary Progressive Aphasia*, *49*, 425–432.

Mesulam, M. M. (2013). Primary progressive aphasia: A dementia of the language network. *Dementia & Neuropsychologia*, *7*(1), 2–9. doi:10.1590/S1980-57642013DN70100002 PMID:24707349

Mesulam, M. M., Weintraub, S., Rogalski, E. J., Wieneke, C., Geula, C., & Bigio, E. H. (2014). Asymmetry and heterogeneity of Alzheimer's and frontotemporal pathology in primary progressive aphasia. *Brain*, *137*(4), 1176–1192. doi:10.1093/brain/awu024 PMID:24574501

Mesulam, M. M., Wieneke, C., Cynthia Thompson, C., Rogalski, E., & Weintraub, S. (2012). Quantitative classification of primary progressive aphasia at early and mild impairment stages. *Brain*, *135*(5), 1537–1553. doi:10.1093/brain/aws080 PMID:22525158

Miotto, E. C., Lúcia, M. C. S., & Scaff, M. (2012). *Neuropsicologia Clínica*. Roca.

Moretti, R., Torre, P., Antonello, R. M., Cattaruzza, T., Cazzato, G., & Bava, A. (2004). Rivastigmine in Frontotemporal Dementia. *Drugs & Aging*, *21*(14), 931–937. doi:10.2165/00002512-200421140-00003 PMID:15554751

Neary, D., Snowden, J. S., Gustafson, L., Passant, U., Stuss, D., Black, S., Freedman, M., Kertesz, A., Robert, P. H., Albert, M., Boone, K., Miller, B. L., Cummings, J., & Benson, D. F. (1998). Frontotemporal lobar degeneration: A consensus on clinical diagnostic criteria. *Neurology*, *51*(6), 1546–1554. doi:10.1212/WNL.51.6.1546 PMID:9855500

Ossenkoppele, R., Jansen, W. J., Rabinovici, G. D., Knol, D. L., van der Flier, W. M., van Berckel, B. N. M., Scheltens, P., Visser, P. J., Verfaillie, S. C. J., Zwan, M. D., Adriaanse, S. M., Lammertsma, A. A., Barkhof, F., Jagust, W. J., Miller, B. L., Rosen, H. J., Landau, S. M., Villemagne, V. L., Rowe, C. C., ... Brooks, D. J. (2015). Prevalence of amyloid PET positivity in dementia syndromes: A meta-analysis. *Journal of the American Medical Association*, *313*(19), 1939–1949. doi:10.1001/jama.2015.4669 PMID:25988463

Ota, S., Kanno, S., Morita, A., Narita, W., Kawakami, N., Kakinuma, K., Saito, Y., Kobayashi, E., Baba, T., Iizuka, O., Nishio, Y., Matsuda, M., Odagiri, H., Endo, K., Takanami, K., Mori, E., & Suzuki, K. (2021). Echolalia in patients with primary progressive aphasia. *European Journal of Neurology*, *28*(4), 1113–1122. doi:10.1111/ene.14673 PMID:33305428

Owens, T.E., Machulda, M.M., & Duffy, J. (2018). Patterns of neuropsychological dysfunction and cortical volume changes in logopenic aphasia. *J Alzheimer's Dis*, *66*(3), 1015-1025. doi:10.3233/JAD-171175

Petroi, D., Duffya, J. R., Borgertb, A., Stranda, E. A., Machuldac, M. M., Senjemd, M. L., Jack, C. R. Jr, Josephsa, K. A., & Whitwelld, J. L. (2020). Neuroanatomical correlates of phonologic errors in logopenic progressive aphasia. *Brain and Language*, *204*, 104773. doi:10.1016/j.bandl.2020.104773 PMID:32114145

Prodan, C. I., Monnot, M., & Ross, E. D. (2009). Behavioural abnormalities associated with rapid deterioration of language functions in semantic dementia respond to sertraline. *Journal of Neurology, Neurosurgery, and Psychiatry*, *80*(12), 1416–1417. doi:10.1136/jnnp.2009.173260 PMID:19917830

Rahul, D.R., & Ponniah, R.J. (2019). Language impairment in primary progressive aphasia and other neurodegenerative diseases. *J Genetics*, *98*(95). https://doi. 12041-019-1139-5 doi:org/10.1007/s

Rogalski, E., Cobia, D., Harrison, T. M., Wieneke, C., Thompson, C. K., Weintraub, S., & Mesulam, M.-M. (2011, March). Anatomy of Language Impairments in Primary Progressive Aphasia. *The Journal of Neuroscience : The Official Journal of the Society for Neuroscience*, *31*(9), 3344–3350. doi:10.1523/JNEUROSCI.5544-10.2011 PMID:21368046

Rogalski, E., Sridhar, J., Rader, B., Martersteck, A., Chen, K., Cobia, D., Thompson, C. K., Weintraub, S., Bigio, E. H., & Mesulam, M.-M. (2016). Aphasic variant of Alzheimer disease. Clinical, anatomic, and genetic features. *Neurology*, *87*(13), 1–7. doi:10.1212/WNL.0000000000003165 PMID:27566743

Rogalski, E. J., & Mesulam, M. M. (2009). Clinical Trajectories and Biological Features of Primary Progressive Aphasia (PPA). *Current Alzheimer Research*, *6*(4), 331–336. doi:10.2174/156720509788929264 PMID:19689231

Roher, J.D., Caso, F., & Mahoney, C. (2013). *Patterns of longitudinal brain atrophy in the logopenic variant of primary progressive aphasia*. Academic Press.

Ropper, A. H., Samuels, M. A., Klein, J. P., & Prasad, S. (2023). *Adams and Victor's Principles of Neurology* (12th ed.). McGrawHill.

Schliebs, R., & Arendt, T. (2006). The significance of the cholinergic system in the brain during aging and in Alzheimer's disease. *Journal of Neural Transmission (Vienna, Austria)*, *113*(11), 1625–1644. doi:10.1007/s00702-006-0579-2 PMID:17039298

Senaha, M. L. H., Caramelli, P., Brucki, S. M. D., Smid, J., Takada, L. T., Porto, C. S., César, K. G., Matioli, M. N. P., Soares, R. T., Mansur, L. L., & Nitrini, R. (2013). Primary progressive aphasia Classification of variants in 100 consecutive Brazilian cases. *Dementia & Neuropsychologia*, *7*(1), 110–1121. doi:10.1590/S1980-57642013DN70100017 PMID:29213827

Small, S.L., Hart, J., Nguyen, T., & Gordon, B. (1995). *Distributed representations of semantic knowledge in the brain*. Academic Press.

Snowed, J. S., Goulding, P. J., & Neary, D. (1989). Semantic Dementia: A Form of Circumscribed Cerebral Atrophy. *Behavioural Neurology*, *2*, 167–182. doi:10.1155/1989/124043

Spinelli, E. G., Caso, F., Agosta, F., Gambina, G., Magnani, G., Canu, E., Blasi, V., Perani, D., Comi, G., Falini, A., Gorno-Tempini, M. L., & Filippi, M. (2015). A multimodal neuroimaging study of a case of crossed nonfluent/agrammatic primary progressive aphasia. *Journal of Neurology*, *262*(10), 2336–2345. doi:10.1007/s00415-015-7845-x PMID:26194195

Spitsyna, G., Warren, J. E., Scoot, S. K., Turkheimer, F. E., & Wise, R. J. S. (2006). Covering Language Streams in the Human Temporal Lobe. *The Journal of Neuroscience : The Official Journal of the Society for Neuroscience*, *26*(28), 7328–7336. doi:10.1523/JNEUROSCI.0559-06.2006 PMID:16837579

Teichmann, M. (2021). The current international consensus criteria can lead to under and over-diagnosis of primary progressive aphasia variants. *Revue Neurologique, 177*, 370–375.

Teichmann, M., Kas, A., Boutet, C., Ferrieux, S., Nogues, M., Samri, D., Rogan, C., Dormont, D., Dubois, B., & Migliaccio, R. (2013). Deciphering logopenic primary progressive aphasia: A clinical, imaging and biomarker investigation. *Brain, 136*(11), 3474–3488. doi:10.1093/brain/awt266 PMID:24108322

Thompson, C. K. (2019). Neurocognitive Recovery of Sentence Processing in Aphasia. *Journal of Speech, Language, and Hearing Research: JSLHR, 62*(11), 3947–3972. doi:10.1044/2019_JSLHR-L-RSNP-19-0219 PMID:31756151

Ulugut, H., Stek, S., & Wageman, L. E. E. (2022). The natural history of primary progressive aphasia: Beyond aphasia. *Journal of Neurology, 269*(3), 1375–1385. doi:10.1007/s00415-021-10689-1 PMID:34216263

Vandenberghe, R. (2016). Classification of the primary progressive aphasias: Principles and review of progress since 2011. *Alzheimer's Research & Therapy, 8*(1), 16. doi:10.1186/s13195-016-0185-y PMID:27097664

Warrington, E. K., & Shallice, T. (1984). Category Specific Semantic Impairments. *Brain, 107*(3), 829–854. doi:10.1093/brain/107.3.829 PMID:6206910

Wilson, S. M., Henry, M. L., Besbris, M., Ogar, J. M., Dronkers, N. F., Jarrold, W., Miller, B. L., & Gorno-Tempini, M. L. (2010). Connected speech production in three variants of primary progressive aphasia. *Brain, 133*(7), 2069–2088. doi:10.1093/brain/awq129 PMID:20542982

Zhao, Y., Ficek, B., Webster, K., Frangakis, C., Caffo, B., Hillis, A. E., Faria, A., & Tsapkini, K. (2021). White-matter integrity predicts electrical stimulation (tDCS) and language therapy effects in primary progressive aphasia. *Neurorehabilitation and Neural Repair, 35*(1), 44–57. doi:10.1177/1545968320971741 PMID:33317422

Chapter 10
Changes in Social Cognition in Dementia

Tiago Figueiredo
Cognus Institute for Research, Brazil

Leonardo Caixeta
https://orcid.org/0000-0002-5736-9058
Federal University of Goiás, Brazil

ABSTRACT

Social cognition is an umbrella term for the brain process underlying social behavior. Neurodegenerative dementia might affect the social cognition domains in different ways and degrees. Therefore, each type of dementia has a distinct profile of social cognitive impairments. This chapter presents a discussion about social cognition in healthy aging and in the most prevalent dementia subtypes, Alzheimer's disease (AD) and frontotemporal dementia (FDT). The available data about mentalizing, emotional contagion, and perception of social cues in AD and FTD variants are analyzed based on the correlation between behavioral and neuroimaging findings.

INTRODUCTION

The prevalence of older adults with cognitive impairment is increasing globally, and this fact has been associated with rising healthcare costs and loss of independence (Prince et al., 2013; Jutkowitz et al., 2017). The current version of the Diagnostic and Statistical Manual of Mental Disorders (DSM-5-TR, APA, 2022) includes dementia syndromes belonging to the group of "Major Neurocognitive Disorders (MND)". It considers that they can occur due to multiple etiologies. In this chapter, we will

DOI: 10.4018/979-8-3693-0851-6.ch010

use the term "dementia" to refer to the group of neurodegenerative disorders that commonly affect older adults.

Dementia is a clinical syndrome characterized by an inexorably progressive deterioration in cognitive domains, associated with several impairments in capacity for independent living. Although young-onset cases are increasingly recognized, dementia typically affects older people (Sousa et al., 2010). Alzheimer's Disease (AD), vascular disease (VD), Lewy Body Disease (DLB), and Frontotemporal Dementia (FTD) are the most prevalent dementia syndromes. The current classification criteria for the dementia subtypes considers which cognitive domains are impaired. Attention, executive functions, learning and memory, language, perceptual-motor abilities, and social cognition are considered by DSM-5-TR as the key domains that may be affected. However, although the pattern of cognitive disturbances helps define the etiological classification, declines in some of these domains can be shared by more than one type of dementia.

Social cognitive deficits were kept in DSM-5-TR as one of the core domains of impaired neurocognitive function (Sachdev et al., 2014). Although the early decline in social cognition is a core feature of FTD, changes in social conduct have been recognized as a significant cognitive and clinical marker for AD and LBD (Gregory et al., 2002; Cosentino et al., 2014; Henry et al., 2016). Considering that comprehending impaired cognitive domains in this population can help promote cognition rehabilitation and healthy aging, this chapter aims to discuss current scientific evidence on social cognition in the main subtypes of dementia. For this, initially, we will give a conceptual definition of social cognition and what science tells us about this cognitive domain in normal aging. Then, a more specific discussion of changes in social cognition in AD and FTD will be presented.

Social Cognition as a Specific Cognitive Domain

Jerry Fodor was an American philosopher and scientist who pioneered describing cognitive architecture in "specific processing modules" (Fodor, 1983). Since then, advances in neuroscience have contributed to polishing this theory, and we currently conceive that the human brain has a level of information processing according to the type and complexity of stimulus (Spunt & Adolphs, 2017; Schulz et al., 2021). Scientific data toward a hierarchical model of brain stimulus-processing shows that subcortical areas are specifically activated during the capture of social versus non-social stimulus, and the superior cortical process involves shared neurocircuits that are engaged in complex processing, regardless of the type of stimulus (Shulz et al., 2021).

Literature involving neuroimage data suggest that processing social and non-social stimuli involves semi-independent neural pathways (Van Overwalle, 2009).

According to these data, the medial prefrontal cortex (mPFC), fusiform gyrus, temporoparietal junction, superior temporal sulcus, and amygdala are brain areas preferentially activated during social information processing. At the same time, the activation of regions of the medial prefrontal cortex occurs whenever the participation of inhibitory control is necessary during the processing of social and non-social information (Spreng et al., 2009; Spunk & Adolphs, 2015). Although there is more robust evidence regarding the specificity of neural networks involved in the social stimulus perception stage, the specificity of activation of regions involved in processing complex social information remains under debate (Spunt & Adolphs, 2017).

There is no universal concept that defines social cognition. However, most experts describe it as an "umbrella" construct, which includes the cognitive functions involved in processing social information (Happé et al., 2017; Green et al., 2016; Fiske & Taylor, 2017). Social information refers to stimuli emitted by another being of the same species, and it is possible to classify them according to their implicit or explicit nature. Although social cognition refers to brain processes involved at the interpersonal functional level, its approach through a concrete guide demands that this construct be conceptualized based on its different levels of analysis and implications. Therefore, the expression of human social behavior should never be reduced to the cerebral cognitive level, and it is mandatory to consider the biological and environmental-contextual interface levels.

Social Cognition and Healthy Aging

Population aging has reached its heights worldwide, and it has become mandatory for all health professionals to understand the concept of healthy aging. Although there is abundant evidence proving the association between factors such as physical activity, diet style, use of licit or illicit substances, and socioeconomic status with health and well-being, in the last decade, some data have highlighted that social relationships play the most crucial role in healthy aging and longevity (Holt-Lunstad et al., 2010; Lövden et al. 2005). These data reinforce the idea that it is urgent to concentrate forces on structuring strategies focusing on the comprehension and rehabilitation of the cognitive abilities involved in social connections.

Research investigating social cognition in elderly from non-clinical populations is still relatively scarce (Henry et al., 2013). Happé et al. (1998) was one of the pioneering studies to investigate changes in social cognition throughout healthy aging. Data provided in this survey shows that the performance of old adults in social inference tasks was equal or even superior to that of young adults in terms of precision of answers. However, the authors signaled those old adults performed with a longer response time.

Most recently, a meta-analytic study including 22 articles investigated the differences in the performance between old and young healthy adults in social paradigms (Henry et al., 2013). Among the social cognitive functions investigated in the studies are the social inference about the mental states of others through the presentation of social vignettes, photographs of eye regions, videos of individuals expressing affective states, identifying false beliefs, and social gaffes (faux pas). The results of this review indicated that healthy old adults performed worse when compared with the group of adults in all instruments. However, these findings were not replicated in later studies (Girardi et al., 2018; Grainger et al., 2018). Thus, considering the currently available empirical data, we identify that some issues still require further clarification. For example, are the neural pathways underlying social cognition affected in a dissociated way from other cognitive pathways (e.g., memory)? How is the average decline of social neural pathways throughout healthy aging? And what is the role of social memory decline and the impairments of social competence?

Beyond the direct relationship between social cognition and social behavior, some authors claim that social competence in older adults is also influenced by psychosocial features, such as the perception of the remaining life span (Carstensen et al., 1999). This complementary perspective would be associated with a profile of selectivity in pursuing the social object, an element also related to social motivation.

In summary, empirical data on the developmental course of social cognition during healthy aging still needs to be discovered and conflicting in the current literature. Future research on this topic is urgent and necessary to help establish the parameters to be used in the assessment and pave the way for structuring new, more specific support strategies.

Social Cognition as an Object of Clinical and Research Investigation: The Importance of the Taxonomy

Based on the assumption that social cognition is a multidimensional cognitive construct, its assessment requires that researchers and clinicians know the targets to be investigated. Over the past 15 years, experts have held task forces focused on defining the taxonomy of social cognition (Happé et al., 2017; Green et al., 2015; Insel et al., 2010). However, there is still no consensus about how subdomains must be investigated.

Green et al. (2015) published a review of findings about social cognition impairments in patients with schizophrenia. They delineated four main subdomains of social cognition: perception of social cues, experience sharing, mentalizing, and emotion experience and regulation.

Perceptions of social cues

Green et al. (2015) describes perceptions of social cues as the ability to perceive verbal and non-verbal social cues, such as cues from others' facial expressions (affective and non-affective), voice prosody, and body movements.

Social signals from others' face processing have been extensively studied in some psychiatric disorders. Non-affective facial stimuli are the individual resources, such as identity, sex, or age. This ability involves activation in the lateral fusiform gyrus, visual extrastriate cortex, lateral occipital gyri, anterior temporal pole, and posterior superior temporal gyrus (Atkinson & Adolphs, 2011; Haxby et al., 2002). On the other side, the processing of facial expressions of emotions encompasses the perception of affective stimuli and involves activation in limbic structures, the medial prefrontal area, and the putamen (Vuilleumier & Pourtuois, 2007; Sabatinelli et al., 2011).

Voice perception refers to discriminating implicit features of speech, such as prosody. Voice prosody conveys information beyond words and plays an important role in understanding social contexts. Non-literal language, also known as metalanguage, refers to the ability to think about language. This skill makes us more efficient communicators, enabling us to use and understand language segments, such as irony, sarcasm, metaphors, and proverbs (Lebrun & Buyssens, 1982). Hence, non-literal language is an important semantic and pragmatic language skill that supports the meaning of words and sentences and their social use.

The effective processing of non-literal language promotes social competence since we need to know how to use and understand this type of language properly to perceive our interlocutors' intentions (Marschark & Nall, 1985). As non-literal language is a linguistic resource widely used in everyday communication, being unable to use or understand these segments can lead to lower social competence and, therefore, to social impairments (Mancopes & Schultz, 2008). This process is associated with activation in the superior temporal gyrus and inferior frontal gyrus.

Experience Sharing

Experience sharing (ES) refers to the neural activity triggered in one's individual from directly observing another person's behavior or emotions (Iacoboni, 2009). Affective experience sharing occurs automatically (non-deliberately) and is central to promoting alignment between mental states during social interaction. Green et al. (2015) classified the ES into two subdomains: motor resonance and affective resonance. Motor resonance occurs from the activation of the brain mirror neuron, and other authors refer to this ability as imitation (Happé et al., 2017). The mirror neuron system involved in motor resonance includes the anterior region of the temporal sulcus, the premotor cortex, and the inferior region of the frontal gyrus (Molenberghs et al., 2012). Affective resonance refers to the ability to be emotionally contained by the affective state of others while observing. The emotional contagion is commonly called "affective empathy" in the literature. The amygdala, the insula's anterior region, and the cingulate cortex's anterior region participate in affective resonance (Molenberghs et al., 2012).

Mentalizing

Mentalizing refers to the ability to understand others' behavior in social situations. This phenomenon requires the integration of social cues from the counterpart and the social context. On top of that, perspective-taking about other mental states is a central ability to promote the Theory of Mind (ToM).

ToM was first referred to by Premack & Woodruff (1978) as the ability to make inferences and theorizations about the elements from others' mental states which are not directly observable. Therefore, it is a cognitive ability that blends metalanguage and social skills (Jolliffe & Baron-Cohen, 1999; Caixeta & Nitrini, 2002). Further, ToM allows people to make attributions and, through clues given by linguistic content, context, and non-linguistic hints, know if the other is telling a joke, being ironic, or is interested in the conversation, for example.

Neuroimaging studies have indicated that there is an activation of the medial regions of the prefrontal cortex, the bilateral temporoparietal junction, and the precuneus when the individual performs perspective-taking (Carrington & Bailey, 2009; Schurz et al., 2014). Neurobiologically, the brain areas involved in mentalization are related to the medial prefrontal cortex, bilateral temporoparietal junctions, and precuneus (Chen et al, 1998; Castelli et al, 2002; Diaz & Hogstrom, 2011).

The Experience and Regulation of Emotions

Green et al. (2015) includes the experience and regulation of emotions as a domain of social cognition due to considering the emotional reactions to others. The experience of emotions refers to the perception of the immediate emotional response in front of the presence of a positive or negative social stimulus. This perception is subjective and based on the affective, motor, and neurophysiological reactions. A large area of the brain is involved in the immediate emotional response: the limbic system (amygdala, anterior region of the hippocampus, anterior part of the insula, and the cingulate gyrus), basal ganglia, thalamus, the ventral striatum, the medial prefrontal cortex, the posterior cingulate cortex, precuneus, lateral temporal cortex, and the temporal pole (Kober et al., 2008; Satpute et al., 2015).

Green et al., (2015) conceptualize emotion regulation according to the influential model elaborated by Gross, 2013. Thus, they conceptualize the control of emotions as a dynamic construct, which has two main processes: the generation and regulation of emotions. The deliberate control of emotions relies on the participation of the dorsolateral and ventrolateral regions of the prefrontal cortex. Experiencing emotions involves the activation of the amygdala (Phelps & LeDoux, 2005).

The taxonomy of the social cognition suggested in the study of Green et al., (2015) can be helpful as a guide in interpreting the enormous variety of data on social cognition in different neuropsychiatric disorders, including dementia syndromes. Such a dimensional approach is aligned with current efforts to understand psychiatric disorders through underlying brain-based dimensions (Cuthbert, 2014).

In a comprehensive and integrative manner, data extracted from functional neuroimaging studies have significantly advanced our understanding of how social information is processed. Schulz et al. (2021) recently published a meta-analytic sizeable survey that investigated the brain areas active during social task performances. Analysis of the data from this meta-analysis allowed the authors to propose a multilevel model of social-cognitive processes, which have cognitive and affective processes forming an opposite pole, and the intermediate cluster bears similarities to both poles. Considering that the analysis of empirical data involves the methodological strategy used by the authors in their studies, these findings are instrumental in guiding current data interpretation according to the methodology used by the authors.

SOCIAL COGNITION IN ALZHEIMER'S DISEASE (AD)

Alzheimer's Disease (AD) is the leading cause of dementia worldwide, with amnestic disturbances as nuclear and initial presenting features (APA, 2022). From a pathophysiological perspective, AD is initially marked by neurodegeneration occurring in the hippocampal regions, the entorhinal cortex, and the posterior cingulate cortex. Later, AD causes degeneration of the temporal, parietal, and frontal cortices (Frisoni et al., 2010). This pattern of deterioration in the neural level led to abnormalities in several cognitive domains in the mild to moderate stage of the illness, including the social cognitive domains.

A growing body of literature has shown that AD compromises some domains of social cognition (Kessels et al., 2021). In general, social cognition impairments become more pronounced with the evolution of the disease, mainly due to neurodegeneration in areas implicated in the ability to perspective-taking (Pick et al., 2019; Rankin et al., 2005). These findings justify why clinically noted behavioral changes appear in more advanced stages of the disorder.

Evidence proves that AD patients showed a deterioration in mentalizing, emotional processing, and social perception cues compared to the healthy controls (Setién-Suero et al., 2022).

Mentalizing in AD

The vast majority of studies in the current literature focus on the investigation of mentalizing (which includes ToM subdomains) in AD patients. Although there is a significant number of studies investigating mentalization in patients with AD, this is the subdomain of social cognition most marked by mixed or conflicting results. In addition to possible selection and methodological application biases, this construct refers to a complex domain of social cognition, which involves the participation of several other non-social cognitive domains (Schulz et al., 2021). In other words, their deficits are recorded in more advanced stages of the disease or, at least, are considered less specific for processing social information.

To this hypothesis, conflicting results tend to be more noticeable as the paradigms assessing mentalization become more complex. For example, the agreement between data from studies that used first-order ToM tasks is higher than that from studies that used second-order ToM tasks (Demichelis et al., 2020; Castelli et al., 2021; Choong & Doody, 2013; Youmans & Bourgeois, 2010). The agreement of data in studies that investigated TOM through second-order false belief paradigms occurred only in groups with moderate/severe stages of AD.

In the same line, the reproducibility of the data is more robust when investigating the sharing of affective experiences (Demichelis et al., 2020; Henry et al., 2009; Laisney et al., 2013). The domain of sharing affective experiences involves subcortical processing (less deliberate) and can be considered more specific to social cognition (Schulz et al., 2021).

The sense of self is a nuclear feature in mentalizing (Fonagy & Bateman, 2008). The self-persistence or diminishment in AD is still under debate (Fazio & Mitchell, 2009; Davis, 2004). This issue is partly complicated by the difficulties inherent in defining the self, with researchers using varying definitions of self as a basis for their investigations. In the social cognition research domain, evidence has highlighted a reduction in the sense of self in the mild stage of AD, with a focus on specific cognitive impairment and self-reflection capacity (Simm et al., 2015). These findings are essential to alert clinicians and researchers about the importance of investigating a domain complementary to the ToM capacity when evaluating mentalization.

Sharing Experiences in AD

Studies investigating the sharing of affective experience abilities in old adults also refer to this social cognition domain through the alternative term "affective empathy" (Baron-Cohen, 2006). The methodological strategies to assess this domain significantly vary in the current literature. Researchers have evaluated the affective domain of empathy through records of autonomic reactions presented by the individual in the face of direct observation of affective events experienced by other people or through psychometric self-response instruments, for example. Both strategies represent significantly different investigative methods, making data reproducibility a challenge.

Qualitative review and meta-analysis studies have indicated no significant difference in healthy controls when investigating the sharing of affective experiences in patients with AD (Demichelis et al., 2020; Fischer et al., 2019). As discussed in the previous topic, the subdomain of social cognition involved in the processing of affective information involves distinct neural substrates that may include more specific elements of social cognition, as it does not require the participation of higher non-social cognitive functions in processing the stimulus (Schulz et al., 2021). Based on this theory, investigating subdomains such as the sharing of affective experiences can offer more specific data about the deterioration of social cognition in AD, including in prodromal stages.

Perception of Social Cues in AD

There is a significant degree of agreement between studies in stating that there is a deterioration in the ability to perceive implicit social cues in patients with AD when compared to healthy older adults (Setién-Suero et al., 2021). Furthermore, evidence suggests that impairments in the ability to process social cues from images of human faces can occur already in prodromal stages (Cadieux & Greve, 1997; Teng et al., 2007).

A longitudinal study with AD patients shows a significant difference in affective cues recognition from the faces of humans from baseline assessment, mainly when the emotions presented were relatively subtle (Torres et al., 2015). These findings highlight the ability to perceive social cues as a marker of disease progression. However, scientific data regarding this topic present mixed results in defining which types of facial expressions are misinterpreted in the prodromal and early stages of the disease (Cadiex & Greve, 1997; Shimokawa et al., 2003; Burnham & Hogervorst, 2004).

Social Cognition in AD Patients: The Summary of the Data

The current literature presents many studies investigating social cognition in patients with AD. However, the heterogeneity of research methods and the profile of patients included in analyses (for example, regarding the stage of the disease) are contributing factors to the low reproducibility of some data and holding back research advances. Since AD is one of the archetypes of cognitive degeneration, tasks that involve complex or polarized social information processing in the cognitive cluster tend to offer fewer specific data about the course of deterioration of social cognition. The evaluation of subdomains, such as the sharing of affective experiences through tasks that measure the reaction pattern in paradigms that directly observe mental states of pain and other emotions, have the potential to extract more specific data.

SOCIAL COGNITION IN FRONTOTEMPORAL DEMENTIA (FTD)

DSM-5-TR currently defines FTD as an insidious onset and gradual progressive neurocognitive disorder, in which the clinical features define FTD into the behavioral (bvFTD) or language (semantic dementia - SD) variants (APA, 2022). Still, according to this Manual, bvFTD is characterized by some social skill impairments as a nuclear clinical feature, such as disinhibition, apathy, and/or loss of empathy or sympathy. The SD is marked by the deterioration of some language domains (e.g., word finding, object naming, and word comprehension), constituting a deterioration in a highly significant communicative element for achieving social competence.

Unlike AD, evidence suggests that typical cognitive screening measures (e.g., Mini-Mental State Examination - MMSE) frequently do not show abnormalities at early stages of illness in FTD patients (Osher et al., 2007. On the other hand, studies indicate that impaired social cognitive processing is characteristic of FTD even in the early stages, mainly bvFTD (Bertoux et al., 2012; Shany-Ur et al., 2012; Torralva et al., 2009). Because of this, screening measures of social cognition are significant steps for identifying individuals with bvFTD or PPA variants and in differential diagnoses.

Studies associating aspects of perspective-taking in FTD began to appear in 1997 with a series of cases of patients with FTD presenting changes in social behavior linked to impairments in ToM tasks (Caixeta et al., 1997). However, despite recent advances, social cognition in bvFTD has not yet received sufficient attention, nor has its association with underlying pathology (Johen & Bertoux, 2019).

The topics below present a discussion of the subdomains of social cognition in both variants of FTD, which will be discussed separately.

Social Cognition in Behavioral Variant Frontotemporal Dementia (bvFTD)

Behavioral variant frontotemporal dementia (bvFTD) is the second leading neurodegenerative disorder in patients under 65 years of age (Hogan et al., 2016). This illness provokes progressive changes in social behaviors, emotional processing, and cognitive functioning (Raskovsky et al., 2011). In neural level, evidence shows a pronounced and relatively focused losses in the regions of bilateral frontal, insular, and anterior temporal cortices in patients with bvFTD (Johen & Bertoux, 2019)

Mentalizing in bvFTD

Most studies focusing on investigating social cognition in FTD patients show results from ToM-related paradigms and a cross-sectional design of the studies (Setién-Suero et al., 2022; Bora et al., 2016). However, unlike the studies with AD, many authors use paradigms that require social inference capacity to evaluate patients with bvFTD.

Huey et al. (2009) demonstrate that the primary deficit presented by patients with bvFTD regarding non-social cognitive functions is the impairment of executive functions, with a secondary memory deficit. Thus, bvFTD patients show significant impairment in abstraction, problem-solving, organization, attention, and mental flexibility abilities. Through neuroimaging studies, the authors linked such deficits to the left prefrontal area atrophy. Some of the patients in the study also had significant difficulty performing language tasks, such as verbal fluency, which was related to the atrophy of the left frontal perisylvian cortex.

Because social cognition and executive functions suffer deterioration in shared neural subtracts (e.g., orbitofrontal cortex and the medial and dorsolateral prefrontal cortex), the investigation of social cognition through tasks that involve metarepresentation abilities presents an enormous challenge in distinguishing both cognitive domains (Hur & Caixeta, 2013; Decety & Sommerville, 2003). It is no coincidence that many authors investigate mentalization abilities in patients with FTD, considering the participants' "social executive" abilities (Sodeno et al., 2016; Custodio et al., 2015).

In summary, current scientific data agrees in proving that there is a significant deterioration in the mentalizing abilities of patients with bvFTD, and highlights that the degree of damage tends to be more significant when compared to other subtypes of dementia (Setién-Suero et al., 2022). However, it is essential to highlight that these results reflect the deterioration of not only the social cognitive domains but also of the executive component, which is necessary for mentalization mainly in ambiguous social contexts (Thompson et al., 2005).

Sharing Experiences in bvFTD

As discussed above, most authors investigate the ability to share affective experiences (emotional contagion) through the alternative term of "affective empathy" in research. Some papers show results agreeing about the deterioration in the affective subdomain of empathy in the bvFTD group (Torralva et al., 2015; Poletti et al., 2012). On top of that, the current data consistently suggests that emotional contagion-related impairments tend to be evident even in the early stages and to progress over time (Torralva et al., 2015; Adenzato et al., 2010).

Perception of Social Cues in bvFTD

Studies investigating the subdomain of perception of social cues in bvFTD patients have primarily focused on perception of facial expressions (Kumfor & Piguet, 2012). The scientific data about this topic point a deterioration in recognizing facial expressions mainly with a negative valence in this clinical population. In correspondence with neural level, several studies highlight the involvement of underlying atrophy in amygdala, orbito-frontal cortex, temporal pole, and insula (Kumfor et al., 2013).

However, the brain regions involved in facial expression recognition have also been associated with perceptions of static and dynamic body expressions in healthy individuals (Gelder et al., 2010). Due to this, Van den Stock et al. (2015) investigated the ability to perceive dynamic and static social cues in bvFTD patients, and the results suggested that this group revealed impaired recognition of static and dynamic bodily and facial expressions while identity processing was intact. Therefore, patients with bvFTD appear to not discriminate between harmful and non-harmful intentions and cannot respond with appropriate facial expressions to other people's emotions, probably because they are unable to perceive social cues (Pick et al. al., 2019).

Social Cognition in bvFTD: The Summary of the Data

Scientific data provides significant evidence that patients with bvFTD present deficits in social cognition tasks compared to healthy controls. Although deficiencies in social cognition can be noted in the early stages of the disease, they may be more specific if cognitive subdomains that do not require the participation of executive functions during the processing of social stimuli are considered. In part because of the closer relationship between social cognition and executive functions during the brain information processes, it was unable to clearly determine the course of deterioration of social cognition in the long term, mainly regarding the cluster of cognitive functions.

Social Cognition in Semantic Dementia (SD)

Semantic Dementia (SD) is characterized by a progressive neurodegenerative disorder that affects the language system. Clinically, SD is associated with circumscribed but profound loss of semantic knowledge, which also may progress to aphasia, causing significant impairment in activities of daily living (Hodges et al., 1992; Gorno-Tempini et al., 2011). At neural levels, SD is initially marked by bilateral and asymmetrical atrophy of the anterior temporal lobe. As the illness progresses, the neural deterioration extends to the posterior and contralateral temporal regions, as well as the frontal lobe (Agosta et al, 2012; Brambati et al., 2009).

Compared to the other subtypes of dementia, there is a significantly smaller number of studies focused on investigating social cognition in SD patients (Setién-Suero et al., 2022). However, the available data provide relevant and consistent findings suggesting that SD patients show a significant deficit in social cognition, which are not exclusive to individuals with bvFTD. However, comparative studies inform that SD patients performs better in social cognitive tasks when compared with bvFTD. Fittipaldi et al., 2019 defend that similarities between SD and bvFTD are explained by the fronto-temporal disturbances present in people with both conditions.

Regarding to the pattern of social cognitive deficits in SD, Hur & Caixeta (2013) reported a series of cases of SD patients that were investigated for deficits in non-literal language skills, using the Alzheimer's disease screening test with proverbs, metaphor test, and irony test. The findings reveal that the patients performed poorly in the tasks applied, mainly in interpreting proverbs. The poor performance was mainly attributed to the semantic alterations characteristic of the disease. Still, frontal symptoms characteristic of other forms of frontotemporal lobar degeneration that interfered with the testing, such as negativism, reduced attentional breadth, concretism, and perseveration, were also observed.

According to these results, Rankin et al. (2009) studied the detection of sarcasm in individuals with different neurodegenerative pathologies. The authors reported that all subjects could correctly identify sincere items, but only DS patients had difficulty detecting sarcasm. In this skill, patients with DS performed statistically significantly worse than patients with bvFTD. In this case, single-word comprehension problems or challenges in naming objects in the early stages of SD might cause poor performance on social tasks that involve verbal stimuli.

However, some studies have provided data suggesting a dissociation between verbal and non-verbal processing impairments in SD (Binney et al., 2016; Irish et al., 2013). For example, in the study by Binney et al. (2016), SD patients varied in recognizing the identities of famous people, although they all failed in semantic processing. This variation was associated with the predominance or absence of involvement of the right brain regions.

Mentalizing in SD

Few studies investigated cognitive abilities implied in mentalizing in SD patients (Bejanin et al., 2017; Hazelton et al., 2017). Available scientific data agree to suggest that SD patients revealed impaired performance in cognitive ToM tasks, even after correcting for mnemic variables. However, neuroimaging findings indicate that cognitive ToM performances correlated with the midline cortical structures, especially the medial prefrontal cortex (Bejanin et al., 2016). Thus, these findings should be interpreted from an integrative view (considering the influence of executive functions in social cognitive tasks) because these neural substrates were part of non-social networks that are topographically close.

Shared Experiences in SD

Some studies have been provided data suggesting that SD patients shows impairment in emotional contagion (Bejanin et al., 2016; Hazelton et al., 2017). These findings were associated with a subgroup of patients that shows degeneration in the amygdala, anterior hippocampus, and parahippocampal gyrus (Bejanin et al., 2016).

Perception of Social Cues in SD

Most studies focusing on investigating social cognition in SD patients provide data from paradigms involving the interpretation of verbal and non-verbal social cues (Setién-Suero et al., 2022). Some studies reported significant impairment in recognizing cues from others' facial expressions and speech, such as prosody tasks (Bertoux et al., 2020).

Due to atrophy of the frontal and anterior temporal lobes, patients with FTD have significant difficulties related to metalanguage and non-verbal language content. Some studies suggest that individuals with FTD fail in linguistic and communicative tasks involving abstraction and pragmatic language use, such as understanding proverbs and metaphors (Eslinger et al., 2005; Kipps et al., 2009; Rapp & Wild, 2011). Anterior temporal areas, especially those in the right hemisphere, are also essential for skills involving abstraction and pragmatic use of language to occur successfully.

Eviatar & Just (2006) performed a relevant study on irony recognition and neuronal activation in healthy individuals. The authors stated that the recognition of ironies exceeds syntactic and semantic knowledge, requiring information about the interlocutor's intentions and beliefs and using this knowledge to perform a non-literal interpretation. Regarding the areas of most significant activation during this task, the authors emphasize the right hemisphere's superior and middle temporal gyrus. Although this study was not carried out with patients with FTD, from the results presented, we can infer that patients with FTD have difficulty in this task, as they have failures in the skills described by the authors as necessary for success in recognizing ironies.

Concerning proverbs, some authors state that the central aspect involved in this interpretation is the individual's familiarity with the proverb in question (Báez et al., 2009). The performance of proverbs involves several cognitive aspects, including declarative memory, categorization of concepts, logical reasoning, and listening comprehension (Santos et al., 2008). The performance in interpreting proverbs can be helpful in the differential diagnosis when differentiating healthy aging from pathological degeneration (Santos et al., 2008).

Proverbio et al. (2009) call attention to the importance of the right hemisphere in understanding such expressions from the observation of brain-damaged subjects. Based on functional neuroimaging tests, the authors report that the limbic areas of both hemispheres are activated in this task, which reveals a crucial emotional connotation of metaphorical expressions.

Social Cognition in SD Patients: Summary of Data

Although fewer studies focused on evaluating social cognition in patients with DS, the available scientific data are of great relevance for understanding changes in social cognition in this condition. Scientific findings indicate that there are social cognitive impairments mainly in interpreting verbal social cues of an implicit nature, which involve interpreting words and concepts. However, these changes are not only secondary to the deterioration of linguistic domains, as patients who mainly present degeneration in neural networks on the right tend to give impairments related to emotional contagion, with this subdomain primarily associated with social cognition.

CONCLUSIONS AND FUTURE PERSPECTIVES

Most of social cognitive domains are affected in patients with neurodegenerative dementia. However, it is possible to identify different degree of social cognition impairments among the dementia subtypes. In the same direction, the pattern of progressive degeneration among patients with neurodegenerative dementia can therefore differentially involve distinctive components of social cognition.

In general, the available scientific data may vary according to two main variables: the stage of the disease and the type of task used to assess social cognition. Regarding the stage of the disease, it is expected that advanced stages are directly associated with a more significant number of affected neural subtracts. Therefore, the findings will be less specific regarding the course of change for social cognition in each of the dementia subtypes.

Instruments commonly used to assess social cognition may vary in the specificity of the domain assessed. Tasks that investigate the Theory of Mind and other subdomains of mentalization require the controlled processing of information, which involves the participation of different domains, such as executive functions and language. Paradigms involving affective stimuli processing tend to involve subcortical neural extracts, which can be considered primary in social cognition. Findings related to affective processing (for example, the perception of social cues and emotional contagion) have been more promising in offering information about changes involving social cognition already in the early stages of dementia.

Finally, the insights gained from reading this chapter help guide future research, especially regarding the need for batteries of more comprehensive tasks that can assess at least most of the domains that make up social cognition. Furthermore, the choice of task type (cognitive versus affective) is fundamental in providing data more specifically related to social cognition.

REFERENCES

Araújo, A. A., & Perissinoto, J. (2004). Desenvolvimento da linguagem na adolescência: Competências Semânticas, sintáticas e pragmáticas. *Pro-Fono*, *16*(3), 251–260. PMID:15609580

Arroyo-Anlló, E. M., Dauphin, S., Fargeau, M. N., Ingrand, P., & Gil, R. (2019). Music and emotion in Alzheimer's disease. *Alzheimer's Research & Therapy*, *11*(1), 69. doi:10.1186/s13195-019-0523-y PMID:31391062

Atkinson, A. P., & Adolphs, R. (2011). The neuropsychology of face perception: Beyond simple dissociations and functional selectivity. *Philosophical Transactions of the Royal Society of London. Series B, Biological Sciences*, *366*(1571), 1726–1738. doi:10.1098/rstb.2010.0349 PMID:21536556

Ávila-Villanueva, M., Gómez-Ramírez, J., Ávila, J., & Fernández-Blázquez, M. A. (2021). Alzheimer's Disease and Empathic Abilities: The Proposed Role of the Cingulate Cortex. *Journal of Alzheimer's Disease Reports*, *5*(1), 345–352. doi:10.3233/ADR-200282 PMID:34189406

Bertoux, M., Delavest, M., de Souza, L. C., Funkiewiez, A., Lépine, J. P., Fossati, P., Dubois, B., & Sarazin, M. (2012). Social Cognition and Emotional Assessment differentiates frontotemporal dementia from depression. *Journal of Neurology, Neurosurgery, and Psychiatry*, *83*(4), 411–416. doi:10.1136/jnnp-2011-301849 PMID:22291219

Budgett, J., Brown, A., Daley, S., Page, T. E., Banerjee, S., Livingston, G., & Sommerlad, A. (2019). The Social Functioning in Dementia Scale (SF-DEM): Exploratory Factor Analysis and Psychometric Properties in Mild, Moderate, and Severe Dementia. *Alzheimer's & Dementia: Diagnosis, Assessment & Disease Monitoring*, *11*(1), 45–52. doi:10.1016/j.dadm.2018.11.001 PMID:30623018

Burnham, H., & Hogervorst, E. (2004). Recognition of facial expressions of emotion by patients with dementia of the Alzheimer type. *Dementia and Geriatric Cognitive Disorders*, *18*(1), 75–79. doi:10.1159/000077813 PMID:15087581

Cadieux, N. L., & Greve, K. W. (1997). Emotion processing in Alzheimer's disease. *Journal of the International Neuropsychological Society*, *3*(5), 411–419. doi:10.1017/S1355617797004116 PMID:9322399

Caixeta, L., & Nitrini, R. (2001). Subtipos clínicos da demência frontotemporal. *Arquivos de Neuro-Psiquiatria*, *59*(3A, n.3A), 577–581. doi:10.1590/S0004-282X2001000400017 PMID:11588638

Carrington, S. J., & Bailey, A. J. (2009). Are there theory of mind regions in the brain? A review of the neuroimaging literature. *Human Brain Mapping*, *30*(8), 2313–2335. doi:10.1002/hbm.20671 PMID:19034900

Castelli, F., Frith, C., Happé, F., & Frith, U. (2002). Autism, Asperger syndrome and brain mechanisms for the attribution of mental states to animated shapes. *Brain*, *125*(Pt 8), 1839–1849. doi:10.1093/brain/awf189 PMID:12135974

Castelli, I., Pini, A., Alberoni, M., Liverta-Sempio, O., Baglio, F., Massaro, D., Marchetti, A., & Nemni, R. (2011). Mapping levels of theory of mind in Alzheimer's disease: A preliminary study. *Aging & Mental Health*, *15*(2), 157–168. doi:10.1080/13607863.2010.513038 PMID:21140304

Chen, E., Widick, P., & Chatterjee, A. (2008). Functional-anatomical organization of predicate metaphor processing. *Brain and Language*, *107*(3), 194–202. doi:10.1016/j.bandl.2008.06.007 PMID:18692890

Choong, C. S., & Doody, G. A. (2013). Can theory of mind deficits be measured reliably in people with mild and moderate Alzheimer's dementia? *BMC Psychology*, *1*(1), 28. doi:10.1186/2050-7283-1-28 PMID:25566376

Cuthbert, B. N. (2014). The RDoC framework: Facilitating transition from ICD/DSM to dimensional approaches that integrate neuroscience and psychopathology. *World Psychiatry; Official Journal of the World Psychiatric Association (WPA)*, *13*(1), 28–35. doi:10.1002/wps.20087 PMID:24497240

Demichelis, O. P., Coundouris, S. P., Grainger, S. A., & Henry, J. D. (2020). Empathy and Theory of Mind in Alzheimer's Disease: A Meta-analysis. *Journal of the International Neuropsychological Society*, *26*(10), 963–977. doi:10.1017/S1355617720000478 PMID:32431261

Diaz, M. T., & Hogstrom, L. J. (2011). The influence of context on hemispheric recruitment during metaphor processing. *Journal of Cognitive Neuroscience*, *23*(11), 3586–3597. doi:10.1162/jocn_a_00053 PMID:21568642

Duval, C., Bejanin, A., Piolino, P., Laisney, M., de La Sayette, V., Belliard, S., Eustache, F., & Desgranges, B. (2012). Theory of mind impairments in patients with semantic dementia. *Brain*, *135*(Pt 1), 228–241. doi:10.1093/brain/awr309 PMID:22232593

Eslinger, P. J., Moore, P., Troiani, V., Antani, S., Cross, K., Kwok, S., & Grossman, M. (2007). Oops! Resolving social dilemmas in frontotemporal dementia. *Journal of Neurology, Neurosurgery, and Psychiatry*, *78*(5), 457–460. doi:10.1136/jnnp.2006.098228 PMID:17012339

Eviatar, Z., & Just, M. A. (2006). Brain correlates of discourse processing: An fMRI investigation of irony and conventional metaphor comprehension. *Neuropsychologia*, *44*(12), 2348–2359. doi:10.1016/j.neuropsychologia.2006.05.007 PMID:16806316

Figueiredo, T. (2022). *Tratado de Cognição Social: uma abordagem multidimensional*. Ed. Ampla.

Fonagy, P., & Bateman, A. W. (2008). The development of borderline personality disorder—A mentalizing model. *Journal of Personality Disorders*, *22*(1), 4–21. doi:10.1521/pedi.2008.22.1.4 PMID:18312120

Frisoni, G. B., Fox, N. C., Jack, C. R. Jr, Scheltens, P., & Thompson, P. M. (2010). The clinical use of structural MRI in Alzheimer disease. *Nature Reviews. Neurology*, *6*(2), 67–77. doi:10.1038/nrneurol.2009.215 PMID:20139996

Gorno-Tempini, M. L., Hillis, A. E., Weintraub, S., Kertesz, A., Mendez, M., Cappa, S. F., Ogar, J. M., Rohrer, J. D., Black, S., Boeve, B. F., Manes, F., Dronkers, N. F., Vandenberghe, R., Rascovsky, K., Patterson, K., Miller, B. L., Knopman, D. S., Hodges, J. R., Mesulam, M. M., & Grossman, M. (2011). Classification of primary progressive aphasia and its variants. *Neurology*, *76*(11), 1006–1014. doi:10.1212/WNL.0b013e31821103e6 PMID:21325651

Green, M. F., Horan, W. P., & Lee, J. (2015). Social cognition in schizophrenia. *Nature Reviews. Neuroscience*, *16*(10), 620–631. doi:10.1038/nrn4005 PMID:26373471

Gregory, C., Lough, S., Stone, V., Erzinclioglu, S., Martin, L., Baron-Cohen, S., & Hodges, J. R. (2002). Theory of mind in patients with frontal variant frontotemporal dementia and Alzheimer's disease: Theoretical and practical implications. *Brain*, *125*(Pt 4), 752–764. doi:10.1093/brain/awf079 PMID:11912109

Happé, F., Cook, J. L., & Bird, G. (2017). The Structure of Social Cognition: In(ter)dependence of Sociocognitive Processes. *Annual Review of Psychology*, *68*(1), 243–267. doi:10.1146/annurev-psych-010416-044046 PMID:27687121

Haxby, J. V., Hoffman, E. A., & Gobbini, M. I. (2002). Human neural systems for face recognition and social communication. *Biological Psychiatry*, *51*(1), 59–67. doi:10.1016/S0006-3223(01)01330-0 PMID:11801231

Henry, J. D., Rendell, P. G., Scicluna, A., Jackson, M., & Phillips, L. H. (2009). Emotion experience, expression, and regulation in Alzheimer's disease. *Psychology and Aging*, *24*(1), 252–257. doi:10.1037/a0014001 PMID:19290761

Hogan, D. B., Jetté, N., Fiest, K. M., Roberts, J. I., Pearson, D., Smith, E. E., Roach, P., Kirk, A., Pringsheim, T., & Maxwell, C. J. (2016). The Prevalence and Incidence of Frontotemporal Dementia: a Systematic Review. *The Canadian Journal of Neurological Sciences. Le journal canadien des sciences neurologiques*, *43*(Suppl 1), S96–S109. doi:10.1017/cjn.2016.25

Hur, M. R., & Caixeta, L. (2013). Non-literal language and semantic dementia. *Dementia & Neuropsychologia*, *7*(4), 424–427. doi:10.1590/S1980-57642013DN74000011 PMID:29213868

Iacoboni, M. (2009). Imitation, empathy, and mirror neurons. *Annual Review of Psychology*, *60*(1), 653–670. doi:10.1146/annurev.psych.60.110707.163604 PMID:18793090

Johnen, A., & Bertoux, M. (2019). Psychological and Cognitive Markers of Behavioral Variant Frontotemporal Dementia-A Clinical Neuropsychologist's View on Diagnostic Criteria and Beyond. *Frontiers in Neurology*, *10*, 594. doi:10.3389/fneur.2019.00594 PMID:31231305

Kessels, R., Waanders-Oude Elferink, M., & van Tilborg, I. (2021). Social cognition and social functioning in patients with amnestic mild cognitive impairment or Alzheimer's dementia. *Journal of Neuropsychology*, *15*(2), 186–203. doi:10.1111/jnp.12223 PMID:32979297

Kipps, C. M., Nestor, P. J., Acosta-Cabronero, J., Arnold, R., & Hodges, J. R. (2009). Understanding social dysfunction in the behavioural variant of frontotemporal dementia: The role of emotion and sarcasm processing. *Brain*, *132*(Pt 3), 592–603. doi:10.1093/brain/awn314 PMID:19126572

Kober, H., Barrett, L. F., Joseph, J., Bliss-Moreau, E., Lindquist, K., & Wager, T. D. (2008). Functional grouping and cortical-subcortical interactions in emotion: A meta-analysis of neuroimaging studies. *NeuroImage*, *42*(2), 998–1031. doi:10.1016/j.neuroimage.2008.03.059 PMID:18579414

Kumfor, F., Irish, M., Hodges, J. R., & Piguet, O. (2013). Discrete Neural Correlates for the Recognition of Negative Emotions: Insights from Frontotemporal Dementia. *PLoS One*, *8*(6), 457. doi:10.1371/journal.pone.0067457 PMID:23805313

Laisney, M., Bon, L., Guiziou, C., Daluzeau, N., Eustache, F., & Desgranges, B. (2013). Cognitive and affective theory of mind in mild to moderate Alzheimer's disease. *Journal of Neuropsychology*, *7*(1), 107–120. doi:10.1111/j.1748-6653.2012.02038.x PMID:23088554

Lansford, J. E., Sharma, C., Malone, P. S., Woodlief, D., Dodge, K. A., Oburu, P., Pastorelli, C., Skinner, A. T., Sorbring, E., Tapanya, S., Tirado, L. M., Zelli, A., Al-Hassan, S. M., Alampay, L. P., Bacchini, D., Bombi, A. S., Bornstein, M. H., Chang, L., Deater-Deckard, K., & Di Giunta, L. (2014). Corporal punishment, maternal warmth, and child adjustment: a longitudinal study in eight countries. *Journal of Clinical Child and Adolescent Psychology*, *43*(4), 670–685. doi:10.1080/15374416.2014.893518

Lebrun, Y., & Buyssens, E. (1982). Metalanguage and speech pathology. *The British Journal of Disorders of Communication*, *17*(1), 21–25. doi:10.3109/13682828209011404 PMID:7044405

Libon, D. J., Xie, S. X., Wang, X., Massimo, L., Moore, P., Vesely, L., Khan, A., Chatterjee, A., Coslett, H. B., Hurtig, H. I., Liang, T. W., & Grossman, M. (2009). Neuropsychological decline in frontotemporal lobar degeneration: A longitudinal analysis. *Neuropsychology, 23*(3), 337–346. doi:10.1037/a0014995 PMID:19413447

Mancopes, R., & Schultz, F. (2008). Processing of metaphors in transcortical motor aphasia. *Dementia & Neuropsychologia, 2*(4), 339–348. doi:10.1590/S1980-57642009DN20400019 PMID:29213596

Marschark, M., & Nall, L. (1985). Metaphoric competence in cognitive and language development. *Advances in Child Development and Behavior, 19*, 49–81. doi:10.1016/S0065-2407(08)60388-2 PMID:2418642

Molenberghs, P., Cunnington, R., & Mattingley, J. B. (2012). Brain regions with mirror properties: A meta-analysis of 125 human fMRI studies. *Neuroscience and Biobehavioral Reviews, 36*(1), 341–349. doi:10.1016/j.neubiorev.2011.07.004 PMID:21782846

Morton, J., & Frith, U. (1995). Causal modelling: a structural approach to developmental psychopathology. In D. Cicchetti & D. Cohen (Eds.), *Developmental Psychopathology* (pp. 357–390). Wiley.

Osher, J. E., Wicklund, A. H., Rademaker, A., Johnson, N., & Weintraub, S. (2007). The mini-mental state examination in behavioral variant frontotemporal dementia and primary progressive aphasia. *American Journal of Alzheimer's Disease and Other Dementias, 22*(6), 468–473. doi:10.1177/1533317507307173 PMID:18166606

Pick, E., Kleinbub, J. R., Mannarini, S., & Palmieri, A. (2019). Empathy In Neurodegenerative Diseases: A Systematic Review. *Neuropsychiatric Disease and Treatment, 15*, 3287–3304. doi:10.2147/NDT.S225920 PMID:31819455

Premack, D., & Woodruff, G. (1978). Does the chimpanzee have a "theory of mind"? *Behavioral and Brain Sciences, 4*(4), 515–526. doi:10.1017/S0140525X00076512

Proverbio, A. M., Crotti, N., Zani, A., & Adorni, R. (2009). The role of left and right hemispheres in the comprehension of idiomatic language: An electrical neuroimaging study. *BMC Neuroscience, 10*(1), 116. doi:10.1186/1471-2202-10-116 PMID:19754932

Rahman, S., Sahakian, B. J., Hodges, J. R., Rogers, R. D., & Robbins, T. W. (1999). Specific cognitive deficits in mild frontal variant frontotemporal dementia. *Brain, 122*(Pt 8), 1469–1493. doi:10.1093/brain/122.8.1469 PMID:10430832

Rankin, K. P., Kramer, J. H., & Miller, B. L. (2005). Patterns of cognitive and emotional empathy in frontotemporal lobar degeneration. *Cognitive and Behavioral Neurology: Official Journal of the Society for Behavioral and Cognitive Neurology, 18*(1), 28–36. https://doi.org/ doi:10.1097/01.wnn.0000152225.05377.ab

Rapp, A. M., & Wild, B. (2011). Nonliteral language in Alzheimer dementia: A review. *Journal of the International Neuropsychological Society, 17*(2), 207–218. doi:10.1017/S1355617710001682 PMID:21241530

Reilly, J., Rodriguez, A. D., Lamy, M., & Neils-Strunjas, J. (2010). Cognition, language, and clinical pathological features of non-Alzheimer's dementias: An overview. *Journal of Communication Disorders, 43*(5), 438–452. doi:10.1016/j.jcomdis.2010.04.011 PMID:20493496

Rousseaux, M., Sève, A., Vallet, M., Pasquier, F., & Mackowiak-Cordoliani, M. A. (2010). An analysis of communication in conversation in patients with dementia. *Neuropsychologia, 48*(13), 3884–3890. doi:10.1016/j.neuropsychologia.2010.09.026 PMID:20888846

Sabatinelli, D., Fortune, E. E., Li, Q., Siddiqui, A., Krafft, C., Oliver, W. T., Beck, S., & Jeffries, J. (2011). Emotional perception: Meta-analyses of face and natural scene processing. *NeuroImage, 54*(3), 2524–2533. doi:10.1016/j.neuroimage.2010.10.011 PMID:20951215

Santos, M. T. F., Sougey, E. B., & Alchieri, J. C. (2009). Validity and reliability of the screening test for Alzheimer's disease with proverbs (STADP) for the elderly. *Arquivos de Neuro-Psiquiatria, 67*(3b, n.3B), 836–842. doi:10.1590/S0004-282X2009000500011 PMID:19838514

Satpute, A. B., Kang, J., Bickart, K. C., Yardley, H., Wager, T. D., & Barrett, L. F. (2015). Involvement of Sensory Regions in Affective Experience: A Meta-Analysis. *Frontiers in Psychology, 6*, 1860. doi:10.3389/fpsyg.2015.01860 PMID:26696928

Schurz, M., Radua, J., Aichhorn, M., Richlan, F., & Perner, J. (2014). Fractionating theory of mind: A meta-analysis of functional brain imaging studies. *Neuroscience and Biobehavioral Reviews, 42*, 9–34. doi:10.1016/j.neubiorev.2014.01.009 PMID:24486722

Setién-Suero, E., Murillo-García, N., Sevilla-Ramos, M., Abreu-Fernández, G., Pozueta, A., & Ayesa-Arriola, R. (2022). Exploring the Relationship Between Deficits in Social Cognition and Neurodegenerative Dementia: A Systematic Review. *Frontiers in Aging Neuroscience, 14*, 778093. doi:10.3389/fnagi.2022.778093 PMID:35572150

Shany-Ur, T., Poorzand, P., Grossman, S. N., Growdon, M. E., Jang, J. Y., Ketelle, R. S., Miller, B. L., & Rankin, K. P. (2012). Comprehension of insincere communication in neurodegenerative disease: Lies, sarcasm, and theory of mind. *Cortex*, *48*(10), 1329–1341. doi:10.1016/j.cortex.2011.08.003 PMID:21978867

Shimokawa, A., Yatomi, N., Anamizu, S., Ashikari, I., Kohno, M., Maki, Y., Torii, S., Isono, H., Sugai, Y., Koyama, N., & Matsuno, Y. (2000). Comprehension of emotions: Comparison between Alzheimer type and vascular type dementias. *Dementia and Geriatric Cognitive Disorders*, *11*(5), 268–274. doi:10.1159/000017249 PMID:10940678

Sousa, R. M., Ferri, C. P., Acosta, D., Albanese, E., Guerra, M., Huang, Y., Jacob, K. S., Jotheeswaran, A. T., Rodriguez, J. J., Pichardo, G. R., Rodriguez, M. C., Salas, A., Sosa, A. L., Williams, J., Zuniga, T., & Prince, M. (2009). Contribution of chronic diseases to disability in elderly people in countries with low and middle incomes: A 10/66 Dementia Research Group population-based survey. *Lancet*, *374*(9704), 1821–1830. doi:10.1016/S0140-6736(09)61829-8 PMID:19944863

Torralva, T., Roca, M., Gleichgerrcht, E., Bekinschtein, T., & Manes, F. (2009). A neuropsychological battery to detect specific executive and social cognitive impairments in early frontotemporal dementia. *Brain*, *132*(Pt 5), 1299–1309. doi:10.1093/brain/awp041 PMID:19336463

Torres, B., Santos, R. L., Barroso de Sousa, M. F., Simoes Neto, J. P., Lima Nogueira, M. M., & Belfort, T. T. (2015). Facial expression recognition in Alzheimer's disease: A longitudinal study. *Arquivos de Neuro-Psiquiatria*, *73*(5), 383–389. doi:10.1590/0004-282X20150009 PMID:26017202

Vuilleumier, P., & Pourtois, G. (2007). Distributed and interactive brain mechanisms during emotion face perception: Evidence from functional neuroimaging. *Neuropsychologia*, *45*(1), 174–194. doi:10.1016/j.neuropsychologia.2006.06.003 PMID:16854439

Youmans, G., & Bourgeois, M. (2010). Theory of mind in individuals with Alzheimer-type dementia. *Aphasiology*, *24*(4), 515–534. doi:10.1080/02687030902801573

Zaki, J., & Ochsner, K. N. (2012). The neuroscience of empathy: Progress, pitfalls and promise. *Nature Neuroscience*, *15*(5), 675–680. doi:10.1038/nn.3085 PMID:22504346

KEY TERMS AND DEFINITIONS

Alzheimer's Disease (AD): The most prevalent cause of dementia, which has as its main characteristic element the loss of episodic memory.

Behavioral Variant of Frontotemporal Dementia (bvFTD): A subtype of FTD frequently associated with social behavioral changes.

Cognition: Refers to the brain functions involved in stimuli processing. Stimuli can be from an internal or external mental state.

Dementia: Generic term used to refer to primary neurodegenerative syndromes that lead to significant functional impairments.

Frontotemporal Dementia (FTD): Prevalent cause of dementia, which has two variants - semantic and behavioral.

Healthy Aging: Refers to the course of physiological aging. In other words, it involves the expected physical and mental deterioration.

Semantic Dementia (SD): A subtype of FTD characterized by language processing losses, can lead to aphasia.

Social Cognition: Is an "umbrella" term used to refer to the brain processes implied in social stimuli, i.e., stimuli emitted by other human beings.

Chapter 11
Sleep Disorders and Neurodegeneration:
Updates and Future Directions

Katie Moraes de Almondes
Federal University of Rio Grande do Norte, Brazi

Teresa Paiva
University of Lisbon, Portugal

ABSTRACT

Dementia affects millions of people around the world each year. At this moment, 55.2 million people are living with dementia in the world, with an estimated social cost of US$1.3 trillion. Dementia associated with Alzheimer's disease and other forms of dementia was ranked as the 7th leading cause of death. One of the risk factors for developing neurodegenerative disorders is sleep disorders. Scientific evidence shows that sleep impairment accelerates cognitive decline by up to three years. Taken together, a better understanding of age-related changes in sleep architecture, their potential impact on cognition, and their underlying mechanisms are essential to developing an efficient treatment of sleep disorders in older people, serving as a promising target for dementia prevention. This chapter will discuss sleep changes and sleep disorders that are predictors of cognitive decline in aging, explaining the pathophysiological mechanisms, and pointing out the gaps that still need to be answered in this area through future research perspectives.

DOI: 10.4018/979-8-3693-0851-6.ch011

SLEEP DISORDERS AND NEURODEGENERATION: A THEORETICAL REVIEW, UPDATES AND FUTURE DIRECTIONS

Recent advances in health technology have significantly increased global life expectancy, transforming disease diagnosis and treatment. Population aging is now a worldwide reality, presenting varied challenges across countries. For example, France had 150 years to adjust to a 20% increase in its elderly population, while Brazil will only have 20 years (WHO, 2018). This situation poses new challenges for healthcare systems globally, as aging brings about changes requiring responsive public policies.

Today, while improved socioeconomic conditions enable a better quality of life for many in old age (World Economic Situation and Prospects as of mid-2023, 2023), there's also a rise in age-related neurodegenerative diseases. Age remains the primary risk factor for dementia, leading to increased dependence and diminished quality of life (Ponjoan et al., 2019).

Dementia is a widespread issue globally, affecting millions annually. According to the Global Action Plan on the Public Health Response to Dementia 2017–2025 and the Global Dementia Observatory (WHO, 2018), approximately 55.2 million people are currently living with dementia worldwide, incurring an estimated social cost of US$1.3 trillion. In 2019, dementia, including Alzheimer's disease and other forms (for example, dementia with Lewy bodies disease, Frontotemporal dementia, Huntington's disease), ranked as the 7th leading cause of death, predominantly affecting women. Projections for Latin America indicate a significant and rapid increase in dementia cases within a short timeframe (Patterson, 2018). This region faces substantial barriers related to diagnosis and treatment, including challenges within healthcare systems, limited access to care, and the necessity to standardize protocols and diagnostic methods (Baez & Ibanez, 2016). Understanding the aging process is crucial for implementing preventive interventions that could potentially alter future projections of dementia prevalence worldwide (Prince et al., 2013).

Sleep is essential for maintaining overall health and well-being, serving as a period for restorative processes, memory consolidation, and regulating bodily functions. Changes in sleep patterns negatively affect cognitive processes such as memory, attention, and executive functions (Almondes, 2017; Castro & Almondes, 2018; Khan & Al-Jahdali, 2023; Lima, Kirov & Almondes, 2022). Individuals with sleep disorders like insomnia or obstructive sleep apnea syndrome, as well as those experiencing partial or acute sleep deprivation, exhibit significant cognitive changes (Almondes, Júnior, Leonardo, & Alves, 2020; Bucks, Olaithe, Rosenzweig, and Morrell, 2017; Seda & Han, 2020).

Persistent sleep disorders, with limited clinical improvement and ongoing cognitive impairment and daytime dysfunction, may indicate a potential risk for developing neurodegenerative conditions (Anghel et al, 2023). Research suggests that sleep disturbances can accelerate cognitive decline by up to three years. Between 18 to 53,4% of individuals with mild cognitive impairment experience pre-existing sleep issues, including low sleep efficiency, increased sleep latency, and shorter sleep duration (Casagrande, Forte, Favieri, & Corbo, 2022). Elderly individuals with sleep disorders have a heightened risk of mild neurocognitive disorder. Approximately 50% of the elderly population experience sleep disorders and related complaints, with rates reaching 70% among those with dementia syndromes (Anghel et al, 2023; Casagrande, Forte, Favieri, & Corbo, 2022). Sleep disorders are recognized as predictors of dementia, notably Alzheimer's disease and vascular dementia (Almondes et al., 2016; Shi et al., 2018; Westwood et al., 2017; Lutsey et al., 2018).

Taken together, a better understanding of aging-related changes in sleep, their potential impact on cognition, and their underlying mechanisms are essential to developing an efficient treatment of sleep disorders in older people, serving as a promising target for dementia prevention. This information is even more crucial to better understand and treat sleep disorders in neurodegenerative diseases, such as Dementia associated with Alzheimer's Disease, for example, where these sleep changes, considered later symptoms of the disease, can be evident years before diagnosis.

Recently, numerous cross-sectional and cohort studies have emerged to investigate how various sleep disorders serve as predictors of neurodegenerative conditions among older adults. However, these studies primarily concentrate on insomnia and obstructive sleep apnea, the most common sleep disorders. The aim is to explore the involvement of various sleep disorders beyond insomnia and obstructive sleep apnea in the progression of neurodegenerative conditions. Additionally, we will delve into the underlying etiopathological mechanisms linking these disorders and discuss updates and future perspectives in this field. However, it is important to note that the robustness of the scientific evidence presented may be affected by sources of bias, such as significant heterogeneity (e.g., differences in population characteristics, sociocultural factors, varying outcome definitions, and sleep-related exposures), as well as well such as the impact of small samples.

Sleep and Aging: Understanding the Process of Deterioration of Biological Clocks

It is essential to know that sleep is controlled by a biological rhythm, that is, a rhythm generated by a neural structure - the suprachiasmatic nuclei - SCN, located in the hypothalamus, which is considered the main biological clock for mammals, with periodicity (in this case of human sleep, happens every 24 hours) and cyclical pattern (occurrence between sleep and wakefulness). More specifically, sleep is generated by the brain, namely by the ventrolateral preoptic nucleus of the hypothalamus (VLPO) which will inhibit the stimuli that put the cortex on alert, which originate in the tubero-mamillary nuclei, lateral hypothalamus, locus coeruleus, dorsal raphe, and tegmental nuclei of the protuberance. In this system, the balance between sleep and wakefulness is modulated by orexin (Brinkman, Reddy, & Sharma, 2023).

The suprachiasmatic nuclei coordinate sleep-wake synchronization with temporal cues such as light-dark (day-night), our work, study, and leisure schedules. The result is sleep occurring during the night and without association with activities. This phenomenon is called External Temporal Order. Furthermore, there is synchronization with other biological rhythms of our body such as blood pressure, heart rate, temperature, and hormones such as cortisol, and melatonin, which results in the Internal Temporal Order. This process is carried out by the regulatory mechanisms of the Circadian Timing System, a system that regulates rhythms close to 24 hours, which is the central characteristic of the sleep-wake rhythm of human beings; associated with homeostatic regulatory mechanisms that drive the system towards sleep propensity depending on the amount of wakefulness during the 24 hours (Brinkman, Reddy, & Sharma, 2023).

Moreover, there are two fundamentally distinct types of sleep regulated by the ultradian mechanism, which controls the alternation between NREM (non-rapid eye movement) sleep and REM (rapid eye movement) sleep throughout the night (Carley & Farabi, 2016).

During aging, older people exhibit a rhythmic pattern of advanced sleep phase (sleep onset and wake times are earlier). This could intensify, with more and more advances in these schedules. This explanation is due to the circadian changes that occur during aging, due to a deterioration of biological oscillators, and degeneration of the suprachiasmatic nucleus (SCN), including a reduction in the amplitude of circadian rhythms, and a worsening in the ability to tolerate phase changes. Other circadian rhythms are also affected, such as a decrease in melatonin secretion levels, which causes a decrease in the propensity to sleep (Morgan et al., 2017).

This process is increased by the loss of the ability to synchronize with photic cues (light-dark) from the environment, by changes in vision, and also by the loss of social and work activities, which are present in some retired elderly people. Over time, the result involves sleep changes such as fragmentation of nighttime sleep, greater difficulty falling asleep (sleep latency), and reduction in the duration and efficiency of nighttime sleep. With the absence of activities and changes in sleep resulting from the deterioration of the Circadian Timing System, the elderly begin to take long daytime naps, which intensifies the difficulty in falling asleep at night, making it easier for sleep to become more fragmented.

In addition to this process, the changes promoted by the degeneration of the SCN bring changes in the NREM-REM sleep cycles, which are shorter, with a pattern of NREM stages 1 and 2 being more overbalanced, making sleep more superficial and less restorative and more frequent. awakenings after the onset of sleep. Older people tend to awaken less from REM sleep and more from NREM sleep, and have a reduced number of sleep spindles and K complexes in NREM (Pace & Spencer, 2011).

Not all elderly people experience the same changes at the same intensity levels. However, it is important to know your sleep pattern throughout your life to identify what is changing and whether the change necessarily means a sleep disorder.

In addition to these sleep changes attributed to aging, many elderly people develop sleep disorders.

It is estimated that sleep disorders affect around 50% of elderly people. Among them, insomnia is the most prevalent, reaching a prevalence of up to 40% of the population aged 60 and over (Patel, Steinberg, & Patel, 2018). Obstructive sleep apnea also has a high prevalence, with a rate of 36% to 84% in the elderly (Senaratna et al., 2017). Restless legs syndrome can affect 10% to 13% of people over 65 years of age; the prevalence of Periodic Sleep Movements also increases significantly with age, determining greater daytime sleepiness if associated with restless legs syndrome (Leary et al, 2018). REM Sleep Behavior Disorder, which occurs in the elderly, has a prevalence ranging from 1.06% (Haba-Rubio et al., 2018) to 1.79% (Sasai-Sakuma et al., 2020).

Finally, the increase in medical, neurological, and psychiatric diseases associated with greater longevity is another conditioning factor for sleep disorders. We highlight, for example, menopausal insomnia (Bonanni et al, 2019), rheumatic diseases (Reis et al, 2018), diabetes (Holingue et al, 2018), obesity, and psychiatric disorders in the elderly (Zimmerman et al, 2013). Around 40% of elderly cancer patients have sleep disorders, which increase the negative impact on Instrumental Activities of Daily Living (OR=1.96, 95% [CI] 1.23-3.13, P = 0.005)(Loh et al, 2017). Neurological diseases, in turn, have a doubly negative impact, as many simultaneously affect sleep and cognition. An example of this is Parkinson's Disease, whose risk of cognitive decline increases significantly in elderly people with sleep disorders, and this change may even be a marker of Parkinson's Dementia (Lysen et al, 2019), while, in turn, Lewy Body Dementia is associated with insomnia, restless legs syndrome, and REM Sleep Behavior Disorder. An identical panorama emerges in strokes, the risk of which increases with insomnia and apnea in the elderly (Liu et al, 2022), which in turn leads to specific sleep changes.

The importance of these sleep disorders in the development, progression, management, and treatment of neurocognitive conditions has not been fully elucidated. Sleep problems may be underreported by the patient or caregiver/family member years before and even in the early stages, but they can lead to behavioral problems that are predictive of worsening dementia.

How to Explain Data on Sleep Changes as Predictors of Neurodegeneration

When there is sleep deprivation or sleep disturbance, there are metabolic and functional changes in certain brain areas that are activated, or not, during certain behaviors or information processing, impacting the optimal functioning of cognition. Scientific data has demonstrated that sleep deprivation affects areas of the prefrontal cortex, in particular, causing an impact on highly complex cognitive processes such as executive functions. Furthermore, memory is consolidated during the phases of night sleep, and if there is a reduction in night sleep in the elderly, there is consequently an impact on this cognitive process (Almondes, 2017; Lima, Kirov & Almondes, 2022).

Furthermore, difficulties in initiating sleep, which can evolve into insomnia, together with changes in the NREM-REM stages, are associated with dementia. Several proteins, including Aβ and tau, are present in the interstitial space around brain cells. Cerebrospinal fluid circulates throughout the brain, and exchanges with interstitial fluid with the function of removing interstitial proteins, including Aβ and tau proteins. During sleep, the clearance of these excess proteins is carried out by slow-wave sleep (or NREM sleep), and the sleep-wake cycle has the function of regulating lymphatic clearance. That is, sleep contributes to the removal of harmful proteins from the brain (Wang & Holtzman, 2020). It is understood that an imbalance between production, clearance, and aggregation of peptides causes Aβ to accumulate, Disrupted sleep patterns may impair this clearance process, leading to the accumulation of these proteins and increasing the risk of neurodegeneration.

Research indicates that sleep disorders can induce changes in both the structure and function of the brain, affecting gray matter volume, white matter integrity, and the connectivity among brain regions. These alterations have been linked to cognitive decline and an elevated susceptibility to neurodegenerative diseases, particularly Alzheimer's disease (Stankeviciute et al, 2023).

Futhermore, sleep disorders are associated with increased levels of neuroinflammation and oxidative stress in the brain. These processes are known to contribute to the development and progression of neurodegenerative diseases such as Alzheimer's disease, Parkinson's disease and dementia. In their study, Palagini et al. (2022) highlight how insomnia, sleep deprivation, and disrupted circadian rhythms can induce chronic inflammation by directly and indirectly triggering neuroinflammation through the activation of microglia and astrocytes. These alterations in sleep patterns serve as neurobiological stressors, hyperactivating the stress system and detrimentally impacting neural plasticity, resulting in neuronal damage. Furthermore, sleep disturbances may facilitate the buildup of neurotoxic proteins, heighten oxidative stress, and compromise neuroprotection, ultimately exacerbating neurodegeneration and aggravating mood disorders.

On the other hand, inflammatory mechanisms are associated with the relationship between sleep disorders, cognitive changes, and dementia. Precisely, sleep deprivation/reduction activates inflammatory mechanisms: factor-κ-B, activator of protein-1 and STAT family proteins that together increase the expression of mRNA in pro-inflammatory cytokines (Irwin & Vitiello, 2019).

It is also important to consider the complex relationships of multimorbidities and brain changes common in aging, which, aggravated by multiple chronic diseases, give rise to anatomical changes in the brain, evident in imaging studies. In vascular dementia, there is a widening of the cerebral sulci, subclinical infarcts, hyperintensities of the white matter, and enlargement of the ventricles, which are aggravated by glucose metabolism disorders in diabetes, or by the cardiovascular consequences of hypertension or sleep apnea. In dementia associated with Alzheimer's disease, decreased cortical thickness and smaller brain and hippocampal volumes are associated with greater insulin resistance and cardiovascular risks. (Holingue et al, 2018). Sleep disorders, in particular, are associated with detrimental metabolic alterations that can impact health and potentially contribute to the onset of various diseases. Extended periods of sleep deprivation elevate the risk of insulin resistance and type 2 diabetes. Additionally, sleep disorders like insomnia, sleep apnea, and disruptions in circadian rhythms are linked to changes in glucose, amino acid, and lipid metabolism. Intriguingly, many of the metabolic changes induced by sleep disturbances resemble those observed in aging and neurodegenerative conditions such as Alzheimer's disease (Carroll & Macauley, 2019).

Finally, numerous psychiatric conditions, including depression, anxiety, bipolar affective disorder, and attention deficit hyperactivity disorder, are linked to disruptions in sleep patterns. These disturbances can worsen cognitive impairment and raise the likelihood of neurodegeneration. Recognizing these connections is vital for devising specific interventions to enhance sleep quality and potentially reduce the risk of neurodegenerative diseases in those impacted (Colwell, 2021; Palagini et al, 2022). In summary, psychiatric conditions associated with sleep disturbances can contribute to neurodegeneration through a myriad of interconnected mechanisms, encompassing inflammation, oxidative stress, impaired protein clearance, neurotransmitter imbalance, hormonal dysregulation, and genetic/environmental factors, as mentioned earlier. Furthermore, the disruption of the sleep-wake cycle rhythm itself significantly contributes to this detrimental association (Colwell, 2021; Palagini et al, 2022)..

Understanding the Role of Different Sleep Disorders in the Development of Neurodegenerative Conditions

Overall, the presence of sleep disorders may serve as an early warning sign or predictor of neurodegeneration, highlighting the importance of addressing sleep disturbances for maintaining brain health and potentially preventing the onset or progression of neurodegenerative diseases.

The most common subtype of sleep apnea is obstructive sleep apnea syndrome (OSAS). OSAS is characterized by recurrent sleep fragmentation and excessive daytime sleepiness, chronic intermittent hypoxia, hypercapnia, hypoventilation, peripheral and central inflammation, and changes in cerebral perfusion and glucose metabolism, with significant impairment in cognitive processes. Research suggests that specific brain damage associated with OSAS may increase the risk of developing dementia (Kim et al, 2011). It has been estimated that the prevalence of OSAS in patients with dementia may reach 60% (Guarnieri et al, 2012). A longitudinal study of elderly women diagnosed with sleep-disordered breathing found an association between the presence of this disorder and a greater risk of developing mild cognitive impairment compared to elderly women who did not have this disorder (Yaffe et al., 2011). In another longitudinal study, patients diagnosed with sleep apnea had an increased risk of developing dementia (Chang et al., 2013). However, it remains unclear whether sleep-disordered breathing precedes cognitive impairment or the opposite. What is known is that there is an association between biomarkers of OSA and Alzheimer's Disease (Kinugawa, 2023).

Insomnia disorder has been associated with an increased risk of developing Mild Neurocognitive Disorder and dementia (Chang et al., 2013; Yaffe et al., 2011; Hung et al., 2018). A systematic review and meta-analysis provided data that insomnia may be a predictor of dementia (de Almondes et al., 2016).

One of the lines of studies to explain this relationship concerns neurodegenerative pathophysiological processes related to tau and Amyloid-β (Aβ) proteins, which indicate a link between insomnia and dementia. These proteins are present in the interstitial space around brain cells (Querfurth & LaFerla, 2010) and it is during sleep, especially in slow-wave sleep (Huang et al., 2012), that the removal of these proteins that harm the brain occurs. Studies have reported that Aβ accumulation in the preclinical stage of Alzheimer's disease (Ju et al., 2013), as well as tauopathies (Lee et al., 2001), are associated with worse sleep quality (sleep deprivation due to insomnia) and support the hypothesis that insomnia is a potential risk factor for subsequent dementia and a pathological marker of severity in Alzheimer's disease.

REM Sleep Behavior Disorder (SRBD), more prevalent in males, may be idiopathic and related to treatments with tricyclics, MAOIs, and SSRIs. However, it is frequently associated with alpha-synucleinopathies, Parkinson's disease, Parkinson's Disease with dementia, Lewy Body Dementia (LBD), and Multisystem Atrophy, preceding the onset of these conditions. These diseases, in addition to neurological changes and sleep disorders, have a characteristic set of symptoms: constipation, urinary and sexual dysfunction, cardiovascular autonomic symptoms (orthostatic hypotension, and reduced heart rate variability). REM Sleep Behavior Disorder precedes Parkinson's Disease: 38% in the first 4 years; and 17% in the following 13 years; the association is greater in forms in which there are no tremors (Kunz et al, 2023).

REM Sleep Behavior Disorder occurs in 77% of LBD cases and its presence is part of the diagnosis. LBD has a progressive evolution, which after diagnosis lasts between 5 and 8 years, but there are cases described between 2 and 20 years of survival. The neuronal accumulation of "Lewy bodies" leads to significant changes in the production of two essential neurotransmitters in memory and motor control: acetylcholine and dopamine. Patients regularly experience visual hallucinations, unpredictable and fluctuating changes in wakefulness, attention, and concentration, and illogical thinking. Memory changes are not usually an initial symptom and appear as the disease progresses (Schumacher et al, 2019). Motor symptoms are important: rigidity, slow gait, frozen posture, balance changes and frequent falls, difficulty swallowing, hypomimia, micrographia, and weak voice. The associated psychiatric symptoms are variable: depression, apathy, anxiety, agitation, false beliefs and opinions, paranoia, and irrational suspicions. In terms of sleep disorders, in addition to REM Sleep Behavior Disorder, the patients may have excessive sleepiness, insomnia, and restless leg syndrome. (Patel et al, 2022).

The risk of dementia in Restless Legs Syndrome has been considered possible taking into account poor sleep quality, depression or anxiety, poor nutrition, microvasculopathy, and hypoxia, which together constitute risk factors for dementia. Recently, a study in South Korea demonstrated that RLS was a risk factor for all forms of dementia (the risk was 1.74 times higher in RLS than in the control group (95% [CI] 1.51–2.02, $p<0.001$). It was specifically a risk for dementia associated with Alzheimer's disease (aHR 1.38, 95% CI 1.11–1.72) and vascular dementia (aHR 1.81, 95% CI 1.30–2.53) (Kim et al, 2023).

Less common sleep disorders, such as circadian rhythm disorders, may also impact cognitive function and contribute to neurodegenerative conditions, although their specific roles are less well understood and require further investigation. It is evident from scientific literature that a disruption in the sleep/wake cycle is a common symptom shared among neurodegenerative diseases, often manifesting early in the disease progression. This disruption frequently includes disturbances in the timing of sleep and rhythmic physiological markers, suggesting the potential involvement of neurodegenerative diseases in disrupting the circadian timing system (Colwell, 2021).

Future Directions

One of the current hot topics for directing research with possibilities for innovative therapeutic management is the role of cognitive reserve as a successful moderating and protective factor for the relationship between sleep and neurodegeneration.

The concept of Cognitive Reserve (CR) was originally defined by Stern (2002) as the ability of the brain to optimize and maximize performance through the differential recruitment of brain networks and the use of alternative cognitive strategies to deal with brain and neurodegenerative pathologies (Stern, 2012). Patients with more cognitive and brain reserve can better deal with neurodegenerative diseases, due to active compensatory mechanisms located throughout the brain network. Lifelong experiences can influence the effectiveness and flexibility of brain networks, promoting neural plasticity. Education, occupational experience, rich and heterogeneous social environments, and leisure activities can modulate the brain, contributing to the construction of cognitive reserve. Thus, individuals with high reserve loads have greater neural resources, requiring more pathological load to reach the critical threshold point necessary to manifest clinical pathological processes, which translates into a delay in the onset of clinical symptoms. (Barulli & Stern, 2013; Stern, 2012).

Some studies on sleep changes in sleep architecture, and sleep disorders in different neurocognitive conditions have tested the association with cognitive reserve. Ourry et al (2023) studied the role of cognitive reserve in cognition, studying slow wave sleep (SWS), as this stage is essential for learning and memory, is reduced in aging, and has an impact on cognitive decline, as we mentioned earlier. The researchers found that those individuals with greater cognitive reserve were able to maintain cognitive performance despite low amounts of SWS, demonstrating that CR appears to protect cognition against the degrading effects of age-related sleep changes. Zavecz et al, (2023) demonstrated that slow-wave sleep is a cognitive reserve factor that protects against memory impairment caused by the high pathological load of amyloid beta plaques in Dementia associated with Alzheimer's Disease, making them conclude that sleep is a modifiable factor and can be seen as an independent cognitive reserve resource. In this sense, the data taken together suggest that an effective intervention for sleep could help preserve cognitive function in the face of neurodegenerative conditions.

D'Este et al (2023), through clinical and neuropsychological assessments with fifty-five patients with REM sleep behavioral disorder, showed that those patients with high levels of cognitive reserve among those with REM sleep behavioral disorder had the lowest percentage of mild cognitive impairment and improved performance in visual-constructive and verbal memory functions, suggesting that cognitive reserve may help deal with the ongoing cognitive decline processes associated with this sleep disorder.

Another area that deserves to be highlighted in clinical research with a future perspective is the design of protocols for the assessment and therapeutic management of sleep disorders, such as respiratory disorders or movement disorders associated with sleep, as they are conditions that increase the risk of neurodegenerative conditions such as Alzheimer's, dementia in Parkinson's and dementia with Lewy bodies. For example, Prete et al (2023) observed a significant negative correlation between sleep difficulties and executive performance only in patients with low cognitive reserve, who may be considered at greater risk of developing cognitive deficits in the presence of sleep difficulties.

Progress in virtual reality technology holds the promise of immersive and interactive sleep therapy experiences in the future. Tailored virtual environments could target specific sleep disturbances associated with various neurological disorders, offering a distinctive and engaging approach to sleep intervention. In addition, innovations in digital sleep monitoring and AI algorithms may pave the way for personalized sleep feedback systems. By integrating sleep data from wearable devices and other sources, AI-driven platforms could provide real-time insights and practical recommendations for enhancing sleep in individuals with neurodegeneration disorders (Al-Sharman, Ismaiel, Khalil, El-Salem, 2021).

In the authors' view of this chapter, the employment of artificial intelligence algorithms for processing vast datasets is noteworthy. Within the domain of sleep and neurodegenerative diseases, AI-driven platforms have the capacity to amalgamate data from wearable devices, electronic health records, genetic profiles, and behavioral patterns. This integration allows for the prediction of the probability of developing such disorders based on sleep disruptions. Identifying individuals at risk early on could enable timely interventions, preventive strategies, and personalized treatment plans. The application of AI-based approaches holds promise in enhancing the precision and efficiency of clinical assessments, ultimately leading to improved patient outcomes.

Some research has demonstrated the relevant association between sleep and physical activity for the prevention of neurodegenerative conditions, especially dementia associated with Alzheimer's disease (Sun, He & Qu, 2023), with a call for a more comprehensive approach to reducing factors risk for these conditions.

Lastly, sleep biomarkers are emerging as valuable tools for the early detection of neurodegenerative diseases such as Parkinson's disease, Alzheimer's disease, dementia with Lewy bodies, among others. These biomarkers provide information about changes in sleep patterns and physiology that may precede the onset of clinical symptoms (Miglis et al, 2021).

There is still a long way to go to clarify many gaps in this field. However, we consider that international cooperation should be considered in light of socio-cultural guidelines, which could help in the construction of clearer, faster, and more effective therapeutic management.

REFERENCES

Al-Sharman, A., Ismaiel, I. A., Khalil, H., & El-Salem, K. (2021). Exploring the Relationship Between Sleep Quality, Sleep-Related Biomarkers, and Motor Skill Acquisition Using Virtual Reality in People With Parkinson's Disease: A Pilot Study. *Frontiers in Neurology*, *12*, 582611. doi:10.3389/fneur.2021.582611 PMID:33737900

Almondes, K. M. (2017). *Neurocognitive impairment of Acute and Chronic Sleep Loss. Williams, Larry Sleep Deprivation: Global prevalence, dangers, and impacts on cognitive performance*. Nova Science.

Anghel, L., Ciubară, A., Nechita, A., Nechita, L., Manole, C., Baroiu, L., Ciubară, A. B., & Mușat, C. L. (2023). Sleep Disorders Associated with Neurodegenerative Diseases. *Diagnostics (Basel)*, *13*(18), 2898. doi:10.3390/diagnostics13182898 PMID:37761265

Baez, S., & Ibáñez, A. (2016). Dementia in Latin America: An emergent silent tsunami. *Frontiers in Aging Neuroscience*, *8*, 253. doi:10.3389/fnagi.2016.00253 PMID:27840605

Barulli, D., & Stern, Y. (2013). Efficiency, capacity, compensation, maintenance, plasticity: Emerging concepts in cognitive reserve. *Trends in Cognitive Sciences*, *17*(10), 502–509. doi:10.1016/j.tics.2013.08.012 PMID:24018144

Bonanni, E., Schirru, A., Di Perri, M. C., Bonuccelli, U., & Maestri, M. (2019). Insomnia and hot flashes. *Maturitas*, *126*, 51–54. doi:10.1016/j.maturitas.2019.05.001 PMID:31239118

BrinkmanJ. E.ReddyV.SharmaS. (2023). *Physiology of Sleep*. Available from: https://www.ncbi.nlm.nih.gov/books/NBK482512/

Bucks, R. S., Olaithe, M., Rosenzweig, I., & Morrell, M. J. (2017). Reviewing the relationship between OSA and cognition: Where do we go from here? *Respirology (Carlton, Vic.)*, *22*(7), 1253–1261. doi:10.1111/resp.13140 PMID:28779504

Carley, D. W., & Farabi, S. S. (2016). Physiology of Sleep. *Diabetes Spectrum*, *29*(1), 5–9. doi:10.2337/diaspect.29.1.5 PMID:26912958

Carroll, C. M., & Macauley, S. L. (2019). The Interaction Between Sleep and Metabolism in Alzheimer's Disease: Cause or Consequence of Disease? *Frontiers in Aging Neuroscience*, *11*, 258. doi:10.3389/fnagi.2019.00258 PMID:31616284

Casagrande, M., Forte, G., Favieri, F., & Corbo, I. (2022). Sleep Quality and Aging: A Systematic Review on Healthy Older People, Mild Cognitive Impairment and Alzheimer's Disease. *International Journal of Environmental Research and Public Health*, *19*(14), 8457. doi:10.3390/ijerph19148457 PMID:35886309

Castro, E. D. A. S., & de Almondes, K. M. (2018). Sleep pattern and decision-making in physicians from mobile emergency care service with 12-h work schedules. *The International Journal of Neuroscience*, *128*(6), 530–539. doi:10.1080/00207454.2017.1400970 PMID:29098917

Chang, W. P., Liu, M. E., Chang, W. C., Yang, A. C., Ku, Y. C., Pai, J. T., Huang, H. L., & Tsai, S. J. (2013). Sleep apnea and the risk of dementia: A population-based 5-year follow-up study in Taiwan. *PLoS One*, *8*(10), e78655. doi:10.1371/journal.pone.0078655 PMID:24205289

Colwell, C. S. (2021). Defining circadian disruption in neurodegenerative disorders. *The Journal of Clinical Investigation*, *131*(19), e148288. doi:10.1172/JCI148288 PMID:34596047

D'Este, G., Berra, F., Carli, G., Leitner, C., Marelli, S., Zucconi, M., Casoni, F., Ferini-Strambi, L., & Galbiati, A. (2023). Cognitive Reserve in Isolated Rapid Eye-Movement Sleep Behavior Disorder. *Brain Sciences*, *13*(2), 176. doi:10.3390/brainsci13020176 PMID:36831719

de Almondes, K. M., Costa, M. V., Malloy-Diniz, L. F., & Diniz, B. S. (2016). Insomnia and risk of dementia in older adults: Systematic review and meta-analysis. *Journal of Psychiatric Research*, *77*, 109–115. doi:10.1016/j.jpsychires.2016.02.021 PMID:27017287

de Almondes, K. M., Júnior, F. W. N. H., Leonardo, M. E. M., & Alves, N. T. (2020). Facial Emotion Recognition and Executive Functions in Insomnia Disorder: An Exploratory Study. *Frontiers in Psychology*, *11*, 502. doi:10.3389/fpsyg.2020.00502 PMID:32362851

Guarnieri, B., Adorni, F., Musicco, M., Appollonio, I., Bonanni, E., Caffarra, P., Caltagirone, C., Cerroni, G., Concari, L., Cosentino, F. I., Ferrara, S., Fermi, S., Ferri, R., Gelosa, G., Lombardi, G., Mazzei, D., Mearelli, S., Morrone, E., Murri, L., ... Sorbi, S. (2012). Prevalence of sleep disturbances in mild cognitive impairment and dementing disorders: A multicenter Italian clinical cross-sectional study on 431 patients. *Dementia and Geriatric Cognitive Disorders*, *33*(1), 50–58. doi:10.1159/000335363 PMID:22415141

Haba-Rubio, J., Frauscher, B., Marques-Vidal, P., Toriel, J., Tobback, N., Andries, D., Preisig, M., Vollenweider, P., Postuma, R., & Heinzer, R. (2018). Prevalence and determinants of rapid eye movement sleep behavior disorder in the general population. *Sleep (Basel)*, *41*(2), zsx197. Advance online publication. doi:10.1093/sleep/zsx197 PMID:29216391

Holingue, C., Wennberg, A., Berger, S., Polotsky, V. Y., & Spira, A. P. (2018). Disturbed sleep and diabetes: A potential nexus of dementia risk. *Metabolism: Clinical and Experimental*, *84*, 85–93. doi:10.1016/j.metabol.2018.01.021 PMID:29409842

Huang, Y., Potter, R., Sigurdson, W., Santacruz, A., Shih, S., Ju, Y. E., ... Bateman, R. J. (2012). Effects of age and amyloid deposition on Aβ dynamics in the human central nervous system. *Archives of Neurology*, *69*(1), 51–58. doi:10.1001/archneurol.2011.235 PMID:21911660

Hung, C. M., Li, Y. C., Chen, H. J., Lu, K., Liang, C. L., Liliang, P. C., Tsai, Y. D., & Wang, K. W. (2018). Risk of dementia in patients with primary insomnia: A nationwide population-based case-control study. *BMC Psychiatry*, *18*(1), 38. doi:10.1186/s12888-018-1623-0 PMID:29415688

Irwin, M. R., & Vitiello, M. V. (2019). Implications of sleep disturbance and inflammation for Alzheimer's disease dementia. *Lancet Neurology*, *18*(3), 296–306. doi:10.1016/S1474-4422(18)30450-2 PMID:30661858

Ju, Y. E. S., McLeland, J. S., Toedebusch, C. D., Xiong, C., Fagan, A. M., Duntley, S. P., Morris, J. C., & Holtzman, D. M. (2013). Sleep quality and preclinical Alzheimer disease. *JAMA Neurology*, *70*(5), 587–593. doi:10.1001/jamaneurol.2013.2334 PMID:23479184

Khan, M. A., & Al-Jahdali, H. (2023). The consequences of sleep deprivation on cognitive performance. *Neurosciences (Riyadh)*, *28*(2), 91–99. doi:10.17712/nsj.2023.2.20220108 PMID:37045455

Kim, K. Y., Kim, E. H., Lee, M., Ha, J., Jung, I., & Kim, E. (2023). Restless leg syndrome and risk of all-cause dementia: A nationwide retrospective cohort study. *Alzheimer's Research & Therapy*, *15*(1), 46. doi:10.1186/s13195-023-01191-z PMID:36879327

Kim, S. J., Lee, J. H., Lee, D. Y., Jhoo, J. H., & Woo, J. I. (2011). Neurocognitive dysfunction associated with sleep quality and sleep apnea in patients with mild cognitive impairment. *The American Journal of Geriatric Psychiatry*, *19*(4), 374–381. doi:10.1097/JGP.0b013e3181e9b976 PMID:20808148

Kinugawa, K. (2023, October). Obstructive sleep apnea and dementia: A role to play? *Revue Neurologique*, *179*(7), 793–803. doi:10.1016/j.neurol.2023.08.006 PMID:37633736

Kunz, D., Stotz, S., de Zeeuw, J., Papakonstantinou, A., Dümchen, S., Haberecht, M., Plotkin, M., & Bes, F. (2023). Prognostic biomarkers in prodromal α-synucleinopathies: DAT binding and REM sleep without atonia. *Journal of Neurology, Neurosurgery, and Psychiatry*, *94*(7), 532–540. doi:10.1136/jnnp-2022-330048 PMID:36725328

Leary, E. B., Moore, H. E. IV, Schneider, L. D., Finn, L. A., Peppard, P. E., & Mignot, E. (2018). Periodic limb movements in sleep: Prevalence and associated sleepiness in the Wisconsin Sleep Cohort. *Clinical Neurophysiology*, *129*(11), 2306–2314. doi:10.1016/j.clinph.2018.08.022 PMID:30243181

Lee, V. M., Goedert, M., & Trojanowski, J. Q. (2001). Neurodegenerative tauopathies. *Annual Review of Neuroscience*, *24*(1), 1121–1159. doi:10.1146/annurev.neuro.24.1.1121 PMID:11520930

Lima, N.C., Kirov, R., & de Almondes, K.M. (2022). Impairment of executive functions due to sleep alterations: An integrative review on the use of P300. *Frontiers in Neuroscience*, *22*, 16:906492. . doi:10.3389/fnins.2022.906492

Liu, L., Yang, J., Wang, J., Nie, M., Wang, Z., Guan, H., Hu, J., & Hong, F. (2022). Relationship between Sleep Duration and Stroke History in Middle-Aged and Elderly in Guiyang: A Cross-Sectional Survey. *Neuropsychiatric Disease and Treatment*, *18*, 243–252. doi:10.2147/NDT.S340834 PMID:35185332

Loh, K. P., Pandya, C., Zittel, J., Kadambi, S., Flannery, M., Reizine, N., Magnuson, A., Braganza, G., Mustian, K., Dale, W., Duberstein, P., & Mohile, S. G. (2017). Associations of sleep disturbance with physical function and cognition in older adults with cancer. *Supportive Care in Cancer*, *25*(10), 3161–3169. doi:10.1007/s00520-017-3724-6 PMID:28455547

Lutsey, P. L., Misialek, J. R., Mosley, T. H., Gottesman, R. F., Punjabi, N. M., Shahar, E., MacLehose, R., Ogilvie, R. P., Knopman, D., & Alonso, A. (2018). Sleep characteristics and risk of dementia and Alzheimer's disease: The Atherosclerosis Risk in Communities Study. *Alzheimer's & Dementia*, *14*(2), 157–166. doi:10.1016/j.jalz.2017.06.2269 PMID:28738188

Lysen, T. S., Darweesh, S. K. L., Ikram, M. K., Luik, A. I., & Ikram, M. A. (2019). Sleep and risk of parkinsonism and Parkinson's disease: A population-based study. *Brain*, *142*(7), 2013–2022. doi:10.1093/brain/awz113 PMID:31038176

Miglis, M. G., Adler, C. H., Antelmi, E., Arnaldi, D., Baldelli, L., Boeve, B. F., Cesari, M., Dall'Antonia, I., Diederich, N. J., Doppler, K., Dušek, P., Ferri, R., Gagnon, J.-F., Gan-Or, Z., Hermann, W., Högl, B., Hu, M. T., Iranzo, A., Janzen, A., ... Oertel, W. H. (2021). Biomarkers of conversion to α-synucleinopathy in isolated rapid-eye-movement sleep behaviour disorder. *Lancet Neurology*, *20*(8), 671–684. doi:10.1016/S1474-4422(21)00176-9 PMID:34302789

Morgan, E., Schumm, L. P., McClintock, M., Waite, L., & Lauderdale, D. S. (2017). Sleep characteristics and daytime cortisol levels in older adults. *Sleep (Basel)*, *40*(5), zsx043. Advance online publication. doi:10.1093/sleep/zsx043 PMID:28329370

Ourry, V., Rehel, S., André, C., Mary, A., Paly, L., Delarue, M., Requier, F., Hendy, A., Collette, F., Marchant, N. L., Felisatti, F., Palix, C., Vivien, D., de la Sayette, V., Chételat, G., Gonneaud, J., & Rauchs, G. (2023). Medit-Ageing Research Group. Effect of cognitive reserve on the association between slow wave sleep and cognition in community-dwelling older adults. *Aging (Albany NY)*, *15*(18), 9275–9292. doi:10.18632/aging.204943 PMID:37770186

Pace-Schott, E. F., & Spencer, R. M. (2011). Age-related changes in the cognitive function of sleep. *Progress in Brain Research*, *191*, 75–89. doi:10.1016/B978-0-444-53752-2.00012-6 PMID:21741545

Palagini, L., Geoffroy, P. A., Miniati, M., Perugi, G., Biggio, G., Marazziti, D., & Riemann, D. (2022). Insomnia, sleep loss, and circadian sleep disturbances in mood disorders: A pathway toward neurodegeneration and neuroprogression? A theoretical review. *CNS Spectrums*, *27*(3), 298–308. doi:10.1017/S1092852921000018 PMID:33427150

Patel, B., Irwin, D. J., Kaufer, D., Boeve, B. F., Taylor, A., & Armstrong, M. J. (2022). Outcome Measures for Dementia With Lewy Body Clinical Trials: A Review. *Alzheimer Disease and Associated Disorders*, *36*(1), 64–72. doi:10.1097/WAD.0000000000000473 PMID:34393189

Patel, D., Steinberg, J., & Patel, P. (2018). Insomnia in the Elderly: A Review. *Journal of Clinical Sleep Medicine, 14*(6), 1017–1024. doi:10.5664/jcsm.7172 PMID:29852897

Patterson, C. (2018). The state of the art of dementia research: New frontiers. *World Alzheimer Report.* https://www.alz.co.uk/research/world-report-2018

Ponjoan, A., Garre-Olmo, J., Blanch, J., Fages, E., Alves-Cabratosa, L., Martí-Lluch, R., Comas-Cufí, M., Parramon, D., Garcia-Gil, M., & Ramos, R. (2019). Epidemiology of dementia: Prevalence and incidence estimates using validated electronic health records from primary care. *Clinical Epidemiology, 11*, 217–228. doi:10.2147/CLEP.S186590 PMID:30881138

Prete, M., Cellini, N., Ronconi, L., & Di Rosa, E. (2023). Cognitive reserve moderates the relationship between sleep difficulties and executive functions in patients with Parkinson's disease. *Sleep Medicine, 111*, 82–85. doi:10.1016/j.sleep.2023.09.009 PMID:37738789

Prince, M., Bryce, R., Albanese, E., Wimo, A., Ribeiro, W., & Ferri, C. P. (2013). The global prevalence of dementia: A systematic review and metaanalysis. *Alzheimer's & Dementia, 9*(1), 63–75. doi:10.1016/j.jalz.2012.11.007 PMID:23305823

Querfurth, H. W., & LaFerla, F. M. (2010). Mechanisms of disease. *The New England Journal of Medicine, 362*(4), 329–344. doi:10.1056/NEJMra0909142 PMID:20107219

Reis, C., Dias, S., Rodrigues, A. M., Sousa, R. D., Gregório, M. J., Branco, J., Canhão, H., & Paiva, T. (2018). Sleep duration, lifestyles and chronic diseases: A cross-sectional population-based study. *Sleep Science, 11*(4), 217–230. doi:10.5935/1984-0063.20180036 PMID:30746039

Sasai-Sakuma, T., Takeuchi, N., Asai, Y., Inoue, Y., & Inoue, Y. (2020). Prevalence and clinical characteristics of REM sleep behavior disorder in Japanese elderly people. *Sleep (Basel), 43*(8), zsaa024. Advance online publication. doi:10.1093/sleep/zsaa024 PMID:32064524

Schumacher, J., Peraza, L. R., Firbank, M., Thomas, A. J., Kaiser, M., Gallagher, P., O'Brien, J. T., Blamire, A. M., & Taylor, J. P. (2019). Dysfunctional brain dynamics and their origin in Lewy body dementia. *Brain, 142*(6), 1767–1782. doi:10.1093/brain/awz069 PMID:30938426

Seda, G., & Han, T. S. (2020). Effect of Obstructive Sleep Apnea on Neurocognitive Performance. *Sleep Medicine Clinics, 15*(1), 77–85. doi:10.1016/j.jsmc.2019.10.001 PMID:32005352

Senaratna, C. V., Perret, J. L., Lodge, C. J., Lowe, A. J., Campbell, B. E., Matheson, M. C., Hamilton, G. S., & Dharmage, S. C. (2017). Prevalence of obstructive sleep apnea in the general population: A systematic review. *Sleep Medicine Reviews*, *34*, 70–81. doi:10.1016/j.smrv.2016.07.002 PMID:27568340

Shi, L., Chen, S.-J., Ma, M.-Y., Bao, Y.-P., Han, Y., Wang, Y.-M., Shi, J., Vitiello, M. V., & Lu, L. (2018). Sleep disturbances increase the risk of dementia: A systematic review and meta-analysis. *Sleep Medicine Reviews*, *40*, 4–16. doi:10.1016/j.smrv.2017.06.010 PMID:28890168

Stankeviciute, L., Falcon, C., Operto, G., Garcia, M., Shekari, M., Iranzo, Á., Niñerola-Baizán, A., Perissinotti, A., Minguillón, C., Fauria, K., Molinuevo, J. L., Zetterberg, H., Blennow, K., Suárez-Calvet, M., Cacciaglia, R., Gispert, J. D., & Grau-Rivera, O. (2023). Differential effects of sleep on brain structure and metabolism at the preclinical stages of AD. *Alzheimer's & Dementia*, *19*(12), 5371–5386. doi:10.1002/alz.13102 PMID:37194734

Stern, Y. (2012). Cognitive reserve in ageing and Alzheimer's disease. *Lancet Neurology*, *11*(11), 1006–1012. doi:10.1016/S1474-4422(12)70191-6 PMID:23079557

Sun, X., He, C., & Qu, H. (2023). Bibliometric review on sleep and Alzheimer disease between 1986 and 2023. *Medicine*, *102*(44), e35764. doi:10.1097/MD.0000000000035764 PMID:37932981

Wang, C., & Holtzman, D. M. (2020). Bidirectional relationship between sleep and Alzheimer's disease: Role of amyloid, tau, and other factors. *Neuropsychopharmacology*, *45*(1), 104–120. doi:10.1038/s41386-019-0478-5 PMID:31408876

Westwood, A. J., Beiser, A., Jain, N., Himali, J. J., DeCarli, C., Auerbach, S. H., Pase, M. P., & Seshadri, S. (2017). Prolonged sleep duration as a marker of early neurodegeneration predicting incident dementia. *Neurology*, *88*(12), 1172–1179. doi:10.1212/WNL.0000000000003732 PMID:28228567

World Economic Situation and Prospects as of mid-2023. (2023). https://www.un.org/development/desa/dpad/publication/world-economic-situation-and-prospects-as-of-mid-2023/

World Health Organization. (2018). *Aging and health*. https://www.who.int/news-room/fact-sheets/detail/ageing-and-health

Yaffe, K., Laffan, A. M., Harrison, S. L., Redline, S., Spira, A. P., Ensrud, K. E., Ancoli-Israel, S., & Stone, K. L. (2011). Sleep-disordered breathing, hypoxia, and risk of mild cognitive impairment and dementia in older women. *Journal of the American Medical Association, 306*(6), 613–619. doi:10.1001/jama.2011.1115 PMID:21828324

Zavecz, Z., Shah, V. D., Murillo, O. G., Vallat, R., Mander, B. A., Winer, J. R., Jagust, W. J., & Walker, M. P. (2023). NREM sleep as a novel protective cognitive reserve factor in the face of Alzheimer's disease pathology. *BMC Medicine, 21*(1), 156. doi:10.1186/s12916-023-02811-z PMID:37138290

Zimmerman, M. E., Bigal, M. E., Katz, M. J., Derby, C. A., & Lipton, R. B. (2013). Are sleep onset/maintenance difficulties associated with medical or psychiatric comorbidities in nondemented community-dwelling older adults? *Journal of Clinical Sleep Medicine, 9*(4), 363–369. doi:10.5664/jcsm.2590 PMID:23585752

Chapter 12
Neuropsychological Manifestations Associated With Cancer Chemotherapy

Marcelo Houat de Brito
Hospital das Clínicas of the Faculty of Medicine, University of São Paulo, Brazil

Sonia Maria Dozzi Brucki
Hospital das Clínicas of the Faculty of Medicine, University of São Paulo, Brazil

Ricardo Nitrini
HHospital das Clínicas of the Faculty of Medicine, University of São Paulo, Brazil

ABSTRACT

In recent years, we have seen a significant increase in the incidence and prevalence of oncological diseases. With the evolution of treatments and the tendency to increase the life expectancy of cancer patients, it is necessary to consider even more: not only survival but also the quality of life in this period, with the individual being able to maintain their family, social, and labor activities. An important and often neglected cause of impact on quality of life, functionality, and work capacity is cancer-related cognitive impairment, which arises from several etiologies, with chemotherapy being one of the most common. This chapter reviews the literature on cognitive dysfunction associated with chemotherapy and how to evaluate and clinically manage patients with this important complication, providing a clear and didactic text on the topic for health professionals who deal with cancer patients in their clinical practice or who are interested in studying and researching the theme.

DOI: 10.4018/979-8-3693-0851-6.ch012

1. INTRODUCTION

In recent years, we have seen a process of demographic transition in the world, with a decrease in birth and mortality rates, leading to an aging population. Likewise, a process of epidemiological transition is taking place, with the replacement of communicable diseases by non-communicable diseases and external causes. These changes have generated a shift in the burden of morbidity and mortality from younger groups to older groups and a transformation from a situation in which mortality predominates to one in which morbidity is dominant. As a result, both the incidence and prevalence of oncological diseases have increased, producing a significant increase in the years lived with disability due to these pathologies, an increase estimated at 40.6% between 2007 and 2017 (James et al., 2018; McCracken & Phillips, 2017).

With the evolution of treatments and the tendency to increase the life expectancy of cancer patients, it is necessary to consider even more not only survival but also the quality of life in this period, with the individual being able to maintain their family, social, and labor activities. An important cause of impact on quality of life, functionality, and work capacity is cancer-related cognitive impairment, which arises from several etiologies, often combined in the same patient: direct effects of cancer (which can occur even in neoplasms outside the central nervous system), effects of comorbidities (such as fatigue, depression, anxiety, insomnia, and pain) and direct effects of cancer treatment; the latter group is divided into the effects of chemotherapy, brain radiotherapy, immunotherapy and hormone therapy (Pendergrass et al., 2018). It is estimated that up to 30% of cancer patients have cognitive changes before treatment, up to 75% have measurable impairment during treatment, and around 35% of cancer survivors will continue to exhibit cognitive changes in the months to years following cancer treatment (Janelsins et al., 2014).

The period in which the highest percentage of cancer patients are affected by cognitive dysfunction is during treatment, with a large part of the cases being directly related to chemotherapy, the subject that we will cover in greater detail in this chapter.

2. CONCEPTS AND EPIDEMIOLOGY

It is already well documented that patients undergoing chemotherapy can suffer cognitive changes, which can be subjective (when there is only the patient's complaint without dysfunction proven by tests) and objective (with documentation of impairment by cognitive tests), mainly in short term memory, attention, executive functions, and processing speed (Ahles et al., 2012; Fleming et al., 2023; Joly et al., 2015; Lange et al., 2019; Pendergrass et al., 2018). In the literature, these changes were called chemotherapy-related cognitive impairment, or colloquially "chemobrain."

Studies demonstrate "chemobrain" in neoplasms such as colorectal cancer, lung cancer, ovarian cancer, testicular cancer, and lymphoma, but this phenomenon has been much more studied in patients with breast cancer (Fayette et al., 2023; Fleming et al., 2023; Lange et al., 2019; Whittaker et al., 2022). Cognitive changes seem more critical during the chemotherapy period, generally with improvement over time, but this improvement is only partial in about a third of cases (Collins et al., 2014). Data from cross-sectional studies indicate rates of 16 to 75% of moderate or severe cognitive impairment during the period of chemotherapy for breast cancer, compared with 4 to 11% in controls (Brezden et al., 2000; Janelsins et al., 2011; Tchen et al., 2003). It is essential to mention that this classification as moderate to severe in these studies was based on scores in cognitive tests and not directly on functionality, having no relation in size to what occurs, for example, in moderate to severe dementias, when even basic activities of daily living tend to be compromised. In the cognitive decline associated with chemotherapy, even in more severe cases, what usually happens are difficulties in instrumental activities of daily living. A more recent meta-analysis on chemotherapy-related cognitive impairment in breast cancer patients estimated its prevalence using self-report, short cognitive screening tools, and neuropsychological test batteries at 44%, 16%, and 21–34%, respectively (Whittaker et al., 2022).

Among the most studied chemotherapy drugs related to cognitive changes are anthracyclines (e.g., doxorubicin), taxanes (e.g., paclitaxel), platinums (e.g., cisplatin), cyclophosphamide, fluorouracil, and methotrexate; the latter mainly when administered intrathecally (Lange et al., 2019). Cognitive effects tend to be more significant the higher the doses of these medications, when there is an association of possible synergistic effects in multidrug regimens (Pendergrass et al., 2018; Shah, 2005), and the longer the duration of chemotherapy (Hodgson et al., 2013).

Some factors seem to indicate more significant risks of patients developing cognitive dysfunction when submitted to chemotherapy, such as the presence of cardiovascular risk factors (e.g., diabetes mellitus and arterial hypertension), some genetic polymorphisms (APOE ε4, COMT, IL-1R1, and BDNF), low education and low premorbid intelligence quotient. Advanced age has conflicting evidence regarding increased risk (Lange et al., 2019).

3. PATHOPHYSIOLOGY AND BIOMARKERS

In addition to clinical and epidemiological data, cognitive impairment related to chemotherapy is also based on experimental models in animals, mainly in rodents, with the use of chemotherapy drugs commonly used in the treatment of cancer, in which changes were demonstrated in behavioral tests that require the use of cognitive functions (Dubois et al., 2014; Fardell et al., 2013, 2015; Seigers et al., 2016; Winocur et al., 2012).

Although the pathophysiology that leads to chemotherapy-related cognitive impairment has not yet been fully elucidated, some of these experimental studies have helped to propose mechanisms. Among the main ones are the decrease in neurogenesis (mainly seen in hippocampal cells), changes in myelin and oligodendrocyte precursors, mitochondrial dysfunction and oxidative stress, increased production of cytokines systemically and locally in the brain, and neuroinflammation with activation of microglia and astrocytes when peripheral toxins cross a disrupted blood-brain barrier (Fleming et al., 2023; Lange et al., 2019). The possibility of an acceleration of the brain aging process by medications (Ahles, 2012) and dysfunction in the dynamics of glial cells (Whalley, 2019) were also proposed.

Studies with advanced neuroimaging techniques also corroborate the demonstration of brain changes secondary to chemotherapy and are interesting methods for better understanding the pathophysiology of the disease and possible clinical use (not yet well established) for diagnosis, evaluation of extension of involvement, and monitoring changes over time. Assessment by brain magnetic resonance imaging (MRI) with diffusion tensor imaging (DTI) demonstrated a decrease in fractional anisotropy in white matter regions, denoting a reduction in their integrity. Studies with voxel-based morphometry of brain MRI demonstrated reduced white and gray matter density. Researches that used functional MRI analyzed the resting state and tasks mainly related to working memory, showing different changes in connectomes and patterns of brain activation in patients undergoing chemotherapy (Sousa et al., 2020). A study with brain MRI using multivoxel spectroscopy demonstrated a reduction in the N-acetyl aspartate (NAA) peak in the posterior cingulate gyrus and in the dorsal thalamus bilaterally, which denotes neuronal depopulation in these regions (Tong et al., 2020). Assessments using positron emission tomography (PET) with glucose marker (FDG) demonstrated a reduction in global brain metabolism and some regions of interest, such as the frontal lobes and the mesial areas of the temporal lobes (Gamal et al., 2021; Oh et al., 2012). Furthermore, there are studies correlating neuroinflammation verified by PET with the 18-kDA translocator protein (TSPO) radioligand, a marker of microglia activation, with systemic chemotherapy in breast cancer patients (Schroyen et al., 2021, 2022).

Some serum biomarkers may be associated with chemotherapy-related cognitive decline, emphasizing biomarkers of systemic inflammation and neuronal injury. One of the studies with the largest sample size that used serum measurement of inflammatory response biomarkers demonstrated an association of increased IL-6 levels with more severe cognitive complaints and the association of higher IL-1 β levels with decreased processing speed in cognitive testing (Cheung et al., 2015). Among the biomarkers of neuronal injury, the most promising appears to be neurofilament light chain (NfL), which has already been extensively studied in other neurological diseases that lead to cognitive decline, such as Alzheimer's disease, frontotemporal dementia, and multiple sclerosis. Recent research found that patients who received chemotherapy had NfL values around 20 times higher than those who did not receive treatment, correlating with findings of cognitive decline (Schroyen et al., 2021).

4. CLINICAL AND NEUROPSYCHOLOGICAL EVALUATION

Being confused or unusually disorganized, feeling mental fogginess, taking longer than usual to complete routine tasks, and having difficulty concentrating, finding the right word, learning new skills, and multitasking are commonly referred to as chemotherapy-related cognitive impairment symptoms. However, it is essential to highlight the importance of those who deal with oncological patients to remain aware of the possibility of cognitive dysfunction with cancer therapies, as patients and their companions often do not notice the signs and symptoms despite suffering consequences from them daily.

As with any other condition, the first step in approaching a cancer patient with potential cognitive decline is to take a good history. Ideally, this should be done with the patient and a companion who lives with him, as cognitive deficits are often not noticed by the patient himself (anosognosia) or the patient is so severe that cannot report the clinical picture well. The anamnesis of these patients must clarify how the symptoms emerged (rapid vs. insidious evolution), whether there is a temporal association with any therapy performed, or even whether the signs and symptoms already existed before any oncological treatment was carried out. In general, the complaint brought by the patient and companion is related to a cognitive domain (e.g., forgetfulness – memory). It is essential to explore through anamnesis whether there are complaints compatible with changes in other domains and the patient's behavior, in addition to evaluating the possibility of symptoms that suggest focal somatic neurological deficits. Some questionnaires can be used to assist in the subjective assessment of cognitive decline, helping to classify and monitor cases better. The Functional Assessment of Cancer Therapy-Cognitive (FACT-Cog) is an interesting instrument for this analysis, as it was specifically designed to assess cognitive impairment related to cancer and its treatment (Wagner et al., 2009).

The presence of common comorbidities in cancer patients, which may be associated with cognitive impairment, must also be assessed in the anamnesis: depression, anxiety, insomnia, pain, and fatigue. Likewise, the patient's chronically used medications must be evaluated, emphasizing drugs with potential cognitive side effects, such as anticholinergic or sedative medications, that can be replaced or withdrawn. Still, in the anamnesis, it is necessary to assess the impact of cognitive complaints on activities of daily living, which can also be done through structured questionnaires, such as the Pfeffer Functional Activities Questionnaire (Pfeffer et al., 1982), the Lawton & Brody Instrumental Activities of Daily Living Scale (Lawton & Brody, 1969), and the Katz Activities of Daily Living Scale (Katz et al., 1970).

Regarding formal cognitive assessment, this can be done by the health professional, among other ways, through screening tests, emphasizing those that assess attention, executive functions, and memory, as these are the main cognitive domains affected by chemotherapy. In this context, the well-known and widely used Montreal Cognitive Assessment (MoCA)(Nasreddine et al., 2005), is a good screening test for patients with more than four years of education and the Brief Cognitive Screening Battery (BCSB) is an interesting test for patients with low education. The BCSB, developed in Brazil, consists of a good memory test through learning and recalling a list of 10 pictures, which is combined with the animal semantic verbal fluency test and the clock drawing test, which evaluates executive functions well (Nitrini et al., 2021). The Mini-Mental State Examination (MMSE) (Folstein et al., 1975), probably the best-known and most widely used cognitive screening test in the world, does not adequately evaluate the majority of cancer patients with suspected cognitive impairment, as the memory test is short (only three words) and relatively easy compared to other tests, in addition to not being a good tool for broader assessment of attention and executive functions. A study compares MMSE and MoCA in the cognitive assessment of geriatric oncology patients, demonstrating better evaluation with the latter (Rambeau et al., 2019).

The most complete way to objectively assess cognition is through a comprehensive neuropsychological exam by a psychologist trained in neuropsychology. Aiming to harmonize studies on cancer-related cognitive impairment, the International Cancer and Cognition Task Force (ICCTF) defined a minimal standard neuropsychological battery for objective assessment of the main cognitive domains affected in this condition (Wefel et al., 2011). This battery includes the Hopkins Verbal Learning Test-Revised (HVLT-R), Trail Making Test (TMT), and Controlled Oral Word Association (COWA) as core assessment (**Table 1**). The HVLT-R is a verbal episodic memory test consisting of three presentations of a list of 12 words, with immediate recall after each presentation, delayed recall after 20-25 minutes, and recognition (Benedict et al., 1998). The TMT consists of part A, which mainly tests attention and processing speed, asking the individual to connect numbers spread out on a sheet of paper in an increasing manner in the shortest possible time, and part B, which assesses executive functions more broadly, asking the individual to connect numbers and letters alternately, in ascending order for numbers and in alphabetical order for letters (Bowie & Harvey, 2006; Tombaugh, 2004). COWA is done through the application of the FAS phonemic verbal fluency test, a mixed test that evaluates executive functions and language, in which the subject must produce words (except proper nouns) that begin with the letters F, A, and S in a period of 60 seconds for each letter (Ruff et al., 1996; Sumerall et al., 1997). Other tests can be carried out in addition to those mentioned in a neuropsychological assessment of a cancer patient, aiming to increase sensitivity for detecting changes. The ICCTF defined the presence of objective cognitive impairment in patients who are below -1.5 standard deviations in 2 tests or below -2 standard deviations, which are characterized through the standardization of tests for the age and education of the person being evaluated (Wefel et al., 2011).

In patients with documented cognitive impairment, it may be interesting to perform some complementary exams to exclude other potential causes of decline that are not related to chemotherapy. This evaluation may include tests of thyroid function, renal function and electrolytes, liver function and enzymes, vitamin B12 and blood count, in addition to serology for syphilis and HIV if in an adequate epidemiology context. Neuroimaging exams (head CT or MRI) may be requested according to the patient's clinical condition, especially if there are findings of focal neurological deficits, presence of epileptic seizures, dementia syndrome or signs of intracranial hypertension on neurological examination. Collection of cerebrospinal fluid (CSF) may be necessary depending on the clinic and neuroimaging study, such as in suspected meningeal carcinomatosis or encephalitis. Occasionally, an electroencephalogram may also be required, as in cases of suspected epileptic seizures.

Table 1. International Cognition and Cancer Task Force (ICCTF) recommendation of cognitive assessments (Pendergrass et al. 2018; Wefel et al. 2011)

<table>
<tr><th>Recommended Cognitive Domain</th><th>Recommended Cognitive Assessment Measures</th><th>Specific Cognitive Skills Evaluated</th><th>Similar Cognitive Assessment Measures</th></tr>
<tr><td colspan="4">Core Assessment</td></tr>
<tr><td>Learning and memory</td><td>Hopkins Verbal Learning Test–Revised (HVLT-R)</td><td>Verbal learning and memory assessed by list learning, immediate recall, delayed recall, and recognition</td><td>California Verbal Learning Test - II (CVLT-II), Rey Auditory Verbal Learning Test (RAVT), Brief Visuospatial Memory–Revised (BVMT-R, Visual Learning and Memory)</td></tr>
<tr><td rowspan="2">Executive function</td><td>Trail Making Test</td><td>Multiple cognitive skills involved in performing task, including attention, working memory, information processing speed, and mental flexibility</td><td>Delis-Kaplan Executive Function Scale (D-KEFS) Trail Making</td></tr>
<tr><td>Controlled Oral Word Association (COWA)</td><td>Spontaneous generation of words</td><td>Delis-Kaplan Executive Function Scale (D-KEFS) Verbal Fluency</td></tr>
<tr><td>Processing speed</td><td>Trail Making Test</td><td>Multiple cognitive skills involved in performing task, including attention, working memory, information processing speed, and mental flexibility</td><td>Delis-Kaplan Executive Function Scale (D-KEFS) Trail Making, Wechscler Adult Intelligence Scale-IV (WAIS-IV) Coding, Symbol Digit Modalities Test (SDMT)</td></tr>
<tr><td colspan="4">Suplemental Assessment</td></tr>
<tr><td rowspan="4">Working memory</td><td>Auditory Consonant Trigrams (ACT)</td><td>Short term or working memory task requiring on-line maintenance of information while performing an interference task during a delay</td><td rowspan="4"></td></tr>
<tr><td>Paced Auditory Serial Attention Test (PASAT)</td><td>Serial attention task assessing working memory, divided attention, and information processing speed</td></tr>
<tr><td>Brief Test of Attention (BTA)</td><td>Auditory divided attention</td></tr>
<tr><td>Wechsler Adult Intelligence Scale -IV (WAIS-IV) Letter Number Sequencing</td><td>Working memory, attention, and mental control</td></tr>
</table>

5. THERAPEUTIC MANAGEMENT

Regarding symptom management, it is essential to adequately assess and treat the presence of pain, anxiety, depression, fatigue, and insomnia, as they can also interfere with the cognitive status of patients undergoing cancer treatment regardless of chemotherapy (Lange et al., 2019; Pendergrass et al., 2018; Pullens et al., 2013). It is also important to try to suspend, change or reduce the doses of any medications that may be harmful to cognition in these patients.

Strategies have been studied with a specific focus on chemotherapy-related cognitive decline. Among the non-pharmacological interventions studied with some evidence of benefit are cognitive rehabilitation (Ercoli et al., 2015; Kesler et al., 2013), cognitive behavioral therapy (Kucherer & Ferguson, 2017), occupational therapy (Player et al., 2014), mindfulness techniques (Bulen Love, 2018), and physical activity (Hartman et al., 2018). There are also initial studies with transcranial direct current electrical stimulation (Gaynor et al., 2020). For pharmacological therapy, there are clinical studies with psychostimulants (methylphenidate and modafinil), ginkgo biloba, erythropoietin, memantine and donepezil. However, with conflicting or not-so-robust results and, to date, there are no guidelines for the use of these medications (Karschnia et al., 2019). The three studies with medications for the symptomatic treatment of "chemobrain" which found some benefit in cognitive assessment, despite negative primary outcomes, are summarized in Table 2.

Table 2. Clinical studies of pharmacotherapy for chemotherapy-related cognitive symptoms with benefit in secondary cognitive outcomes

Study	Pharmacological Therapy	Study Design	Study Population	Duration	Cognitive Assessment With Evidence of Benefit
Kohli et al., 2009	Modafinil 200mg/day	Phase 1: open-label single arm; phase 2: RCT	Adults (n=82): breast cancer	4 weeks	COGDRAS (speed and quality of memory, and attention)
Escalante et al., 2014	Methylphenidate 10-20mg/day	N-of-1 crossover	Adults (n=33): breast cancer	4 weeks	WAIS-III digit span test (attention)
Lawrence et al., 2016	Donepezil 10mg/day	Double-blind RCT	Adults (n=62): breast cancer	24 weeks	HVLT-R (memory)

RCT: Randomized controlled trial; COGDRAS: Cognitive Drug Research Computerized Assessment System; WAIS-III: Wechsler Adult Intelligence Scale, Third edition; HVLT-R: Hopkins Verbal Learning Test–Revised

6. CURRENT ISSUES AND FUTURE DIRECTIONS

Although cognitive decline is a common complication of chemotherapy, unfortunately, it is still largely neglected in clinical practice, and there is still much to be studied. One of the main problems in the literature on the subject is the lack of uniformity in the use of cognitive tests and the objective criteria to establish the diagnosis homogeneously between studies. The ICCTF's efforts were very important in defining a minimum neuropsychological battery that should be used in all studies on the subject, as well as an objective definition of what cognitive impairment would be in this population based on the results of these tests, which were published in 2011 and already cited more fully in the clinical and neuropsychological evaluation section of this chapter (Wefel et al., 2011). It is essential to stimulate new epidemiological studies on the condition using these criteria defined by the ICCTF, especially in neoplasms other than breast cancer.

Another critical problem in the literature on the subject is the lack of studies focused on the treatment of cognitive impairment associated with chemotherapy. With a better understanding of the pathophysiology of the condition through studies on animal models and biomarkers to define therapeutic targets and a greater awareness of the subject, new studies will probably emerge trying to find specific pharmacological treatments for the condition since, to date, most therapies studies were with stimulants or with repositioning of medications used for symptomatic treatment of neurodegenerative dementia (Fleming et al., 2023). It is also vital to encourage new research into non-pharmacological interventions, increasing the level of scientific evidence to recommend some interventions that have already been studied, as well as evaluating new potential interventions.

Finally, although there is a growing number of research on "chemobrain" biomarkers, there has yet to be a direction for their use in clinical practice outside the research area, which could help in the diagnosis and, thus, in the management of the condition. With the advancement of understanding epidemiology, pathophysiology, diagnosis, and treatment of the condition, it will be possible to create a complete guideline on the subject, which would further help in recognizing and managing this complication of chemotherapy in practice.

7. CONCLUSION

Chemotherapy-related cognitive impairment can have a significant impact on the quality of life of this population. Although there is currently no formal recommendation for routine cognitive assessment in cancer patients without complaints, one should always be alert to signs and symptoms that may indicate cognitive impairment.

The evaluation of these patients involves a multidisciplinary approach, which may include neurologists, oncologists, hematologists, psychiatrists, psychologists, and occupational therapists. Therefore, these professionals must know how to recognize this condition and evaluate its differential diagnoses, in addition to carrying out appropriate clinical management, including guidance, consideration of suspending or changing oncological therapies, and proposing pharmacological and non-pharmacological treatments for the improvement of these patients' symptoms.

REFERENCES

Ahles, T. A. (2012). Brain vulnerability to chemotherapy toxicities. Psycho-Oncology. doi:10.1002/pon.3196

Ahles, T. A., Root, J. C., & Ryan, E. L. (2012). Cancer- and cancer treatment-associated cognitive change: An update on the state of the science. *Journal of Clinical Oncology*, *30*(30), 3675–3686. doi:10.1200/JCO.2012.43.0116 PMID:23008308

Benedict, R. H. B., Schretlen, D., Groninger, L., & Brandt, J. (1998). Hopkins verbal learning test - Revised: Normative data and analysis of inter-form and test-retest reliability. *The Clinical Neuropsychologist*, *12*(1), 43–55. Advance online publication. doi:10.1076/clin.12.1.43.1726

Bowie, C. R., & Harvey, P. D. (2006). Administration and interpretation of the Trail Making Test. *Nature Protocols*, *1*(5), 2277–2281. Advance online publication. doi:10.1038/nprot.2006.390 PMID:17406468

Brezden, C. B., Phillips, K. A., Abdolell, M., Bunston, T., & Tannock, I. F. (2000). Cognitive function in breast cancer patients receiving adjuvant chemotherapy. *Journal of Clinical Oncology*, *18*(14), 2695–2701. Advance online publication. doi:10.1200/JCO.2000.18.14.2695 PMID:10894868

Bulen Love, D. (2018). Mindfulness to Improve Cognitive Function in Cancer Survivors. *Oncology Times*, *40*(8), 28. doi:10.1097/01.COT.0000532447.18584.01

Cheung, Y. T., Ng, T., Shwe, M. K., Ho, H., Foo, K. M., Cham, M. T., Lee, J. A., Fan, G., Tan, Y. P., Yong, W. S., Ng, R., & Chan, A. (2015). Association of proinflammatory cytokines and chemotherapy-associated cognitive impairment in breast cancer patients: A multi-centered, prospective, cohort study. *Annals of Oncology : Official Journal of the European Society for Medical Oncology*, *26*(7), 1446–1451. doi:10.1093/annonc/mdv206 PMID:25922060

Collins, B., MacKenzie, J., Tasca, G. A., Scherling, C., & Smith, A. (2014). Persistent cognitive changes in breast cancer patients 1 year following completion of chemotherapy. *Journal of the International Neuropsychological Society*, *20*(4), 370–379. doi:10.1017/S1355617713001215 PMID:24229809

Dubois, M., Lapinte, N., Villier, V., Lecointre, C., Roy, V., Tonon, M.-C., Gandolfo, P., Joly, F., Hilber, P., & Castel, H. (2014). Chemotherapy-induced long-term alteration of executive functions and hippocampal cell proliferation: Role of glucose as adjuvant. *Neuropharmacology*, *79*, 234–248. doi:10.1016/j.neuropharm.2013.11.012 PMID:24291465

Ercoli, L. M., Petersen, L., Hunter, A. M., Castellon, S. A., Kwan, L., Kahn-Mills, B. A., Embree, L. M., Cernin, P. A., Leuchter, A. F., & Ganz, P. A. (2015). Cognitive rehabilitation group intervention for breast cancer survivors: Results of a randomized clinical trial. *Psycho-Oncology*, *24*(11), 1360–1367. doi:10.1002/pon.3769 PMID:25759235

Escalante, C. P., Meyers, C., Reuben, J. M., Wang, X., Qiao, W., Manzullo, E., Alvarez, R. H., Morrow, P. K., Gonzalez-Angulo, A. M., Wang, X. S., Mendoza, T., Liu, W., Holmes, H., Hwang, J., Pisters, K., Overman, M., & Cleeland, C. (2014). A Randomized, Double-blind, 2-Period, Placebo-Controlled Crossover Trial of a Sustained-Release Methylphenidate in the Treatment of Fatigue in Cancer Patients. *Cancer Journal (Sudbury, Mass.)*, *20*(1), 8–14. doi:10.1097/PPO.0000000000000018 PMID:24445757

Fardell, J. E., Vardy, J., & Johnston, I. N. (2013). The short and long term effects of docetaxel chemotherapy on rodent object recognition and spatial reference memory. *Life Sciences*, *93*(17), 596–604. doi:10.1016/j.lfs.2013.05.006 PMID:23693082

Fardell, J. E., Vardy, J., Monds, L. A., & Johnston, I. N. (2015). The long-term impact of oxaliplatin chemotherapy on rodent cognition and peripheral neuropathy. *Behavioural Brain Research*, *291*, 80–88. doi:10.1016/j.bbr.2015.04.038 PMID:25934489

Fayette, D., Juríčková, V., Kozák, T., Mociková, H., Gaherová, L., Fajnerová, I., & Horáček, J. (2023). Cognitive impairment associated with Hodgkin's lymphoma and chemotherapy. *Neuroscience Letters*, *797*, 137082. doi:10.1016/j.neulet.2023.137082 PMID:36693557

Fleming, B., Edison, P., & Kenny, L. (2023). Cognitive impairment after cancer treatment: Mechanisms, clinical characterization, and management. *BMJ (Clinical Research Ed.)*, *071726*, e071726. Advance online publication. doi:10.1136/bmj-2022-071726 PMID:36921926

Folstein, M. F., Folstein, S. E., & McHugh, P. R. (1975). Mini-mental state. *Journal of Psychiatric Research, 12*(3), 189–198. doi:10.1016/0022-3956(75)90026-6 PMID:1202204

Gamal, S. M. T., Azab, A. O. M. A., El Refaei, S. M., & Houseni, M. (2021). The role of 18-FDG PET/CT assessment of functional brain metabolism in cancer patients after chemotherapy. *The Egyptian Journal of Radiology and Nuclear Medicine, 52*(1), 18. doi:10.1186/s43055-020-00403-y

Gaynor, A. M., Pergolizzi, D., Alici, Y., Ryan, E., McNeal, K., Ahles, T. A., & Root, J. C. (2020). Impact of transcranial direct current stimulation on sustained attention in breast cancer survivors: Evidence for feasibility, tolerability, and initial efficacy. *Brain Stimulation, 13*(4), 1108–1116. doi:10.1016/j.brs.2020.04.013 PMID:32353419

Hartman, S. J., Nelson, S. H., Myers, E., Natarajan, L., Sears, D. D., Palmer, B. W., Weiner, L. S., Parker, B. A., & Patterson, R. E. (2018). Randomized controlled trial of increasing physical activity on objectively measured and self-reported cognitive functioning among breast cancer survivors: The memory & motion study. *Cancer, 124*(1), 192–202. Advance online publication. doi:10.1002/cncr.30987 PMID:28926676

Hodgson, K. D., Hutchinson, A. D., Wilson, C. J., & Nettelbeck, T. (2013). A meta-analysis of the effects of chemotherapy on cognition in patients with cancer. *Cancer Treatment Reviews, 39*(3), 297–304. doi:10.1016/j.ctrv.2012.11.001 PMID:23219452

James, S. L., Abate, D., Abate, K. H., Abay, S. M., Abbafati, C., Abbasi, N., Abbastabar, H., Abd-Allah, F., Abdela, J., Abdelalim, A., Abdollahpour, I., Abdulkader, R. S., Abebe, Z., Abera, S. F., Abil, O. Z., Abraha, H. N., Abu-Raddad, L. J., Abu-Rmeileh, N. M. E., Accrombessi, M. M. K., ... Murray, C. J. L. (2018). Global, regional, and national incidence, prevalence, and years lived with disability for 354 Diseases and Injuries for 195 countries and territories, 1990-2017: A systematic analysis for the Global Burden of Disease Study 2017. *Lancet, 392*(10159), 1789–1858. doi:10.1016/S0140-6736(18)32279-7 PMID:30496104

Janelsins, M. C., Kesler, S. R., Ahles, T. A., & Morrow, G. R. (2014). Prevalence, mechanisms, and management of cancer-related cognitive impairment. *International Review of Psychiatry (Abingdon, England), 26*(1), 102–113. doi:10.3109/09540261.2013.864260 PMID:24716504

Janelsins, M. C., Kohli, S., Mohile, S. G., Usuki, K., Ahles, T. A., & Morrow, G. R. (2011). An update on cancer- and chemotherapy-related cognitive dysfunction: Current status. *Seminars in Oncology*, *38*(3), 431–438. Advance online publication. doi:10.1053/j.seminoncol.2011.03.014 PMID:21600374

Joly, F., Giffard, B., Rigal, O., De Ruiter, M. B., Small, B. J., Dubois, M., Lefel, J., Schagen, S. B., Ahles, T. A., Wefel, J. S., Lange, M., & Castel, H. (2015). Impact of Cancer and Its Treatments on Cognitive Function: Advances in Research from the Paris International Cognition and Cancer Task Force Symposium and Update since 2012. *Journal of Pain and Symptom Management*, *50*(6), 830–841. doi:10.1016/j.jpainsymman.2015.06.019 PMID:26344551

Karschnia, P., Parsons, M. W., & Dietrich, J. (2019). Pharmacologic management of cognitive impairment induced by cancer therapy. *The Lancet. Oncology*, *20*(2), e92–e102. doi:10.1016/S1470-2045(18)30938-0 PMID:30723041

Katz, S., Downs, T. D., Cash, H. R., & Grotz, R. C. (1970). Progress in Development of the Index of ADL. *The Gerontologist*, *10*(1 Part 1), 20–30. doi:10.1093/geront/10.1_Part_1.20 PMID:5420677

Kesler, S., Hadi Hosseini, S. M., Heckler, C., Janelsins, M., Palesh, O., Mustian, K., & Morrow, G. (2013). Cognitive training for improving executive function in chemotherapy-treated breast cancer survivors. *Clinical Breast Cancer*, *13*(4), 299–306. Advance online publication. doi:10.1016/j.clbc.2013.02.004 PMID:23647804

Kohli, S., Fisher, S. G., Tra, Y., Adams, M. J., Mapstone, M. E., Wesnes, K. A., Roscoe, J. A., & Morrow, G. R. (2009). The effect of modafinil on cognitive function in breast cancer survivors. *Cancer*, *115*(12), 2605–2616. doi:10.1002/cncr.24287 PMID:19309747

Kucherer, S., & Ferguson, R. J. (2017). Cognitive behavioral therapy for cancer-related cognitive dysfunction. *Current Opinion in Supportive and Palliative Care*, *11*(1), 46–51. doi:10.1097/SPC.0000000000000247 PMID:27898511

Lange, M., Joly, F., Vardy, J., Ahles, T., Dubois, M., Tron, L., Winocur, G., De Ruiter, M. B., & Castel, H. (2019). Cancer-related cognitive impairment: An update on state of the art, detection, and management strategies in cancer survivors. *Annals of Oncology : Official Journal of the European Society for Medical Oncology*, *30*(12), 1925–1940. doi:10.1093/annonc/mdz410 PMID:31617564

Lawrence, J. A., Griffin, L., Balcueva, E. P., Groteluschen, D. L., Samuel, T. A., Lesser, G. J., Naughton, M. J., Case, L. D., Shaw, E. G., & Rapp, S. R. (2016). A study of donepezil in female breast cancer survivors with self-reported cognitive dysfunction 1 to 5 years following adjuvant chemotherapy. *Journal of Cancer Survivorship: Research and Practice*, *10*(1), 176–184. doi:10.1007/s11764-015-0463-x PMID:26130292

Lawton, M. P., & Brody, E. M. (1969). Assessment of older people: Self-maintaining and instrumental activities of daily living. *The Gerontologist*, *9*(3), 179–186. doi:10.1093/geront/9.3_Part_1.179 PMID:5349366

McCracken, K., & Phillips, D. R. (2017). Demographic and Epidemiological Transition. *International Encyclopedia of Geography: People, the Earth. Environmental Technology*, 1–8. doi:10.1002/9781118786352.wbieg0063

Nasreddine, Z. S., Phillips, N. A., Bédirian, V., Charbonneau, S., Whitehead, V., Collin, I., Cummings, J. L., & Chertkow, H. (2005). The Montreal Cognitive Assessment, MoCA: A Brief Screening Tool For Mild Cognitive Impairment. *Journal of the American Geriatrics Society*, *53*(4), 695–699. doi:10.1111/j.1532-5415.2005.53221.x PMID:15817019

Nitrini, R., Brucki, S. M. D., Yassuda, M. S., Fichman, H. C., & Caramelli, P. (2021). The Figure Memory Test: Diagnosis of memory impairment in populations with heterogeneous educational background. *Dementia & Neuropsychologia*, *15*(2), 173–185. doi:10.1590/1980-57642021dn15-020004 PMID:34345358

Oh, S. Y., Jeong, E., Pahk, K., Cheon, G. J., Park, K. H., & Choe, J. G. (2012). FDG PET-CT of chemobrain: An SPM analysis in patients with breast cancer before and after neoadjuvant chemotherapy. *Journal of Nuclear Medicine*, *53*(supplement 1), 1922–1922. https://jnm.snmjournals.org/cgi/content/short/53/supplement_1/1922

Pendergrass, J. C., Targum, S. D., & Harrison, J. E. (2018). Cognitive Impairment Associated with Cancer: A Brief Review. *Innovations in Clinical Neuroscience*, *15*(1–2), 36–44. https://www.ncbi.nlm.nih.gov/pubmed/29497579 PMID:29497579

Pfeffer, R. I., Kurosaki, T. T., Harrah, C. H., Chance, J. M., & Filos, S. (1982). Measurement of Functional Activities in Older Adults in the Community. *Journal of Gerontology*, *37*(3), 323–329. doi:10.1093/geronj/37.3.323 PMID:7069156

Player, L., Mackenzie, L., Willis, K., & Loh, S. Y. (2014). Women's experiences of cognitive changes or 'chemobrain' following treatment for breast cancer: A role for occupational therapy? *Australian Occupational Therapy Journal*, *61*(4), 230–240. doi:10.1111/1440-1630.12113 PMID:24499127

Pullens, M. J. J., De Vries, J., Van Warmerdam, L. J. C., Van De Wal, M. A., & Roukema, J. A. (2013). Chemotherapy and cognitive complaints in women with breast cancer. *Psycho-Oncology*, *22*(8), 1783–1789. doi:10.1002/pon.3214 PMID:23109296

Rambeau, A., Beauplet, B., Laviec, H., Licaj, I., Leconte, A., Chatel, C., Le Bon, P., Denhaerynck, J., Clarisse, B., Frenkiel, N., Lange, M., & Joly, F. (2019). Prospective comparison of the Montreal Cognitive Assessment (MoCA) and the Mini Mental State Examination (MMSE) in geriatric oncology. *Journal of Geriatric Oncology*, *10*(2), 235–240. doi:10.1016/j.jgo.2018.08.003 PMID:30150019

Ruff, R. M., Light, R. H., Parker, S. B., & Levin, H. S. (1996). Benton controlled Oral Word Association Test: Reliability and updated norms. *Archives of Clinical Neuropsychology*, *11*(4), 329–338. doi:10.1093/arclin/11.4.329 PMID:14588937

Schroyen, G., Blommaert, J., van Weehaeghe, D., Sleurs, C., Vandenbulcke, M., Dedoncker, N., Hatse, S., Goris, A., Koole, M., Smeets, A., van Laere, K., Sunaert, S., & Deprez, S. (2021). Neuroinflammation and Its Association with Cognition, Neuronal Markers and Peripheral Inflammation after Chemotherapy for Breast Cancer. *Cancers (Basel)*, *13*(16), 4198. doi:10.3390/cancers13164198 PMID:34439351

Schroyen, G., Sleurs, C., Bartsoen, E., Smeets, D., van Weehaeghe, D., Van Laere, K., Smeets, A., Deprez, S., & Sunaert, S. (2022). Neuroinflammation as potential precursor of leukoencephalopathy in early-stage breast cancer patients: A cross-sectional PET-MRI study. *The Breast*, *62*, 61–68. doi:10.1016/j.breast.2022.02.001 PMID:35131644

Seigers, R., Loos, M., Van Tellingen, O., Boogerd, W., Smit, A. B., & Schagen, S. B. (2016). Neurobiological changes by cytotoxic agents in mice. *Behavioural Brain Research*, *299*, 19–26. doi:10.1016/j.bbr.2015.10.057 PMID:26602283

Shah, R. R. (2005). Mechanistic basis of adverse drug reactions: The perils of inappropriate dose schedules. Expert Opinion on Drug Safety. doi:10.1517/14740338.4.1.103

Sousa, H., Almeida, S., Bessa, J., & Pereira, M. G. (2020). The Developmental Trajectory of Cancer-Related Cognitive Impairment in Breast Cancer Patients: A Systematic Review of Longitudinal Neuroimaging Studies. *Neuropsychology Review*, *30*(3), 287–309. doi:10.1007/s11065-020-09441-9 PMID:32607817

Sumerall, S. W., Timmons, P. L., James, A. L., Ewing, M. J. M., & Oehlert, M. E. (1997). Expanded norms for the controlled oral word association test. *Journal of Clinical Psychology*, *53*(5), 517–521. Advance online publication. doi:10.1002/(SICI)1097-4679(199708)53:5<517::AID-JCLP14>3.0.CO;2-H PMID:9257231

Tchen, N., Juffs, H. G., Downie, F. P., Yi, Q. L., Hu, H., Chemerynsky, I., Clemons, M., Crump, M., Goss, P. E., Warr, D., Tweedale, M. E., & Tannock, I. F. (2003). Cognitive function, fatigue, and menopausal symptoms in women receiving adjuvant chemotherapy for breast cancer. *Journal of Clinical Oncology*, *21*(22), 4175–4183. Advance online publication. doi:10.1200/JCO.2003.01.119 PMID:14615445

Tombaugh, T. N. (2004). Trail Making Test A and B: Normative data stratified by age and education. *Archives of Clinical Neuropsychology*, *19*(2), 203–214. Advance online publication. doi:10.1016/S0887-6177(03)00039-8 PMID:15010086

Tong, T., Lu, H., Zong, J., Lv, Q., & Chu, X. (2020). Chemotherapy-related cognitive impairment in patients with breast cancer based on MRS and DTI analysis. *Breast Cancer (Tokyo, Japan)*, *27*(5), 893–902. doi:10.1007/s12282-020-01094-z PMID:32301096

Wagner, L. I., Sweet, J., Butt, Z., Lai, J., & Cella, D. (2009). Measuring patient self-reported cognitive function: Development of the Functional Assessment of Cancer Therapy – Cognitive Function instrument. *The Journal of Supportive Oncology*.

Wefel, J. S., Vardy, J., Ahles, T., & Schagen, S. B. (2011). International Cognition and Cancer Task Force recommendations to harmonise studies of cognitive function in patients with cancer. *The Lancet. Oncology*, *12*(7), 703–708. doi:10.1016/S1470-2045(10)70294-1 PMID:21354373

Whalley, K. (2019). A glial contribution to 'chemobrain.'. *Nature Reviews. Neuroscience*, *20*(2), 67–67. doi:10.1038/s41583-018-0115-z PMID:30602722

Whittaker, A. L., George, R. P., & O'Malley, L. (2022). Prevalence of cognitive impairment following chemotherapy treatment for breast cancer: A systematic review and meta-analysis. *Scientific Reports*, *12*(1), 2135. doi:10.1038/s41598-022-05682-1 PMID:35136066

Winocur, G., Henkelman, M., Wojtowicz, J. M., Zhang, H., Binns, M. A., & Tannock, I. F. (2012). The effects of chemotherapy on cognitive function in a mouse model: A prospective study. *Clinical Cancer Research*, *18*(11), 3112–3121. doi:10.1158/1078-0432.CCR-12-0060 PMID:22467680

Compilation of References

A New Drug Development Tool for Parkinson's Disease and Dementia with Lewy Bodies: The Neuronal Synuclein Disease Integrated Staging System. (n.d.). Retrieved September 20, 2023, from https://www.michaeljfox.org/StagingManuscript

Addington, J., Heinssen, R.K., Robinson, D.G., Schooler, N.R., Marcy, P., Brunette, M.F., Correll, C.U., Estroff, S., Mueser, K.T., Penn, D., Robinson, J.A., Rosenheck, R.A., Azrin, S.T., Goldstein, A.B., Severe, J., & Kane, J.M. (2015). Duration of Untreated Psychosis in Community Treatment Settings in the United States. *Psychiatr Serv., 66*(7), 753-6. doi:10.1176/appi.ps.201400124

Afkhami, R. G., Azarnia, G., & Tinati, M. A. (2016). Cardiac arrhythmia classification using statistical and mixture modeling features of ecg signals. *Pattern Recognition Letters*, *70*, 45–51. doi:10.1016/j.patrec.2015.11.018

Agosta, F., Henry, R.G., & Migliaccio, R. (2010). *Language networks in semantic dementia.* Academic Press.

Aharon-Peretz, J., Rosenbaum, H., & Gershoni-Baruch, R. (2004). Mutations in the glucocerebrosidase gene and Parkinson's disease in Ashkenazi Jews. *The New England Journal of Medicine*, *351*(19), 1972–1977. doi:10.1056/NEJMoa033277 PMID:15525722

Ahles, T. A. (2012). Brain vulnerability to chemotherapy toxicities. Psycho-Oncology. doi:10.1002/pon.3196

Ahles, T. A., Root, J. C., & Ryan, E. L. (2012). Cancer- and cancer treatment-associated cognitive change: An update on the state of the science. *Journal of Clinical Oncology*, *30*(30), 3675–3686. doi:10.1200/JCO.2012.43.0116 PMID:23008308

Alarcón-Arís, D., Recasens, A., Galofré, M., Carballo-Carbajal, I., Zacchi, N., Ruiz-Bronchal, E., Pavia-Collado, R., Chica, R., Ferrés-Coy, A., Santos, M., Revilla, R., Montefeltro, A., Fariñas, I., Artigas, F., Vila, M., & Bortolozzi, A. (2018). Selective α-Synuclein Knockdown in Monoamine Neurons by Intranasal Oligonucleotide Delivery: Potential Therapy for Parkinson's Disease. *Molecular Therapy*, *26*(2), 550–567. doi:10.1016/j.ymthe.2017.11.015 PMID:29273501

Albert, N., Melau, M., Jensen, H., Emborg, C., Jepsen, J.R., Fagerlund, B., Gluud, C., Mors, O., Hjorthøj, C., & Nordentoft, M. (2017). Five years of specialised early intervention versus two years of specialised early intervention followed by three years of standard treatment for patients with a first episode psychosis: randomised, superiority, parallel group trial in Denmark (OPUS II). *BMJ, 12*(356).

Al-Medfa, M. K., Al-Ansari, A. M. S., Darwish, A. H., Qreeballa, T. A., & Jahrami, H. (2023). Physicians' attitudes and knowledge toward artificial intelligence in medicine: Benefits and drawbacks. *Heliyon*, *9*(4), e14744. doi:10.1016/j.heliyon.2023.e14744 PMID:37035387

Almondes, K. M. (2017). *Neurocognitive impairment of Acute and Chronic Sleep Loss. Williams, Larry Sleep Deprivation: Global prevalence, dangers, and impacts on cognitive performance*. Nova Science.

Al-Sharman, A., Ismaiel, I. A., Khalil, H., & El-Salem, K. (2021). Exploring the Relationship Between Sleep Quality, Sleep-Related Biomarkers, and Motor Skill Acquisition Using Virtual Reality in People With Parkinson's Disease: A Pilot Study. *Frontiers in Neurology*, *12*, 582611. doi:10.3389/fneur.2021.582611 PMID:33737900

Alvarez-Jimenez, M., O'Donoghue, B., Thompson, A., Gleeson, J.F., Bendall, S., Gonzalez-Blanch, C., Killackey, E., Wunderink, L., & McGorry, P.D. (2016). Beyond Clinical Remission in First Episode Psychosis: Thoughts on Antipsychotic Maintenance vs. Guided Discontinuation in the Functional Recovery Era. *CNS Drugs, 30*(5), 357-68. . doi:10.1007/s40263-016-0331-x

Alves, G., Casali, M., Veras, A., Carrilho, C., Bruno, C., Rodrigues, V., & Dourado, M. (2020). A Systematic Review of Home-Setting Psychoeducation Interventions for Behavioral Changes in Dementia: Some Lessons for the COVID-19 Pandemic and Post-Pandemic Assistance. *Frontiers in Psychiatry*, *11*, 577871. doi:10.3389/fpsyt.2020.577871 PMID:33132937

American Psychiatric Association. (2013). Diagnostic and statistical manual of mental disorders: DSM-5 (5th ed). American Psychiatric Association.

Amici, S., Gorno-Tempini, M. L., Ogara, J. M., Dronkers, H. F., & Millera, B. L. (2006). An overview on Primary Progressive Aphasia and its variants. *Behavioural Neurology*, *17*(2), 77–87. doi:10.1155/2006/260734 PMID:16873918

Anghel, L., Ciubară, A., Nechita, A., Nechita, L., Manole, C., Baroiu, L., Ciubară, A. B., & Muşat, C. L. (2023). Sleep Disorders Associated with Neurodegenerative Diseases. *Diagnostics (Basel)*, *13*(18), 2898. doi:10.3390/diagnostics13182898 PMID:37761265

Araújo, A. A., & Perissinoto, J. (2004). Desenvolvimento da linguagem na adolescência: Competências Semânticas, sintáticas e pragmáticas. *Pro-Fono*, *16*(3), 251–260. PMID:15609580

Ardila, A. (2020). Gerstmann Syndrome. *Current Neurology and Neuroscience Reports*, *20*(11), 48. Advance online publication. doi:10.1007/s11910-020-01069-9 PMID:32852667

Arroyo-Anlló, E. M., Dauphin, S., Fargeau, M. N., Ingrand, P., & Gil, R. (2019). Music and emotion in Alzheimer's disease. *Alzheimer's Research & Therapy*, *11*(1), 69. doi:10.1186/s13195-019-0523-y PMID:31391062

Ascherio, A., & Schwarzschild, M. A. (2016). The epidemiology of Parkinson's disease: Risk factors and prevention. *Lancet Neurology*, *15*(12), 1257–1272. doi:10.1016/S1474-4422(16)30230-7 PMID:27751556

Atkinson, A. P., & Adolphs, R. (2011). The neuropsychology of face perception: Beyond simple dissociations and functional selectivity. *Philosophical Transactions of the Royal Society of London. Series B, Biological Sciences*, *366*(1571), 1726–1738. doi:10.1098/rstb.2010.0349 PMID:21536556

Ávila-Villanueva, M., Gómez-Ramírez, J., Ávila, J., & Fernández-Blázquez, M. A. (2021). Alzheimer's Disease and Empathic Abilities: The Proposed Role of the Cingulate Cortex. *Journal of Alzheimer's Disease Reports*, *5*(1), 345–352. doi:10.3233/ADR-200282 PMID:34189406

Ayres, A. M., Busatto, G., Menezes, P., Schaufelberger, M., Coutinho, L., Murray, R., McGuire, P., Rushe, T., & Scazufca, M. (2007). Cognitive deficits in first-episode psychosis: A population-based study in São Paulo, Brazil. *Schizophrenia Research*, *90*(1-3), 338–343. doi:10.1016/j.schres.2006.09.026 PMID:17123787

Baez, S., & Ibáñez, A. (2016). Dementia in Latin America: An emergent silent tsunami. *Frontiers in Aging Neuroscience*, *8*, 253. doi:10.3389/fnagi.2016.00253 PMID:27840605

Bandeira, D., Andrade, J., & Peixoto, E. (2021). O uso de testes psicológicos: Formação, avaliação e critérios de restrição. *Psicologia (Conselho Federal de Psicologia)*, *41*(spe1), 41. doi:10.1590/1982-3703003252970

Bang, J., Spina, S., & Miller, B. (2015). Frontotemporal dementia. *Lancet*, *386*(10004), 1672–1682. doi:10.1016/S0140-6736(15)00461-4 PMID:26595641

Barulli, D., & Stern, Y. (2013). Efficiency, capacity, compensation, maintenance, plasticity: Emerging concepts in cognitive reserve. *Trends in Cognitive Sciences*, *17*(10), 502–509. doi:10.1016/j.tics.2013.08.012 PMID:24018144

Bassett, A. S., & Chow, E. W. C. (2008). Schizophrenia and 22q11.2 deletion syndrome. *Current Psychiatry Reports*, *10*(2), 148–157. doi:10.1007/s11920-008-0026-1 PMID:18474208

Beber, B. C., Berbert, M. C. B., Grawer, R. S., & Cardoso, M. C. A. F. (2018, March). Rate and rhythm control strategies for apraxia of speech in nonfluent primary progressive aphasia. *Dementia & Neuropsychologia*, *12*(1), 80–84. doi:10.1590/1980-57642018dn12-010012 PMID:29682238

Bechdolf, A., Müller, H., Hellmich, M., de Millas, W., Falkai, P., Gaebel, W., Gallinat, J., Hasan, A., Heinz, A., Janssen, B., Juckel, G., Karow, A., Krüger-Özgürdal, S., Lambert, M., Maier, W., Meyer-Lindenberg, A., Pützfeld, V., Rausch, F., Schneider, F., ... Klosterkötter, J. (2023). Prevention of First-Episode Psychosis in People at Clinical High Risk: A Randomized Controlled, Multicentre Trial Comparing Cognitive-Behavioral Therapy and Clinical Management Plus Low-Dose Aripiprazole or Placebo (PREVENT). *Schizophrenia Bulletin*, *49*(4), 1055–1066. doi:10.1093/schbul/sbad029 PMID:37021666

Benedict, R. H. B., Schretlen, D., Groninger, L., & Brandt, J. (1998). Hopkins verbal learning test - Revised: Normative data and analysis of inter-form and test-retest reliability. *The Clinical Neuropsychologist*, *12*(1), 43–55. Advance online publication. doi:10.1076/clin.12.1.43.1726

Bergh, S., Hjorthøj C., Sørensen H.J., Fagerlund B., Austin S., Secher R.G., Jepsen J.R., & Nordentoft M. (2016). Predictors and longitudinal course of cognitive functioning in schizophrenia spectrumdisorders, 10 years after baseline: The OPUS Study. *Schizophr. Res.*, *175*(1-3), 57-63. . doi:10.1016/j.schres.2016.03.025

Bertola, L. (2019). *Psicometria e Estatística Aplicadas à Neuropsicologia Clínica. Editora*. Editora Pearson Clinical Brasil.

Bertoux, M., Delavest, M., de Souza, L. C., Funkiewiez, A., Lépine, J. P., Fossati, P., Dubois, B., & Sarazin, M. (2012). Social Cognition and Emotional Assessment differentiates frontotemporal dementia from depression. *Journal of Neurology, Neurosurgery, and Psychiatry*, *83*(4), 411–416. doi:10.1136/jnnp-2011-301849 PMID:22291219

Bessi, V., Piaceri, I., & Padiglioni, S. (2018). Crossed aphasia in nonfluent variant of primary progressive aphasia carrying a GRN mutation. *J Neurological Sciences*. Doi: . 2018.06.026 doi:10.1016 / j.jns

Bhavsar, V., McGuire, P., MacCabe, J., Oliver, D., & Fusar-Poli, P. (2018). A systematic review and meta-analysis of mental health service use in people who report psychotic experiences. *Early Intervention in Psychiatry*, *12*(3), 275–285. doi:10.1111/eip.12464 PMID:28805304

Biermann, M., Arruda, C., & Holanda, L. (2022). Machine Learning e Processos Diagnósticos com Avaliação Neuropsicológica: Uma Revisão Integrativa. *Revista de Psicologia da IMED*, *14*(1), 102–120. doi:10.18256/2175-5027.2022.v14i1.4568

Binder, J R. (2015). The Wernicke area. *Neurology*, *85*(24), 2170–2175. doi:.00000000 00002219 doi:10.1212/wnl

Bird, V., Premkumar, P., Kendall, T., Whittington, C., Mitchell, J., & Kuipers, E. (2010). Early intervention services, cognitive-behavioural therapy and family intervention in early psychosis: Systematic review. *Br J Psychiatry*, *197*(5), 350-6. . doi:10.1192/bjp.bp.109.074526

Bjorklund, G., Stejskal, V., Urbina, M. A., Dadar, M., Chirumbolo, S., & Mutter, J. (2018). Metals and Parkinson's Disease: Mechanisms and Biochemical Processes. *Current Medicinal Chemistry*, *25*(19), 2198–2214. doi:10.2174/0929867325666171129124616 PMID:29189118

Bombassaro, T., Carrilho, C. G., Peixoto, C., Alves, G. S., Kahn, J. P., Nardi, A. E., & Veras, A. B. (2023). Cognition in Schizophrenia: A Systematic Review of Wechsler Adult Intelligence Scale Studies. *The Primary Care Companion for CNS Disorders, 25*(5). doi:10.4088/PCC.22r03456 PMID:37857289

Bonanni, E., Schirru, A., Di Perri, M. C., Bonuccelli, U., & Maestri, M. (2019). Insomnia and hot flashes. *Maturitas, 126*, 51–54. doi:10.1016/j.maturitas.2019.05.001 PMID:31239118

Bora, E., & Murray, R.M. (2014). Meta-analysis of cognitive deficits in ultra-high risk to psychosis and first-episode psychosis: do the cognitive deficits progress over, or after, the onset of psychosis? *Schizophr Bull., 40*(4), 744-55. . doi:10.1093/schbul/sbt085

Bosnjak Kuharic, D., Kekin, I., Hew, J., Rojnic Kuzman, M., & Puljak, L. (2019). Interventions for prodromal stage of psychosis. *Cochrane Database of Systematic Reviews, 2019*(11). doi:10.1002/14651858.CD012236.pub2

Botha, H., & Josephs, K.A. (2019). Primary progressive aphasias and apraxia of speech. *Continuum, 25*(1), 101–127.

Bowie, C. R., & Harvey, P. D. (2006). Administration and interpretation of the Trail Making Test. *Nature Protocols, 1*(5), 2277–2281. Advance online publication. doi:10.1038/nprot.2006.390 PMID:17406468

Boxer, A. L., Lipton, A. M., Womack, K., Merrilees, J., Neuhaus, J., Pavlic, D., Gandhi, A., Red, D., Martin-Cook, K., Svetlik, D., & Miller, B. L. (2009). An open-label study of memantine treatment in 3 subtypes of frontotemporal lobar degeneration. *Alzheimer Disease and Associated Disorders, 23*(3), 211–217. doi:10.1097/WAD.0b013e318197852f PMID:19812461

Boydell, J., van Os, J., McKenzie, K., Allardyce, J., Goel, R., McCreadie, R. G., & Murray, R. M. (2001). Incidence of schizophrenia in ethnic minorities in London: Ecological study into interactions with environment. *BMJ (Clinical Research Ed.), 323*(7325), 1336–1336. doi:10.1136/bmj.323.7325.1336 PMID:11739218

Braak, H., Del Tredici, K., Bratzke, H., Hamm-Clement, J., Sandmann-Keil, D., & Rüb, U. (2002). Staging of the intracerebral inclusion body pathology associated with idiopathic Parkinson's disease (preclinical and clinical stages). *Journal of Neurology, 249*(3), 1-5.

Braak, H., & Del Tredici, K. (2008). Invited Article: Nervous system pathology in sporadic Parkinson disease. *Neurology, 70*(20), 1916–1925. doi:10.1212/01.wnl.0000312279.49272.9f PMID:18474848

Braak, H., Ghebremedhin, E., Rüb, U., Bratzke, H., & Del Tredici, K. (2004). Stages in the development of Parkinson's disease-related pathology. *Cell and Tissue Research, 318*(1), 121–134. doi:10.1007/s00441-004-0956-9 PMID:15338272

Breitborde, N. J. K., Srihari, V. H., & Woods, S. W. (2009). Review of the operational definition for first-episode psychosis. *Early Intervention in Psychiatry, 3*(4), 259–265. doi:10.1111/j.1751-7893.2009.00148.x PMID:22642728

Brezden, C. B., Phillips, K. A., Abdolell, M., Bunston, T., & Tannock, I. F. (2000). Cognitive function in breast cancer patients receiving adjuvant chemotherapy. *Journal of Clinical Oncology*, *18*(14), 2695–2701. Advance online publication. doi:10.1200/JCO.2000.18.14.2695 PMID:10894868

Briceño, D. F., Santangeli, P., Frankel, D. S., Liang, J. J., Shirai, Y., Markman, T., Enriquez, A., Walsh, K., Riley, M. P., Nazarian, S., Lin, D., Kumareswaran, R., Arkles, J. S., Hyman, M. C., Deo, R., Supple, G. E., Garcia, F. C., Dixit, S., Epstein, A. E., . . . Schaller, R. D. (2020). Qrs morphology in lead v1 for the rapid localization of idiopathic ventricular arrhythmias originating from the left ventricular papillary muscles: A novel electrocardiographic criterion. *Heart Rhythm*, *17*(10), 1711 – 1718. https://www.sciencedirect.com/science/article/pii/S1547527120304550

BrinkmanJ. E.ReddyV.SharmaS. (2023). *Physiology of Sleep*. Available from: https://www.ncbi.nlm.nih.gov/books/NBK482512/

Brito-Marques, P.R., Vieira-Mello, R., & Montenegro, L. (2001). *Classic Pick´s disease type with ubiquitine positive and tau negative inclusions.* Academic Press.

Brito-Marques, P. R. (2002). Frontotemporal cortical atrophy with gliosis in the Gray mather of cerebral cortex. Case report. *Arquivos de Neuro-Psiquiatria*, *60*, 462–468. doi:10.1590/S0004-282X2002000300023 PMID:12131951

Brito-Marques, P. R. (2018). *Desordens Cognitivas e Demências: Diagnóstico Diferencial e Tratamento*. EDUPE.

Brito-Marques, P. R., & Brito-Marques, J. M. A. M. (2023). A left-handed woman with a non-fluent variant of primary progressive aphasia with a left asymmetric brain SPECT. *Dementia & Neuropsychologia*, *17*, e2022009. doi:10.1590/1980-5764-dn-2022-0095

Brito-Marques, P. R., & Cabral-Filho, J. E. (2004). The role of education on in Mini-Mental State Examination: A study in Northeast Brazil. *Arquivos de Neuro-Psiquiatria*, *62*(2a), 206–211. doi:10.1590/S0004-282X2004000200003 PMID:15235718

Brito-MarquesP. R.Vieira-MelloR. (1999). Amyotrophic lateral Sclerosis with dementia. *Arq. Neuro-Psiquiatr.*, *57*(2A). https://doi.org/ doi:10.1590/S0004-282X1999000 2000 18

Brito-Marques, P. R., Vieira-Mello, R. J., Montenegro, M. F. V., & Aragão, M. F. V. (2011). Clinicopathologic analysis of progressive non-fluent aphasia and corticobasal degeneration, Case report and review. *Dementia & Neuropsychologia*, *5*(2), 135–141. doi:10.1590/S1980-57642011DN05020013 PMID:29213736

Brod, M., Pohlman, B., Lasser, R., & Hodgkins, P. (2012). Comparison of the burden of illness for adults with ADHD across seven countries: A qualitative study. *Health and Quality of Life Outcomes*, *10*(1), 47. doi:10.1186/1477-7525-10-47 PMID:22583562

Brod, M., Schmitt, E., Goodwin, M., Hodgkins, P., & Niebler, G. (2012). ADHD burden of illness in older adults: A life course perspective. *Quality of Life Research: An International Journal of Quality of Life Aspects of Treatment, Care and Rehabilitation*, *21*(5), 795–799. doi:10.1007/s11136-011-9981-9 PMID:21805205

Brooks, B., & Sherman, E. (2012). Computerized neuropsychological testing to rapidly evaluate cognition in pediatric patients with neurologic disorders. *Journal of Child Neurology*, *27*(8), 982–991. doi:10.1177/0883073811430863 PMID:22290863

Browndyke, J., Albert, A., Malone, W., Schatz, P., Paul, R., Cohen, R., & Gouvier, K. (2002). Computer-related anxiety: Examining the impact of technology-specific affect on the performance of a computerized neuropsychological assessment measure. *Applied Neuropsychology*, *9*(4), 210–218. doi:10.1207/S15324826AN0904_3 PMID:12584075

Brunette, M.F., Mueser, K.T., Babbin, S., Meyer-Kalos, P., Rosenheck, R., Correll, C.U., Cather, C., Robinson, D.G., Schooler, N.R., Penn, D.L., Addington, J., Estroff, S.E., Gottlieb, J., Glynn, S.M., Marcy, P., Robinson, J., & Kane, J.M. (2018). Demographic and clinical correlates of substance use disorders in first episode psychosis. *Schizophr Res., 194,* 4-12. doi:10.1016/j.schres.2017.06.039

Bucks, R. S., Olaithe, M., Rosenzweig, I., & Morrell, M. J. (2017). Reviewing the relationship between OSA and cognition: Where do we go from here? *Respirology (Carlton, Vic.)*, *22*(7), 1253–1261. doi:10.1111/resp.13140 PMID:28779504

Budgett, J., Brown, A., Daley, S., Page, T. E., Banerjee, S., Livingston, G., & Sommerlad, A. (2019). The Social Functioning in Dementia Scale (SF-DEM): Exploratory Factor Analysis and Psychometric Properties in Mild, Moderate, and Severe Dementia. *Alzheimer's & Dementia: Diagnosis, Assessment & Disease Monitoring*, *11*(1), 45–52. doi:10.1016/j.dadm.2018.11.001 PMID:30623018

Bulen Love, D. (2018). Mindfulness to Improve Cognitive Function in Cancer Survivors. *Oncology Times*, *40*(8), 28. doi:10.1097/01.COT.0000532447.18584.01

Burke, R. E., Dauer, W. T., & Vonsattel, J. P. G. (2008). A critical evaluation of the Braak staging scheme for Parkinson's disease. *Annals of Neurology*, *64*(5), 485–491. doi:10.1002/ana.21541 PMID:19067353

Burnham, H., & Hogervorst, E. (2004). Recognition of facial expressions of emotion by patients with dementia of the Alzheimer type. *Dementia and Geriatric Cognitive Disorders*, *18*(1), 75–79. doi:10.1159/000077813 PMID:15087581

Cabrera-Martin, M. N., Matías-Guiub, J. A., & Yus-Fuertes, M. (2016). PET-TC 18F-FDG y RM funcional en un caso de afasia progresiva primaria logopénica cruzada. *Revista Española de Medicina Nuclear e Imagen Molecular*, *35*(6), 394–397. Advance online publication. doi:10.1016/j.remn.2016.05.002 PMID:27321169

Cadieux, N. L., & Greve, K. W. (1997). Emotion processing in Alzheimer's disease. *Journal of the International Neuropsychological Society*, *3*(5), 411–419. doi:10.1017/S1355617797004116 PMID:9322399

Caixeta, L., & Nitrini, R. (2001). Subtipos clínicos da demência frontotemporal. *Arquivos de Neuro-Psiquiatria*, *59*(3A, n.3A), 577–581. doi:10.1590/S0004-282X2001000400017 PMID:11588638

Callahan, B. L., Bierstone, D., Stuss, D. T., & Black, S. E. (2017). Adult ADHD: Risk Factor for Dementia or Phenotypic Mimic? *Frontiers in Aging Neuroscience*, *9*, 260. doi:10.3389/fnagi.2017.00260 PMID:28824421

Camacho-Gomez, M., & Castellvi, P. (2020). Effectiveness of Family Intervention for Preventing Relapse in First-Episode Psychosis Until 24 Months of Follow-up: A Systematic Review With Meta-analysis of Randomized Controlled Trials. *Schizophr Bull., 46*(1), 98-109.

Cannon, M., Jones, P. B., & Murray, R. M. (2002). Obstetric Complications and Schizophrenia: Historical and Meta-Analytic Review. *The American Journal of Psychiatry*, *159*(7), 1080–1092. doi:10.1176/appi.ajp.159.7.1080 PMID:12091183

Cantor-Graae, E., & Selten, J.-P. (2005). Schizophrenia and Migration: A Meta-Analysis and Review. *The American Journal of Psychiatry*, *162*(1), 12–24. doi:10.1176/appi.ajp.162.1.12 PMID:15625195

Canu, E., Bessi, V., Leocadi, M., Padiglioni, S., Nacmias, B., Sorbi, S., Filippi, M., & Agosta, F. (2019). Crossed aphasia confrmed by fMRI in a case with nonfuent variant of primary progressive aphasia carrying a GRN mutation. *Journal of Neurology*, *266*(5), 1274–1279. Advance online publication. doi:10.1007/s00415-019-09298-w PMID:30923936

Caramazza, A., Berndt, R.S., & Brownell, H.H. (1982). *The semantic deficit hypothesis: perceptual parsing and object classification by aphasic patients*. Academic Press.

Carley, D. W., & Farabi, S. S. (2016). Physiology of Sleep. *Diabetes Spectrum*, *29*(1), 5–9. doi:10.2337/diaspect.29.1.5 PMID:26912958

Carlsson, A., Lindqvist, M., Magnusson, T., & Waldeck, B. (1958). On the presence of 3-hydroxytyramine in brain. *Science*, *127*(3296), 471. doi:10.1126/science.127.3296.471.a PMID:13529006

Carrillo, F., Sigman, M., Slezak, D., Ashton, P., Fitzgerald, L., Stroud, J., Nutt, D., & Carhart-Harris, R. (2018). Natural speech algorithms applied to baseline interview data can predict which patients will respond to psilocybin for treatment-resistant depression. *Journal of Affective Disorders*, *230*, 84–86. doi:10.1016/j.jad.2018.01.006 PMID:29407543

Carrington, S. J., & Bailey, A. J. (2009). Are there theory of mind regions in the brain? A review of the neuroimaging literature. *Human Brain Mapping*, *30*(8), 2313–2335. doi:10.1002/hbm.20671 PMID:19034900

Carroll, C. M., & Macauley, S. L. (2019). The Interaction Between Sleep and Metabolism in Alzheimer's Disease: Cause or Consequence of Disease? *Frontiers in Aging Neuroscience*, *11*, 258. doi:10.3389/fnagi.2019.00258 PMID:31616284

Casagrande, M., Forte, G., Favieri, F., & Corbo, I. (2022). Sleep Quality and Aging: A Systematic Review on Healthy Older People, Mild Cognitive Impairment and Alzheimer's Disease. *International Journal of Environmental Research and Public Health*, *19*(14), 8457. doi:10.3390/ijerph19148457 PMID:35886309

Castelli, F., Frith, C., Happé, F., & Frith, U. (2002). Autism, Asperger syndrome and brain mechanisms for the attribution of mental states to animated shapes. *Brain*, *125*(Pt 8), 1839–1849. doi:10.1093/brain/awf189 PMID:12135974

Castelli, I., Pini, A., Alberoni, M., Liverta-Sempio, O., Baglio, F., Massaro, D., Marchetti, A., & Nemni, R. (2011). Mapping levels of theory of mind in Alzheimer's disease: A preliminary study. *Aging & Mental Health*, *15*(2), 157–168. doi:10.1080/13607863.2010.513038 PMID:21140304

Castro, E. D. A. S., & de Almondes, K. M. (2018). Sleep pattern and decision-making in physicians from mobile emergency care service with 12-h work schedules. *The International Journal of Neuroscience*, *128*(6), 530–539. doi:10.1080/00207454.2017.1400970 PMID:29098917

Cavalcante, D.A, Coutinho, L.S., Ortiz, B.B., Noto, M.N., Cordeiro, Q., Ota, V.K., Belangeiro, S.I., Bressan, R.A., Gadelha, A., & Noto, C. (2020). Impact of duration of untreated psychosis in short-term response to treatment and outcome in antipsychotic naïve first-episode psychosis. *Early Interv Psychiatry.*, *14*(6), 677-683. . doi:10.1111/eip.12889

Chahine, L. M., Merchant, K., Siderowf, A., Sherer, T., Tanner, C., Marek, K., & Simuni, T. (2023). Proposal for a Biologic Staging System of Parkinson's Disease. *Journal of Parkinson's Disease*, *13*(3), 297–309. doi:10.3233/JPD-225111 PMID:37066922

Chang, W. P., Liu, M. E., Chang, W. C., Yang, A. C., Ku, Y. C., Pai, J. T., Huang, H. L., & Tsai, S. J. (2013). Sleep apnea and the risk of dementia: A population-based 5-year follow-up study in Taiwan. *PLoS One*, *8*(10), e78655. doi:10.1371/journal.pone.0078655 PMID:24205289

Chan, J. Y. C., Yau, S. T. Y., Kwok, T. C. Y., & Tsoi, K. K. F. (2021). Diagnostic performance of digital cognitive tests for the identification of MCI and dementia: A systematic review. *Ageing Research Reviews*, *72*, 101506. doi:10.1016/j.arr.2021.101506 PMID:34744026

Chartier-Harlin, M.-C., Kachergus, J., Roumier, C., Mouroux, V., Douay, X., Lincoln, S., Levecque, C., Larvor, L., Andrieux, J., Hulihan, M., Waucquier, N., Defebvre, L., Amouyel, P., Farrer, M., & Destée, A. (2004). Alpha-synuclein locus duplication as a cause of familial Parkinson's disease. *Lancet*, *364*(9440), 1167–1169. doi:10.1016/S0140-6736(04)17103-1 PMID:15451224

Chen, E., Widick, P., & Chatterjee, A. (2008). Functional-anatomical organization of predicate metaphor processing. *Brain and Language*, *107*(3), 194–202. doi:10.1016/j.bandl.2008.06.007 PMID:18692890

Chen, L., Selvendra, A., Stewart, A., & Castle, D. (2018). Risk factors in early and late onset schizophrenia. *Comprehensive Psychiatry*, *80*, 155–162. doi:10.1016/j.comppsych.2017.09.009 PMID:29096206

Chen, T.-M., Huang, C.-H., Shih, E. S., Hu, Y.-F., & Hwang, M.-J. (2020). Detection and classification of cardiac arrhyth- mias by a challenge-best deep learning neural network model. *iScience*, *23*(3), 100886. doi:10.1016/j.isci.2020.100886 PMID:32062420

Cheung, Y. T., Ng, T., Shwe, M. K., Ho, H., Foo, K. M., Cham, M. T., Lee, J. A., Fan, G., Tan, Y. P., Yong, W. S., Ng, R., & Chan, A. (2015). Association of proinflammatory cytokines and chemotherapy-associated cognitive impairment in breast cancer patients: A multi-centered, prospective, cohort study. *Annals of Oncology : Official Journal of the European Society for Medical Oncology*, *26*(7), 1446–1451. doi:10.1093/annonc/mdv206 PMID:25922060

Choong, C. S., & Doody, G. A. (2013). Can theory of mind deficits be measured reliably in people with mild and moderate Alzheimer's dementia? *BMC Psychology*, *1*(1), 28. doi:10.1186/2050-7283-1-28 PMID:25566376

Collins, B., MacKenzie, J., Tasca, G. A., Scherling, C., & Smith, A. (2014). Persistent cognitive changes in breast cancer patients 1 year following completion of chemotherapy. *Journal of the International Neuropsychological Society*, *20*(4), 370–379. doi:10.1017/S1355617713001215 PMID:24229809

Colwell, C. S. (2021). Defining circadian disruption in neurodegenerative disorders. *The Journal of Clinical Investigation*, *131*(19), e148288. doi:10.1172/JCI148288 PMID:34596047

Compta, Y., Parkkinen, L., O'Sullivan, S. S., Vandrovcova, J., Holton, J. L., Collins, C., Lashley, T., Kallis, C., Williams, D. R., de Silva, R., Lees, A. J., & Revesz, T. (2011). Lewy- and Alzheimer-type pathologies in Parkinson's disease dementia: Which is more important? *Brain*, *134*(5), 1493–1505. doi:10.1093/brain/awr031 PMID:21596773

Conklin, H., Ashford, J., Di Pinto, M., Vaughan, C., Gioia, G., Merchant, T., Ogg, R., Santana, V., & Wu, S. (2013). Computerized assessment of cognitive late effects among adolescent brain tumor survivors. *Journal of Neuro-Oncology*, *113*(2), 333–340. doi:10.1007/s11060-013-1123-5 PMID:23525951

Connell, M., Betts, K., McGrath, J. J., Alati, R., Najman, J., Clavarino, A., Mamun, A., Williams, G., & Scott, J. G. (2016). Hallucinations in adolescents and risk for mental disorders and suicidal behaviour in adulthood: Prospective evidence from the MUSP birth cohort study. *Schizophrenia Research*, *176*(2), 546–551. doi:10.1016/j.schres.2016.06.009 PMID:27374323

Corrêa-Oliveira, G. E., Scarabelot, L. F., Araújo, J. M., Boin, A. C., de Paula Pessoa, R. M., Leal, L. R., & Del-Ben, C. M. (2021). Early intervention in psychosis in emerging countries: Findings from a first-episode psychosis programme in the Ribeirão Preto catchment area, southeastern Brazil. *Early Intervention in Psychiatry*, *16*(7), 800–807. doi:10.1111/eip.13252 PMID:34794209

Cosgrave, J., Purple, R. J., Haines, R., Porcheret, K., van Heugten-van der Kloet, D., Johns, L., Alexander, I., Goodwin, G. M., Foster, R. G., & Wulff, K. (2021). Do environmental risk factors for the development of psychosis distribute differently across dimensionally assessed psychotic experiences? *Translational Psychiatry*, *1*(1), 226. Advance online publication. doi:10.1038/s41398-021-01265-2 PMID:33875641

Croft, J., Heron, J., Teufel, C., Cannon, M., Wolke, D., Thompson, A., Houtepen, L., & Zammit, S. (2019). Association of Trauma Type, Age of Exposure, and Frequency in Childhood and Adolescence With Psychotic Experiences in Early Adulthood. *JAMA Psychiatry*, *76*(1), 79. doi:10.1001/jamapsychiatry.2018.3155 PMID:30477014

Cuthbert, B. N. (2014). The RDoC framework: Facilitating transition from ICD/DSM to dimensional approaches that integrate neuroscience and psychopathology. *World Psychiatry; Official Journal of the World Psychiatric Association (WPA)*, *13*(1), 28–35. doi:10.1002/wps.20087 PMID:24497240

D'Este, G., Berra, F., Carli, G., Leitner, C., Marelli, S., Zucconi, M., Casoni, F., Ferini-Strambi, L., & Galbiati, A. (2023). Cognitive Reserve in Isolated Rapid Eye-Movement Sleep Behavior Disorder. *Brain Sciences*, *13*(2), 176. doi:10.3390/brainsci13020176 PMID:36831719

Damasio, A., & Damasio, H. (1992). Grain and Language. Mind and Brain, 89-95.

Das, D., Cherbuin, N., Easteal, S., & Anstey, K. J. (2014). Attention Deficit/Hyperactivity Disorder Symptoms and Cognitive Abilities in the Late-Life Cohort of the PATH through Life Study. *PLoS One*, *9*(1), e86552. doi:10.1371/journal.pone.0086552 PMID:24489743

de Almondes, K. M., Costa, M. V., Malloy-Diniz, L. F., & Diniz, B. S. (2016). Insomnia and risk of dementia in older adults: Systematic review and meta-analysis. *Journal of Psychiatric Research*, *77*, 109–115. doi:10.1016/j.jpsychires.2016.02.021 PMID:27017287

de Almondes, K. M., Júnior, F. W. N. H., Leonardo, M. E. M., & Alves, N. T. (2020). Facial Emotion Recognition and Executive Functions in Insomnia Disorder: An Exploratory Study. *Frontiers in Psychology*, *11*, 502. doi:10.3389/fpsyg.2020.00502 PMID:32362851

De Loore, E., Gunther, N., Drukker, M., Feron, F., Sabbe, B., Deboutte, D., & van Os, J. (2011). Myin-Germeys I. Persistence and outcome of auditory hallucinations in adolescence: a longitudinal general population study of 1800 individuals. *Schizophr Res.*, *127*(1-3), 252-6. doi:10.1016/j.schres.2011.01.015

de Mello Ayres, A., Scazufca, M., Menezes, P.R., Nakano, E.Y., Regina, A.C., Schaufelberger, M.S., Murray, R.M., McGuire, P.K., Rushe, T., & Busatto, G.F. (2010). Cognitive functioning in subjects with recent-onset psychosis from a low-middle-income environment: multiple-domain deficits and longitudinal evaluation. *Psychiatry Res.*, *179*(2), 157-64. doi:10.1016/j.psychres.2009.11.001

de Oliveira, P. A., Scivoletto, S., & Cunha, P. J. (2010). Estudos neuropsicológicos e de neuroimagem associados ao estresse emocional na infância e adolescência. *Archives of Clinical Psychiatry*, *37*(6), 271–279. doi:10.1590/S0101-60832010000600004

de Vos, C., Thompson, A., Amminger, P., Bendall, S., de Haan, L., Hartmann, J., Lavoie, S., Markulev, C., McGorry, P., Nieman, D., Phillips, L., Reininghaus, U., Yuen, H. P., Yung, A., & Nelson, B. (2019). The relationship between childhood trauma and clinical characteristics in ultrahigh risk for psychosis youth. *Psychosis*, *11*(1), 28–41. doi:10.1080/17522439.2019.1582686

Del-Ben, C. M., Shuhama, R., Loureiro, C. M., Ragazzi, T. C., Zanatta, D. P., Tenan, S. H., Ferreira Santos, J. L., Louzada-Junior, P., dos Santos, A. C., Morgan, C., & Menezes, P. R. (2019). Urbanicity and risk of first-episode psychosis: Incidence Study in Brazil. *The British Journal of Psychiatry*, *215*(6), 726–729. doi:10.1192/bjp.2019.110 PMID:31099323

Demichelis, O. P., Coundouris, S. P., Grainger, S. A., & Henry, J. D. (2020). Empathy and Theory of Mind in Alzheimer's Disease: A Meta-analysis. *Journal of the International Neuropsychological Society*, *26*(10), 963–977. doi:10.1017/S1355617720000478 PMID:32431261

Demjaha, A., Lappin, J.M., Stahl, D., Patel, M.X., MacCabe, J.H., & Howes, O.D. (2017). Antipsychotic treatment resistance in first-episode psychosis: prevalence, subtypes and predictors. *Psychol Med.*, *47*(11), 1981–9.

Den Heijer, A. E., Groen, Y., Tucha, L., Fuermaier, A. B. M., Koerts, J., Lange, K. W., Thome, J., & Tucha, O. (2017). Sweat it out? The effects of physical exercise on cognition and behavior in children and adults with ADHD: A systematic literature review. *Journal of Neural Transmission (Vienna, Austria)*, *124*(S1), 3–26. doi:10.1007/s00702-016-1593-7 PMID:27400928

Di Forti, M., Quattrone, D., Freeman, T. P., Tripoli, G., Gayer-Anderson, C., Quigley, H., Rodriguez, V., Jongsma, H. E., Ferraro, L., La Cascia, C., La Barbera, D., Tarricone, I., Berardi, D., Szöke, A., Arango, C., Tortelli, A., Velthorst, E., Bernardo, M., Del-Ben, C. M., ... van der Ven, E. (2019). The contribution of cannabis use to variation in the incidence of psychotic disorder across Europe (EU-GEI): A multicentre case-control study. *The Lancet. Psychiatry*, *6*(5), 427–436. doi:10.1016/S2215-0366(19)30048-3 PMID:30902669

Diaz, M. T., & Hogstrom, L. J. (2011). The influence of context on hemispheric recruitment during metaphor processing. *Journal of Cognitive Neuroscience*, *23*(11), 3586–3597. doi:10.1162/jocn_a_00053 PMID:21568642

Dickman, M. S. (2001). Von Economo encephalitis. *Archives of Neurology*, *58*(10), 1696–1698. doi:10.1001/archneur.58.10.1696 PMID:11594935

Dillinger, R.L., & Kersun, J.M. (2020). Caring for caregivers: Understanding and meeting their needs in coping with first episode psychosis. *Early Interv Psychiatry*, *14*(5), 528-534. doi:10.1111/eip.12870

Dinu, L. M., Singh, S. N., Baker, N. S., Georgescu, A. L., Singer, B. F., Overton, P. G., & Dommett, E. J. (2023). The Effects of Different Exercise Approaches on Attention Deficit Hyperactivity Disorder in Adults: A Randomised Controlled Trial. *Behavioral Sciences (Basel, Switzerland)*, *13*(2), 129. doi:10.3390/bs13020129 PMID:36829357

Dobrosavljevic, M., Solares, C., Cortese, S., Andershed, H., & Larsson, H. (2020). Prevalence of attention-deficit/hyperactivity disorder in older adults: A systematic review and meta-analysis. *Neuroscience and Biobehavioral Reviews*, *118*, 282–289. doi:10.1016/j.neubiorev.2020.07.042 PMID:32798966

Dobrosavljevic, M., Zhang, L., Garcia-Argibay, M., Du Rietz, E., Andershed, H., Chang, Z., Faraone, S., & Larsson, H. (2022). Attention-deficit/hyperactivity disorder as a risk factor for dementia and mild cognitive impairment: A population-based register study. *European Psychiatry*, *65*(1), e3. doi:10.1192/j.eurpsy.2021.2261 PMID:34924079

Dokur, Z., & Ömez, T. (2001). Ecg beat classification by a novel hybrid neural network. *Computer Methods and Programs in Biomedicine*, *66*(2), 167–181. doi:10.1016/S0169-2607(00)00133-4 PMID:11551391

Dominguez, M.D., Wichers, M., Lieb, R., Wittchen, H.U., & van Os, J. (2011). Evidence that onset of clinical psychosis is an outcome of progressively more persistent subclinical psychotic experiences: an 8-year cohort study. *Schizophr Bull.*, *37*(1), 84-93. doi:10.1093/schbul/sbp022

Dubois, M., Lapinte, N., Villier, V., Lecointre, C., Roy, V., Tonon, M.-C., Gandolfo, P., Joly, F., Hilber, P., & Castel, H. (2014). Chemotherapy-induced long-term alteration of executive functions and hippocampal cell proliferation: Role of glucose as adjuvant. *Neuropharmacology*, *79*, 234–248. doi:10.1016/j.neuropharm.2013.11.012 PMID:24291465

Dunning, C. J. R., Reyes, J. F., Steiner, J. A., & Brundin, P. (2012). Can Parkinson's disease pathology be propagated from one neuron to another? *Progress in Neurobiology*, *97*(2), 205–219. doi:10.1016/j.pneurobio.2011.11.003 PMID:22115849

Duval, C., Bejanin, A., Piolino, P., Laisney, M., de La Sayette, V., Belliard, S., Eustache, F., & Desgranges, B. (2012). Theory of mind impairments in patients with semantic dementia. *Brain*, *135*(Pt 1), 228–241. doi:10.1093/brain/awr309 PMID:22232593

Eakin, L., Minde, K., Hechtman, L., Ochs, E., Krane, E., Bouffard, R., Greenfield, B., & Looper, K. (2004). The marital and family functioning of adults with ADHD and their spouses. *Journal of Attention Disorders*, *8*(1), 1–10. doi:10.1177/108705470400800101 PMID:15669597

Ebrahimzadeh, E., Kalantari, M., Joulani, M., Shahraki, R. S., Fayaz, F., & Ahmadi, F. (2018). Prediction of paroxysmal atrial fibrillation: A machine learning based approach using combined feature vector and mixture of expert classification on hrv signal. *Computer Methods and Programs in Biomedicine*, *165*, 53–67. doi:10.1016/j.cmpb.2018.07.014 PMID:30337081

Edavally, S., Miller, D. D., & Youssef, N. A. (2021). Artificial intelligence to aid detection and diagnostic accuracy of mood disorders and predict suicide risk: A systematic review. *Annals of Clinical Psychiatry*, *33*(4), 270–281. PMID:34672929

Elhaj, F. A., Salim, N., Harris, A. R., Swee, T. T., & Ahmed, T. (2016). Arrhythmia recognition and classification using combined linear and nonlinear features of ecg signals. *Computer Methods and Programs in Biomedicine*, *127*, 52–63. doi:10.1016/j.cmpb.2015.12.024 PMID:27000289

Elyoseph, Z., & Levkovich, I. (2023). Beyond human expertise: The promise and limitations of ChatGPT in suicide risk assessment. *Frontiers in Psychiatry*, *14*, 1–7. doi:10.3389/fpsyt.2023.1213141 PMID:37593450

Emmert-Streib, F., Dehmer, M., & Shi, Y. (2016). Fifty years of graph matching, network alignment and network comparison. *Information Sciences, 346–347*, 180–197. doi:10.1016/j.ins.2016.01.074

Emsley, R., Oosthuizen, P.P., Koen, L., Niehaus, D.J., & Martinez, G. (2012). Symptom recurrence following intermittent treatment in first-episode schizophrenia successfully treated for 2 years: a 3-year open-label clinical study. *J Clin Psychiatry, 73*(4), e541-7. doi:10.4088/JCP.11m07138

Ercoli, L. M., Petersen, L., Hunter, A. M., Castellon, S. A., Kwan, L., Kahn-Mills, B. A., Embree, L. M., Cernin, P. A., Leuchter, A. F., & Ganz, P. A. (2015). Cognitive rehabilitation group intervention for breast cancer survivors: Results of a randomized clinical trial. *Psycho-Oncology, 24*(11), 1360–1367. doi:10.1002/pon.3769 PMID:25759235

Erkkinen, M.G., Kim, M.O., & Geschwind, M.D. (2018). Clinical Neurology and Epidemiology of the Major Neurodegenerative Diseases. *Cold Springer Harbor Perspectives in Biol, 10*(4).

Erlenmeyer-Kimling, L., Rock, D., Roberts, S. A., Janal, M., Kestenbaum, C., Cornblatt, B., Adamo, U. H., & Gottesman, I. I. (2000). Attention, Memory, and Motor Skills as Childhood Predictors of Schizophrenia-Related Psychoses: The New York High-Risk Project. *The American Journal of Psychiatry, 157*(9), 1416–1422. doi:10.1176/appi.ajp.157.9.1416 PMID:10964857

Escalante, C. P., Meyers, C., Reuben, J. M., Wang, X., Qiao, W., Manzullo, E., Alvarez, R. H., Morrow, P. K., Gonzalez-Angulo, A. M., Wang, X. S., Mendoza, T., Liu, W., Holmes, H., Hwang, J., Pisters, K., Overman, M., & Cleeland, C. (2014). A Randomized, Double-blind, 2-Period, Placebo-Controlled Crossover Trial of a Sustained-Release Methylphenidate in the Treatment of Fatigue in Cancer Patients. *Cancer Journal (Sudbury, Mass.), 20*(1), 8–14. doi:10.1097/PPO.0000000000000018 PMID:24445757

Eslinger, P. J., Moore, P., Troiani, V., Antani, S., Cross, K., Kwok, S., & Grossman, M. (2007). Oops! Resolving social dilemmas in frontotemporal dementia. *Journal of Neurology, Neurosurgery, and Psychiatry, 78*(5), 457–460. doi:10.1136/jnnp.2006.098228 PMID:17012339

Eviatar, Z., & Just, M. A. (2006). Brain correlates of discourse processing: An fMRI investigation of irony and conventional metaphor comprehension. *Neuropsychologia, 44*(12), 2348–2359. doi:10.1016/j.neuropsychologia.2006.05.007 PMID:16806316

Eysenbach, G. (2023). The Role of ChatGPT, Generative Language Models, and Artificial Intelligence in Medical Education: A Conversation With ChatGPT and a Call for Papers. *JMIR Medical Education, 9*, e46885. doi:10.2196/46885 PMID:36863937

Fardell, J. E., Vardy, J., & Johnston, I. N. (2013). The short and long term effects of docetaxel chemotherapy on rodent object recognition and spatial reference memory. *Life Sciences, 93*(17), 596–604. doi:10.1016/j.lfs.2013.05.006 PMID:23693082

Fardell, J. E., Vardy, J., Monds, L. A., & Johnston, I. N. (2015). The long-term impact of oxaliplatin chemotherapy on rodent cognition and peripheral neuropathy. *Behavioural Brain Research, 291*, 80–88. doi:10.1016/j.bbr.2015.04.038 PMID:25934489

Fayette, D., Juríčková, V., Kozák, T., Mociková, H., Gaherová, L., Fajnerová, I., & Horáček, J. (2023). Cognitive impairment associated with Hodgkin's lymphoma and chemotherapy. *Neuroscience Letters*, *797*, 137082. doi:10.1016/j.neulet.2023.137082 PMID:36693557

Fayyad, J., Sampson, N. A., Hwang, I., Adamowski, T., Aguilar-Gaxiola, S., Al-Hamzawi, A., Andrade, L. H. S. G., Borges, G., De Girolamo, G., Florescu, S., Gureje, O., Haro, J. M., Hu, C., Karam, E. G., Lee, S., Navarro-Mateu, F., O'Neill, S., Pennell, B.-E., Piazza, M., ... Kessler, R. C. (2017). The descriptive epidemiology of DSM-IV Adult ADHD in the World Health Organization World Mental Health Surveys. *Attention Deficit and Hyperactivity Disorders*, *9*(1), 47–65. doi:10.1007/s12402-016-0208-3 PMID:27866355

Fearnley, J. M., & Lees, A. J. (1991). Ageing and Parkinson's disease: Substantia nigra regional selectivity. *Brain*, *114*(5), 2283–2301. doi:10.1093/brain/114.5.2283 PMID:1933245

Feber, L., Peter, N., Schneider-Thoma, J., Siafis, S., Bighelli, I., Hansen, W.-P., Prates Baldez, D., Salanti, G., Keefe, R. S. E., Engel, R. R., & Leucht, S. (2023). Antipsychotic drugs and their effects on cognitive function: Protocol for a systematic review, pairwise, and network meta-analysis. *Systematic Reviews*, *12*(1), 54. doi:10.1186/s13643-023-02213-5 PMID:36959619

Ferguson, H. J., Brunsdon, V. E. A., & Bradford, E. E. F. (2021). The developmental trajectories of executive function from adolescence to old age. *Scientific Reports*, *11*(1), 1382. doi:10.1038/s41598-020-80866-1 PMID:33446798

Fergusson, D. M., Boden, J. M., & Horwood, L. J. (2008). Exposure to childhood sexual and physical abuse and adjustment in early adulthood. *Child Abuse & Neglect*, *32*(6), 607–619. doi:10.1016/j.chiabu.2006.12.018 PMID:18565580

Figueiredo, T. (2022). *Tratado de Cognição Social: uma abordagem multidimensional*. Ed. Ampla.

Fittipaldi, S., Ibanez, A., Baez, S., Manesa, F., Sedeno, L., & Adolfo, A. M. (2019). More than words: Social cognition across variants of primary progressive aphasia. *Neuroscience and Biobehavioral Reviews*, *100*, 263–284. doi:10.1016/j.neubiorev.2019.02.020 PMID:30876954

Fleischmann, A., & Fleischmann, R. H. (2012). Advantages of an ADHD Diagnosis in Adulthood: Evidence From Online Narratives. *Qualitative Health Research*, *22*(11), 1486–1496. doi:10.1177/1049732312457468 PMID:22910591

Fleming, B., Edison, P., & Kenny, L. (2023). Cognitive impairment after cancer treatment: Mechanisms, clinical characterization, and management. *BMJ (Clinical Research Ed.)*, *071726*, e071726. Advance online publication. doi:10.1136/bmj-2022-071726 PMID:36921926

Folstein, M. F., Folstein, S. E., & McHugh, P. R. (1975). Mini-mental state. *Journal of Psychiatric Research*, *12*(3), 189–198. doi:10.1016/0022-3956(75)90026-6 PMID:1202204

Fonagy, P., & Bateman, A. W. (2008). The development of borderline personality disorder—A mentalizing model. *Journal of Personality Disorders*, *22*(1), 4–21. doi:10.1521/pedi.2008.22.1.4 PMID:18312120

Freudenreich, O., Schulz, S.C., & Goff, D.C. (2009). Initial medical work-up of first-episode psychosis: a conceptual review. *Early Interv Psychiatry, 3*(1), 10-8. . doi:10.1111/j.1751-7893.2008.00105.x

Freudenreich, O. (2020). *Psychotic disorders: A practical guideline* (2nd ed.). Springer International Publishing. doi:10.1007/978-3-030-29450-2

Friedman, J., Hastie, T., & Tibshirani, R. (2001). The elements of statistical learning (Vol. 1). *Springer series in statistics, 1*(10).

Frisoni, G. B., Fox, N. C., Jack, C. R. Jr, Scheltens, P., & Thompson, P. M. (2010). The clinical use of structural MRI in Alzheimer disease. *Nature Reviews. Neurology, 6*(2), 67–77. doi:10.1038/nrneurol.2009.215 PMID:20139996

From, H. H. (2019). Onset and Prodromal Stage to a Life-Long Course of Schizophrenia and Its Symptom Dimensions: How Sex, Age, and Other Risk Factors Influence Incidence and Course of Illness. *Psychiatry Journal, 2019*, 9804836. PMID:31139639

Fuchs, J., Nilsson, C., Kachergus, J., Munz, M., Larsson, E.-M., Schüle, B., Langston, J. W., Middleton, F. A., Ross, O. A., Hulihan, M., Gasser, T., & Farrer, M. J. (2007). Phenotypic variation in a large Swedish pedigree due to SNCA duplication and triplication. *Neurology, 68*(12), 916–922. doi:10.1212/01.wnl.0000254458.17630.c5 PMID:17251522

Fusar-Poli, P., Nelson, B., Valmaggia, L., Yung, A.R., & McGuire, P.K. (2014). Comorbid depressive and anxiety disorders in 509 individuals with an at-risk mental state: impact on psychopathology and transition to psychosis. *Schizophr Bull, 40*(1), 120-131.

Fusar-Poli, P., Salazar de Pablo, G., Correll, C.U., Meyer-Lindenberg, A., Millan, M.J., Borgwardt, S., Galderisi, S., Bechdolf, A., Pfennig, A., Kessing, L.V., van Amelsvoort, T., Nieman, D.H., Domschke, K., Krebs, M.O., Koutsouleris, N., McGuire, P., Do, K.Q., & Arango, C. (2020). Prevention of Psychosis: Advances in Detection, Prognosis, and Intervention. *JAMA Psychiatry, 77*(7), 755-765.

Fusar-Poli, P., Cappucciati, M., De Micheli, A., Rutigliano, G., Bonoldi, I., Tognin, S., Ramella-Cravaro, V., Castagnini, A., & McGuire, P. (2017). Diagnostic and Prognostic Significance of Brief Limited Intermittent Psychotic Symptoms (BLIPS) in Individuals at Ultra High Risk. *Schizophrenia Bulletin, 43*(1), 48–56. doi:10.1093/schbul/sbw151 PMID:28053130

Fusar-Poli, P., De Micheli, A., Cappucciati, M., Rutigliano, G., Davies, C., Ramella-Cravaro, V., Oliver, D., Bonoldi, I., Rocchetti, M., Gavaghan, L., Patel, R., & McGuire, P. (2018). Diagnostic and Prognostic Significance of DSM-5 Attenuated Psychosis Syndrome in Services for Individuals at Ultra High Risk for Psychosis. *Schizophrenia Bulletin, 44*(2), 264–275. doi:10.1093/schbul/sbx055 PMID:28521060

Fusar-Poli, P., Rutigliano, G., Stahl, D., Davies, C., De Micheli, A., Ramella-Cravaro, V., Bonoldi, I., & McGuire, P. (2017). Long-term validity of the At Risk Mental State (ARMS) for predicting psychotic and non-psychotic mental disorders. *European Psychiatry, 42*, 49–54. doi:10.1016/j.eurpsy.2016.11.010 PMID:28212505

Gaebel, W., & Zielasek, J. (2015). Focus on psychosis. *Dialogues in Clinical Neuroscience*, *17*(1), 9–18. doi:10.31887/DCNS.2015.17.1/wgaebel PMID:25987859

Gagnon Shaigetz, V., Proulx, C., Cabral, A., Choudhury, N., Hewko, M., Kohlenberg, E., Segado, M., Smith, M. S. D., & Debergue, P. (2021). An Immersive and Interactive Platform for Cognitive Assessment and Rehabilitation (bWell): Design and Iterative Development Process. *JMIR Rehabilitation and Assistive Technologies*, *8*(4), e26629. doi:10.2196/26629 PMID:34730536

Galletly, C., Castle, D., Dark, F., Humberstone, V., Jablensky, A., & Killackey, E. (2016). Royal Australian and New Zealand College of Psychiatrists clinical practice guidelines for the management of schizophrenia and related disorders. *Aust N Z J Psychiatry*, *50*(5), 410–72.

Galvin, J. E., Howard, D. H., Denny, S. S., Dickinson, S., & Tatton, N. (2017). The social and economic burden of frontotemporal degeneration. *Neurology*, *89*(20), 2049–2056. doi:10.1212/WNL.0000000000004614 PMID:28978658

Gamal, S. M. T., Azab, A. O. M. A., El Refaei, S. M., & Houseni, M. (2021). The role of 18-FDG PET/CT assessment of functional brain metabolism in cancer patients after chemotherapy. *The Egyptian Journal of Radiology and Nuclear Medicine*, *52*(1), 18. doi:10.1186/s43055-020-00403-y

Gan-Or, Z., Liong, C., & Alcalay, R. N. (2018). GBA-Associated Parkinson's Disease and Other Synucleinopathies. *Current Neurology and Neuroscience Reports*, *18*(8), 44. doi:10.1007/s11910-018-0860-4 PMID:29884970

Gaynor, A. M., Pergolizzi, D., Alici, Y., Ryan, E., McNeal, K., Ahles, T. A., & Root, J. C. (2020). Impact of transcranial direct current stimulation on sustained attention in breast cancer survivors: Evidence for feasibility, tolerability, and initial efficacy. *Brain Stimulation*, *13*(4), 1108–1116. doi:10.1016/j.brs.2020.04.013 PMID:32353419

Germine, L., Reinecke, K., & Chaytor, N. S. (2019). Digital neuropsychology: Challenges and opportunities at the intersection of science and software. *The Clinical Neuropsychologist*, *33*(2), 271–286. doi:10.1080/13854046.2018.1535662 PMID:30614374

Ginsberg, Y., Beusterien, K. M., Amos, K., Jousselin, C., & Asherson, P. (2014). The unmet needs of all adults with ADHD are not the same: A focus on Europe. *Expert Review of Neurotherapeutics*, *14*(7), 799–812. doi:10.1586/14737175.2014.926220 PMID:24894408

Goetz, C. G., Fahn, S., Martinez-Martin, P., Poewe, W., Sampaio, C., Stebbins, G. T., Stern, M. B., Tilley, B. C., Dodel, R., Dubois, B., Holloway, R., Jankovic, J., Kulisevsky, J., Lang, A. E., Lees, A., Leurgans, S., LeWitt, P. A., Nyenhuis, D., Olanow, C. W., ... LaPelle, N. (2007). Movement Disorder Society-sponsored revision of the Unified Parkinson's Disease Rating Scale (MDS-UPDRS): Process, format, and clinimetric testing plan. *Movement Disorders*, *22*(1), 41–47. doi:10.1002/mds.21198 PMID:17115387

Goldberger, A. L., Amaral, L. A., Glass, L., Hausdorff, J. M., Ivanov, P. C., Mark, R. G., Mietus, J. E., Moody, G. B., Peng, C.-K., & Stanley, H. E. (2000a). Physiobank, physiotoolkit, and physionet components of a new research resource for complex physiologic signals (Vol. 101). *Circulation*, *101*(23), 215–220. doi:10.1161/01.CIR.101.23.e215 PMID:10851218

Gong, Q., Hu, X., Pettersson-Yeo, W., Xu, X., Lui, S., Crossley, N., Wu, M., Zhu, H., & Mechelli, A. (2017). Network-Level Dysconnectivity in Drug-Naïve First-Episode Psychosis: Dissociating Transdiagnostic and Diagnosis-Specific Alterations. *Neuropsychopharmacology*, *42*(4), 933–940. doi:10.1038/npp.2016.247 PMID:27782128

Goodman, D. W., Mitchell, S., Rhodewalt, L., & Surman, C. B. H. (2016). Clinical Presentation, Diagnosis and Treatment of Attention-Deficit Hyperactivity Disorder (ADHD) in Older Adults: A Review of the Evidence and its Implications for Clinical Care. *Drugs & Aging*, *33*(1), 27–36. doi:10.1007/s40266-015-0327-0 PMID:26659731

Gorestein, C., Wang, Y., & Hungerbuhler, I. (2016). Instrumentos de avaliação em saúde mental. *The Art of Medication.*

Gorno-Tempini, M.L., Brambati, S.M., & Ginex, V. (2008). *The logopenic – phonological variant of primary progressive aphasia.* Academic Press.

Gorno-Tempini, M. L., Dronkers, N. F., Rankin, K. P., Ogar, J. M., Phengrasamy, L., Rosen, H. J., Johnson, J. K., Weiner, M. W., & Miller, B. L. (2004, March). Cognition and Anatomy in Three Variants of Primary Progressive Aphasia. *Annals of Neurology*, *55*(3), 335–346. doi:10.1002/ana.10825 PMID:14991811

Gorno-Tempini, M. L., Hills, A. E., & Weintraub, S. (2011). Classification of primary progressive aphasia and its variants. *Neurology*, *76*(11), 1006–1014. doi:10.1212/WNL.0b013e31821103e6 PMID:21325651

Goulart, A. A., Lucatelli, A., Silveira, P. S. P., Siqueira, J. O., Pereira, V. F. A., Carmona, M. J. C., Valentin, L. S. S., & Vieira, J. E. (2022). Comparison of digital games as a cognitive function assessment tool for current standardized neuropsychological tests. *Brazilian Journal of Anesthesiology*, *72*(1), 13–20. doi:10.1016/j.bjane.2021.06.027 PMID:34411626

Graham, S. A., Lee, E. E., Jeste, D. V., Van Patten, R., Twamley, E. W., Nebeker, C., Yamada, Y., Kim, H. C., & Depp, C. A. (2020). Artificial intelligence approaches to predicting and detecting cognitive decline in older adults: A conceptual review. *Psychiatry Research*, *284*, 112732. doi:10.1016/j.psychres.2019.112732 PMID:31978628

Green, M. F., Horan, W. P., & Lee, J. (2015). Social cognition in schizophrenia. *Nature Reviews. Neuroscience*, *16*(10), 620–631. doi:10.1038/nrn4005 PMID:26373471

Gregory, C., Lough, S., Stone, V., Erzinclioglu, S., Martin, L., Baron-Cohen, S., & Hodges, J. R. (2002). Theory of mind in patients with frontal variant frontotemporal dementia and Alzheimer's disease: Theoretical and practical implications. *Brain*, *125*(Pt 4), 752–764. doi:10.1093/brain/awf079 PMID:11912109

Grossman, M., & Irwin, D. J. (2018). Primary progressive aphasia and stroke aphasia. *Continuum*, *24*(3), 745–767. PMID:29851876

Guarnieri, B., Adorni, F., Musicco, M., Appollonio, I., Bonanni, E., Caffarra, P., Caltagirone, C., Cerroni, G., Concari, L., Cosentino, F. I., Ferrara, S., Fermi, S., Ferri, R., Gelosa, G., Lombardi, G., Mazzei, D., Mearelli, S., Morrone, E., Murri, L., ... Sorbi, S. (2012). Prevalence of sleep disturbances in mild cognitive impairment and dementing disorders: A multicenter Italian clinical cross-sectional study on 431 patients. *Dementia and Geriatric Cognitive Disorders*, *33*(1), 50–58. doi:10.1159/000335363 PMID:22415141

Guldberg-Kjär, T., & Johansson, B. (2009). Old people reporting childhood AD/HD symptoms: Retrospectively self-rated AD/HD symptoms in a population-based Swedish sample aged 65–80. *Nordic Journal of Psychiatry*, *63*(5), 375–382. doi:10.1080/08039480902818238 PMID:19308795

Guldberg-Kjär, T., Sehlin, S., & Johansson, B. (2013). ADHD symptoms across the lifespan in a population-based Swedish sample aged 65 to 80. *International Psychogeriatrics*, *25*(4), 667–675. doi:10.1017/S1041610212002050 PMID:23237190

Haba-Rubio, J., Frauscher, B., Marques-Vidal, P., Toriel, J., Tobback, N., Andries, D., Preisig, M., Vollenweider, P., Postuma, R., & Heinzer, R. (2018). Prevalence and determinants of rapid eye movement sleep behavior disorder in the general population. *Sleep (Basel)*, *41*(2), zsx197. Advance online publication. doi:10.1093/sleep/zsx197 PMID:29216391

Hafner, H. (1998). Onset and course of the first schizophrenic episode. *Kaohsiung J Med Sci.*, *14*(7), 413–31.

Halliday, G. M., Leverenz, J. B., Schneider, J. S., & Adler, C. H. (2014). The neurobiological basis of cognitive impairment in Parkinson's disease. *Movement Disorders*, *29*(5), 634–650. doi:10.1002/mds.25857 PMID:24757112

Halliday, G. M., Li, Y. W., Blumbergs, P. C., Joh, T. H., Cotton, R. G. H., Howe, P. R. C., Blessing, W. W., & Geffen, L. B. (1990). Neuropathology of immunohistochemically identified brainstem neurons in Parkinson's disease. *Annals of Neurology*, *27*(4), 373–385. doi:10.1002/ana.410270405 PMID:1972319

Halliday, G., McCann, H., & Shepherd, C. (2012). Evaluation of the Braak hypothesis: How far can it explain the pathogenesis of Parkinson's disease? *Expert Review of Neurotherapeutics*, *12*(6), 673–686. doi:10.1586/ern.12.47 PMID:22650170

Hall, J. E. (2015). *Guyton and Hall Textbook of Medical Physiology* (13th ed.). Elsevier Health Sciences. https://books.google.com.br/books?id=krLSCQAAQBAJhttp://circ.ahajournals.org/content/101/23/e215http://www.scielo.br/scielo.phpscript=sciarttextpid=S0066-782X2016004000020nrm=isohttps://apps.who.int/iris/bitstream/handle/10665/43007/9786009317059-fas.pdf

Hanssen, M., Bak, M., Bijl, R., Vollebergh, W., & van Os, J. (2005). The incidence and outcome of subclinical psychotic experiences in the general population. *Br J Clin Psychol.*, *44*(Pt 2), 181-91. doi:10.1348/014466505X29611

Hansson Halleröd, S. L., Anckarsäter, H., Råstam, M., & Hansson Scherman, M. (2015). Experienced consequences of being diagnosed with ADHD as an adult – a qualitative study. *BMC Psychiatry*, *15*(1), 31. doi:10.1186/s12888-015-0410-4 PMID:25884685

Happé, F., Cook, J. L., & Bird, G. (2017). The Structure of Social Cognition: In(ter)dependence of Sociocognitive Processes. *Annual Review of Psychology*, *68*(1), 243–267. doi:10.1146/annurev-psych-010416-044046 PMID:27687121

Hartman, S. J., Nelson, S. H., Myers, E., Natarajan, L., Sears, D. D., Palmer, B. W., Weiner, L. S., Parker, B. A., & Patterson, R. E. (2018). Randomized controlled trial of increasing physical activity on objectively measured and self-reported cognitive functioning among breast cancer survivors: The memory & motion study. *Cancer*, *124*(1), 192–202. Advance online publication. doi:10.1002/cncr.30987 PMID:28926676

Hastrup, L.H., Kronborg, C., Bertelsen, M., Jeppesen, P., Jorgensen, P., Petersen, L., Thorup, A., Simonsen, E., & Nordentoft, M. (2013). Cost-effectiveness of early intervention in first-episode psychosis: economic evaluation of a randomised controlled trial (the OPUS study). *Br J Psychiatry, 202*(1), 35-41.

Haug, E., Øie, M., Andreassen, O. A., Bratlien, U., Nelson, B., Aas, M., Møller, P., & Melle, I. (2015). Anomalous self-experience and childhood trauma in first-episode schizophrenia. *Comprehensive Psychiatry*, *56*, 35–41. doi:10.1016/j.comppsych.2014.10.005 PMID:25458477

Havers, L., Stumm, S. von, Cardno, A.G., Freeman, D., & Ronald, A. (2023). Psychotic experiences and negative symptoms from adolescence to emerging adulthood: Developmental trajectories and associations with polygenic scores and childhood characteristics. *Psychol Med.*, *53*(12), 5685–97.

Hawkes, C. H., Del Tredici, K., & Braak, H. (2007). Parkinson's disease: A dual-hit hypothesis. *Neuropathology and Applied Neurobiology*, *33*(6), 599–614. doi:10.1111/j.1365-2990.2007.00874.x PMID:17961138

Hawkes, C. H., Del Tredici, K., & Braak, H. (2009). Parkinson's disease: The dual hit theory revisited. *Annals of the New York Academy of Sciences*, *1170*(1), 615–622. doi:10.1111/j.1749-6632.2009.04365.x PMID:19686202

Haxby, J. V., Hoffman, E. A., & Gobbini, M. I. (2002). Human neural systems for face recognition and social communication. *Biological Psychiatry*, *51*(1), 59–67. doi:10.1016/S0006-3223(01)01330-0 PMID:11801231

Healy, C., Brannigan, R., Dooley, N., Coughlan, H., Clarke, M., Kelleher, I., & Cannon, M. (2019). Childhood and adolescent psychotic experiences and risk of mental disorder: A systematic review and meta-analysis. *Psychological Medicine*, *49*(10), 1589–1599. doi:10.1017/S0033291719000485 PMID:31088578

Heins, M., Simons, C., Lataster, T., Pfeifer, S., Versmissen, D., Lardinois, M., Marcelis, M., Delespaul, P., Krabbendam, L., van Os, J., & Myin-Germeys, I. (2011). Childhood trauma and psychosis: A case–control and case-sibling comparison across different levels of genetic liability, psychopathology, and type of trauma. *The American Journal of Psychiatry*, *168*(12), 1286–1294. doi:10.1176/appi.ajp.2011.10101531 PMID:21955935

Henry, M.L., & Gorno-Tempini, M.L. (2010). *The logopenic variant of primary progressive aphasia.* Academic Press.

Henry, J. D., Rendell, P. G., Scicluna, A., Jackson, M., & Phillips, L. H. (2009). Emotion experience, expression, and regulation in Alzheimer's disease. *Psychology and Aging*, *24*(1), 252–257. doi:10.1037/a0014001 PMID:19290761

Hilker, R., Helenius, D., Fagerlund, B., Skytthe, A., Christensen, K., Werge, T. M., Nordentoft, M., & Glenthøj, B. (2018). Heritability of Schizophrenia and Schizophrenia Spectrum Based on the Nationwide Danish Twin Register. *Biological Psychiatry*, *83*(6), 492–498. doi:10.1016/j.biopsych.2017.08.017 PMID:28987712

Hodges, J. R. (2001). Frontotemporal dementia (Pick's disease): Clinical features and assessment. *Neurology*, *56*(Suppl 4), S6–S10. doi:10.1212/WNL.56.suppl_4.S6 PMID:11402143

Hodges, J. R., Patterson, K., Oxbury, S., & Funnell, E. (1992). Semantic dementia. Progressive fluent aphasia with temporal lobe atrophy. *Brain*, *115*(6), 1783–1806. doi:10.1093/brain/115.6.1783 PMID:1486461

Hodgson, K. D., Hutchinson, A. D., Wilson, C. J., & Nettelbeck, T. (2013). A meta-analysis of the effects of chemotherapy on cognition in patients with cancer. *Cancer Treatment Reviews*, *39*(3), 297–304. doi:10.1016/j.ctrv.2012.11.001 PMID:23219452

Hogan, D. B., Jetté, N., Fiest, K. M., Roberts, J. I., Pearson, D., Smith, E. E., Roach, P., Kirk, A., Pringsheim, T., & Maxwell, C. J. (2016). The Prevalence and Incidence of Frontotemporal Dementia: a Systematic Review. *The Canadian Journal of Neurological Sciences. Le journal canadien des sciences neurologiques*, *43*(Suppl 1), S96–S109. doi:10.1017/cjn.2016.25

HöglingerG. U.AdlerC. H.BergD.KleinC.OuteiroT. F.PoeweW.PostumaR.StoesslJ.LangA. E. (2023). *Towards a Biological Definition of Parkinson's Disease* (No. 2023040108). Preprints. doi:10.20944/preprints202304.0108.v1

Holingue, C., Wennberg, A., Berger, S., Polotsky, V. Y., & Spira, A. P. (2018). Disturbed sleep and diabetes: A potential nexus of dementia risk. *Metabolism: Clinical and Experimental*, *84*, 85–93. doi:10.1016/j.metabol.2018.01.021 PMID:29409842

Höllerhage, M., Klietz, M., & Höglinger, G. U. (2022). Disease modification in Parkinsonism: Obstacles and ways forward. *Journal of Neural Transmission (Vienna, Austria)*, *129*(9), 1133–1153. doi:10.1007/s00702-022-02520-6 PMID:35695938

Homaeinezhad, M., Atyabi, S., Tavakkoli, E., Toosi, H., Ghaffari, A., & Ebrahimpour, R. (2012). Ecg arrhythmia recognition via a neuro-svm–knn hybrid classifier with virtual qrs image- based geometrical features. *Expert Systems with Applications*, *39*(2), 2047–2058. doi:10.1016/j.eswa.2011.08.025

Hornykiewicz, O. (2002). Dopamine miracle: From brain homogenate to dopamine replacement. *Movement Disorders*, *17*(3), 501–508. doi:10.1002/mds.10115 PMID:12112197

Huang, C., Ye, S., Chen, H., Li, D., He, F., & Tu, Y. (2011). A novel method for detection of the transition between atrial fibrillation and sinus rhythm. *IEEE Transactions on Biomedical Engineering*, *58*(4), 1113–1119. doi:10.1109/TBME.2010.2096506 PMID:21134807

Huang, Y., Potter, R., Sigurdson, W., Santacruz, A., Shih, S., Ju, Y. E., ... Bateman, R. J. (2012). Effects of age and amyloid deposition on Aβ dynamics in the human central nervous system. *Archives of Neurology*, *69*(1), 51–58. doi:10.1001/archneurol.2011.235 PMID:21911660

Hung, C. M., Li, Y. C., Chen, H. J., Lu, K., Liang, C. L., Liliang, P. C., Tsai, Y. D., & Wang, K. W. (2018). Risk of dementia in patients with primary insomnia: A nationwide population-based case-control study. *BMC Psychiatry*, *18*(1), 38. doi:10.1186/s12888-018-1623-0 PMID:29415688

Hur, M. R., & Caixeta, L. (2013). Non-literal language and semantic dementia. *Dementia & Neuropsychologia*, *7*(4), 424–427. doi:10.1590/S1980-57642013DN74000011 PMID:29213868

Hutchinson, G., Takei, N., Fahy, T. A., Bhugra, D., Gilvarry, C., Moran, P., Mallett, R., Sham, P., Leff, J., & Murray, R. M. (1996). Morbid Risk of Schizophrenia in First-Degree Relatives of White and African–Caribbean Patients with Psychosis. *The British Journal of Psychiatry*, *169*(6), 776–780. doi:10.1192/bjp.169.6.776 PMID:8968638

Iacoboni, M. (2009). Imitation, empathy, and mirror neurons. *Annual Review of Psychology*, *60*(1), 653–670. doi:10.1146/annurev.psych.60.110707.163604 PMID:18793090

Iliescu, D., Greiff, S., Ziegler, M., & Fokkema, M. (2022). Artificial Intelligence, Machine Learning, and Other Demons. *European Journal of Psychological Assessment*, *38*(3), 163–164. doi:10.1027/1015-5759/a000713

Irwin, M. R., & Vitiello, M. V. (2019). Implications of sleep disturbance and inflammation for Alzheimer's disease dementia. *Lancet Neurology*, *18*(3), 296–306. doi:10.1016/S1474-4422(18)30450-2 PMID:30661858

Isvoranu, A.-M., van Borkulo, C. D., Boyette, L.-L., Wigman, J. T. W., Vinkers, C. H., & Borsboom, D. (2017). A Network Approach to Psychosis: Pathways Between Childhood Trauma and Psychotic Symptoms. *Schizophrenia Bulletin*, *43*(1), 187–196. doi:10.1093/schbul/sbw055 PMID:27165690

Jaaro-Peled, H., & Sawa, A. (2020). Neurodevelopmental Factors in Schizophrenia. *The Psychiatric Clinics of North America*, *43*(2), 263–274. doi:10.1016/j.psc.2020.02.010 PMID:32439021

Jackson, H. J., & McGorry, P. D. (2009). *The recognition and management of early psychosis a preventive approach*. Cambridge University Press. doi:10.1017/CBO9780511576287

Jager, C.A., Haigh, A.M.F., & Garrad, P. (2012). *Logopenic aphasia in Alzheimer's disease: clinical variant or clinical feature?* Academic Press.

Jameson, C., Boulton, K. A., Silove, N., Nanan, R., & Guastella, A. J. (2023). Ectodermal origins of the skin-brain axis: A novel model for the developing brain, inflammation, and neurodevelopmental conditions. *Molecular Psychiatry*, *28*(1), 108–117. doi:10.1038/s41380-022-01829-8 PMID:36284159

James, S. L., Abate, D., Abate, K. H., Abay, S. M., Abbafati, C., Abbasi, N., Abbastabar, H., Abd-Allah, F., Abdela, J., Abdelalim, A., Abdollahpour, I., Abdulkader, R. S., Abebe, Z., Abera, S. F., Abil, O. Z., Abraha, H. N., Abu-Raddad, L. J., Abu-Rmeileh, N. M. E., Accrombessi, M. M. K., ... Murray, C. J. L. (2018). Global, regional, and national incidence, prevalence, and years lived with disability for 354 Diseases and Injuries for 195 countries and territories, 1990-2017: A systematic analysis for the Global Burden of Disease Study 2017. *Lancet*, *392*(10159), 1789–1858. doi:10.1016/S0140-6736(18)32279-7 PMID:30496104

Janelsins, M. C., Kesler, S. R., Ahles, T. A., & Morrow, G. R. (2014). Prevalence, mechanisms, and management of cancer-related cognitive impairment. *International Review of Psychiatry (Abingdon, England)*, *26*(1), 102–113. doi:10.3109/09540261.2013.864260 PMID:24716504

Janelsins, M. C., Kohli, S., Mohile, S. G., Usuki, K., Ahles, T. A., & Morrow, G. R. (2011). An update on cancer- and chemotherapy-related cognitive dysfunction: Current status. *Seminars in Oncology*, *38*(3), 431–438. Advance online publication. doi:10.1053/j.seminoncol.2011.03.014 PMID:21600374

Jenkins, T. A. (2013). Perinatal complications and schizophrenia: Involvement of the immune system. *Frontiers in Neuroscience*, *7*. doi:10.3389/fnins.2013.00110 PMID:23805069

Johnen, A., & Bertoux, M. (2019). Psychological and Cognitive Markers of Behavioral Variant Frontotemporal Dementia-A Clinical Neuropsychologist's View on Diagnostic Criteria and Beyond. *Frontiers in Neurology*, *10*, 594. doi:10.3389/fneur.2019.00594 PMID:31231305

Joly, F., Giffard, B., Rigal, O., De Ruiter, M. B., Small, B. J., Dubois, M., Lefel, J., Schagen, S. B., Ahles, T. A., Wefel, J. S., Lange, M., & Castel, H. (2015). Impact of Cancer and Its Treatments on Cognitive Function: Advances in Research from the Paris International Cognition and Cancer Task Force Symposium and Update since 2012. *Journal of Pain and Symptom Management*, *50*(6), 830–841. doi:10.1016/j.jpainsymman.2015.06.019 PMID:26344551

Joly, M., Welter, G., Martins, R., Marini, J., Montiel, J., Lopes, F., & de Carvalho, M. (2005). Sistema de avaliação para testes informatizados (SAPI): Estudo preliminar. *Psic: Revista da Vetor Editora*, *6*(2), 51–60.

Josephs, K. A., Duffy, J. R., Strand, E. A., Machulda, M. M., Senjem, M. L., Gunter, J. L., Schwarz, C. G., Reid, R. I., Spychalla, A. J., Lowe, V. J., Jack, C. R. Jr, & Whitwell, J. L. (2014). The evolution of primary progressive apraxia of speech. *Brain*, *137*(10), 2783–2795. doi:10.1093/brain/awu223 PMID:25113789

Josephs, K. A., Duffy, J. R., Strand, E. A., Machulda, M. M., Senjem, M. L., Master, A. V., Lowe, V. J., Jack, C. R. Jr, & Whitwell, J. L. (2012). Characterizing a neurodegenerative syndrome: Primary progressive apraxia of speech. *Brain, 135*(pt5), 1522–1536. doi:10.1093/brain/aws032 PMID:22382356

Jovic, A., & Jovic, F. (2017). Classification of cardiac arrhythmias based on alphabet entropy of heart rate variability time series. *Biomedical Signal Processing and Control, 31*, 217–230. doi:10.1016/j.bspc.2016.08.010

Júnior, J., Neto, R., Gusmão, V., Menezes, N., Silva, M., Santos, L., & Reinoso, L. (2023). O futuro da aprendizagem com a inteligência artificial aplicada à educação 4.0. *Revista Educação, Humanidades e Ciências Sociais.*

Ju, Y. E. S., McLeland, J. S., Toedebusch, C. D., Xiong, C., Fagan, A. M., Duntley, S. P., Morris, J. C., & Holtzman, D. M. (2013). Sleep quality and preclinical Alzheimer disease. *JAMA Neurology, 70*(5), 587–593. doi:10.1001/jamaneurol.2013.2334 PMID:23479184

Kahn, R. S., & Sommer, I. E. (2015). The neurobiology and treatment of first-episode schizophrenia. *Molecular Psychiatry, 20*(1), 84–97. doi:10.1038/mp.2014.66 PMID:25048005

Kalia, L. V., & Lang, A. E. (2015). Parkinson's disease. *Lancet, 386*(9996), 896–912. doi:10.1016/S0140-6736(14)61393-3 PMID:25904081

Kaltsas, A., Moustakli, E., Zikopoulos, A., Georgiou, I., Dimitriadis, F., Symeonidis, E. N., Markou, E., Michaelidis, T. M., Tien, D. M. B., Giannakis, I., Ioannidou, E. M., Papatsoris, A., Tsounapi, P., Takenaka, A., Sofikitis, N., & Zachariou, A. (2023). Impact of Advanced Paternal Age on Fertility and Risks of Genetic Disorders in Offspring. *Genes, 14*(2), 486. doi:10.3390/genes14020486 PMID:36833413

Karschnia, P., Parsons, M. W., & Dietrich, J. (2019). Pharmacologic management of cognitive impairment induced by cancer therapy. *The Lancet. Oncology, 20*(2), e92–e102. doi:10.1016/S1470-2045(18)30938-0 PMID:30723041

Katz, S., Downs, T. D., Cash, H. R., & Grotz, R. C. (1970). Progress in Development of the Index of ADL. *The Gerontologist, 10*(1 Part 1), 20–30. doi:10.1093/geront/10.1_Part_1.20 PMID:5420677

Kaymaz, N., Drukker, M., Lieb, R., Wittchen, H.U., Werbeloff, N., Weiser, M., Lataster, T., & van Os, J. (2012). Do subthreshold psychotic experiences predict clinical outcomes in unselected non-help-seeking population-based samples? A systematic review and meta-analysis, enriched with new results. *Psychol Med., 42*(11), 2239-53. doi:10.1017/S0033291711002911

Kelly, B. C., & Vuolo, M. (2023). Trends in Psychotropic Drug–Implicated Cardiovascular Mortality: Patterns in U.S. Mortality, 1999–2020. *American Journal of Preventive Medicine, 65*(3), 377–384. doi:10.1016/j.amepre.2023.02.016 PMID:36894483

Kertesz, A., Davidson, W., & Munoz, D. G. (1999). Clinical and pathological overlap between frontotemporal dementia, primary progressive aphasia and corticobasal degeneration: The Pick complex. *Dementia and Geriatric Cognitive Disorders, 10*(Suppl. 1), 46–49. doi:10.1159/000051212 PMID:10436340

Kertesz, A., Morlog, D., Light, M., Blair, M., Davidson, W., Jesso, S., & Brashear, R. (2008). Galantamine in Frontotemporal Dementia and Primary Progressive Aphasia. *Dementia and Geriatric Cognitive Disorders, 25*(2), 178–185. doi:10.1159/000113034 PMID:18196898

Kesler, S., Hadi Hosseini, S. M., Heckler, C., Janelsins, M., Palesh, O., Mustian, K., & Morrow, G. (2013). Cognitive training for improving executive function in chemotherapy-treated breast cancer survivors. *Clinical Breast Cancer, 13*(4), 299–306. Advance online publication. doi:10.1016/j.clbc.2013.02.004 PMID:23647804

Kessels, R., Waanders-Oude Elferink, M., & van Tilborg, I. (2021). Social cognition and social functioning in patients with amnestic mild cognitive impairment or Alzheimer's dementia. *Journal of Neuropsychology, 15*(2), 186–203. doi:10.1111/jnp.12223 PMID:32979297

Khan, M. A., & Al-Jahdali, H. (2023). The consequences of sleep deprivation on cognitive performance. *Neurosciences (Riyadh), 28*(2), 91–99. doi:10.17712/nsj.2023.2.20220108 PMID:37045455

Kieburtz, K., Tilley, B. C., Elm, J. J., Babcock, D., Hauser, R., Ross, G. W., Augustine, A. H., Augustine, E. U., Aminoff, M. J., Bodis-Wollner, I. G., Boyd, J., Cambi, F., Chou, K., Christine, C. W., Cines, M., Dahodwala, N., Derwent, L., Dewey, R. B. Jr, Hawthorne, K., ... Wills, A.-M. (2015). Effect of creatine monohydrate on clinical progression in patients with Parkinson disease: A randomized clinical trial. *Journal of the American Medical Association, 313*(6), 584–593. doi:10.1001/jama.2015.120 PMID:25668262

Kim, E., Han, J., Choi, H., Prié, Y., Vigier, T., Bulteau, S., & Kwon, G. H. (2021). Examining the Academic Trends in Neuropsychological Tests for Executive Functions Using Virtual Reality: Systematic Literature Review. *JMIR Serious Games, 9*(4), e30249. doi:10.2196/30249 PMID:34822341

Kim, K. Y., Kim, E. H., Lee, M., Ha, J., Jung, I., & Kim, E. (2023). Restless leg syndrome and risk of all-cause dementia: A nationwide retrospective cohort study. *Alzheimer's Research & Therapy, 15*(1), 46. doi:10.1186/s13195-023-01191-z PMID:36879327

Kim, S. J., Lee, J. H., Lee, D. Y., Jhoo, J. H., & Woo, J. I. (2011). Neurocognitive dysfunction associated with sleep quality and sleep apnea in patients with mild cognitive impairment. *The American Journal of Geriatric Psychiatry, 19*(4), 374–381. doi:10.1097/JGP.0b013e3181e9b976 PMID:20808148

Kim, Y. J., Heo, J., Park, K. S., & Kim, S. (2016). Proposition of novel classification approach and features for improved real-time arrhythmia monitoring. *Computers in Biology and Medicine, 75*, 190–202. doi:10.1016/j.compbiomed.2016.06.009 PMID:27318329

Kinugawa, K. (2023, October). Obstructive sleep apnea and dementia: A role to play? *Revue Neurologique*, *179*(7), 793–803. doi:10.1016/j.neurol.2023.08.006 PMID:37633736

Kipps, C. M., Nestor, P. J., Acosta-Cabronero, J., Arnold, R., & Hodges, J. R. (2009). Understanding social dysfunction in the behavioural variant of frontotemporal dementia: The role of emotion and sarcasm processing. *Brain*, *132*(Pt 3), 592–603. doi:10.1093/brain/awn314 PMID:19126572

Kirkbride, J.B., Fearon, P., Morgan, C., Dazzan, P., Morgan, K., & Tarrant, J. (2006). Heterogeneity in incidence rates of schizophrenia and other psychotic syndromes: findings from the 3-center AeSOP study. *Arch Gen Psychiatry*, *63*(3), 250–8.

Kirkbride, J. B., Fearon, P., Morgan, C., Dazzan, P., Morgan, K., Murray, R. M., & Jones, P. B. (2007). Neighborhood variation in the incidence of psychotic disorders in Southeast London. *Social Psychiatry and Psychiatric Epidemiology*, *42*(6), 438–445. doi:10.1007/s00127-007-0193-0 PMID:17473901

Kirkbride, J. B., Fearon, P., Morgan, C., Dazzan, P., Morgan, K., Tarrant, J., Lloyd, T., Holloway, J., Hutchinson, G., Leff, J. P., Mallett, R. M., Harrison, G. L., Murray, R. M., & Jones, P. B. (2006). Heterogeneity in Incidence Rates of Schizophrenia and Other Psychotic Syndromes: Findings From the 3-Center ÆSOP Study. *Archives of General Psychiatry*, *63*(3), 250. doi:10.1001/archpsyc.63.3.250 PMID:16520429

Knecht, L., Folke, J., Dodel, R., Ross, J. A., & Albus, A. (2022). Alpha-synuclein Immunization Strategies for Synucleinopathies in Clinical Studies: A Biological Perspective. *Neurotherapeutics; the Journal of the American Society for Experimental NeuroTherapeutics*, *19*(5), 1489–1502. doi:10.1007/s13311-022-01288-7 PMID:36083395

Knibb, J. A., Xereb, J. H., Patterson, K., & Hodgges, J. R. (2006). Clinical and Pathological Characterization of Progressive Aphasia. *Annals of Neurology*, *59*(1), 156–165. doi:10.1002/ana.20700 PMID:16374817

Kober, H., Barrett, L. F., Joseph, J., Bliss-Moreau, E., Lindquist, K., & Wager, T. D. (2008). Functional grouping and cortical-subcortical interactions in emotion: A meta-analysis of neuroimaging studies. *NeuroImage*, *42*(2), 998–1031. doi:10.1016/j.neuroimage.2008.03.059 PMID:18579414

Kohli, S., Fisher, S. G., Tra, Y., Adams, M. J., Mapstone, M. E., Wesnes, K. A., Roscoe, J. A., & Morrow, G. R. (2009). The effect of modafinil on cognitive function in breast cancer survivors. *Cancer*, *115*(12), 2605–2616. doi:10.1002/cncr.24287 PMID:19309747

Kooij, J. J. S., Bijlenga, D., & Michielsen, M. (2019). Assessment and Treatment of ADHD in People Over 60. *The ADHD Report*, *27*(4), 1–7, 15–16. doi:10.1521/adhd.2019.27.4.1

Kooij, J. J. S., Bijlenga, D., Salerno, L., Jaeschke, R., Bitter, I., Balázs, J., Thome, J., Dom, G., Kasper, S., Nunes Filipe, C., Stes, S., Mohr, P., Leppämäki, S., Casas, M., Bobes, J., Mccarthy, J. M., Richarte, V., Kjems Philipsen, A., Pehlivanidis, A., ... Asherson, P. (2019). Updated European Consensus Statement on diagnosis and treatment of adult ADHD. *European Psychiatry*, *56*(1), 14–34. doi:10.1016/j.eurpsy.2018.11.001 PMID:30453134

Kooij, J. J. S., Michielsen, M., Kruithof, H., & Bijlenga, D. (2016). ADHD in old age: A review of the literature and proposal for assessment and treatment. *Expert Review of Neurotherapeutics*, *16*(12), 1371–1381. doi:10.1080/14737175.2016.1204914 PMID:27334252

Kooij, J. S., & Bijlenga, D. (2013). The circadian rhythm in adult attention-deficit/hyperactivity disorder: Current state of affairs. *Expert Review of Neurotherapeutics*, *13*(10), 1107–1116. doi: 10.1586/14737175.2013.836301 PMID:24117273

Krüger, R., Kuhn, W., Müller, T., Woitalla, D., Graeber, M., Kösel, S., Przuntek, H., Epplen, J. T., Schöls, L., & Riess, O. (1998). Ala30Pro mutation in the gene encoding alpha-synuclein in Parkinson's disease. *Nature Genetics*, *18*(2), 106–108. doi:10.1038/ng0298-106 PMID:9462735

Kucherer, S., & Ferguson, R. J. (2017). Cognitive behavioral therapy for cancer-related cognitive dysfunction. *Current Opinion in Supportive and Palliative Care*, *11*(1), 46–51. doi:10.1097/SPC.0000000000000247 PMID:27898511

Kuhtz-Buschbeck, J. P., Schaefer, J., Wilder, N., & Wolze, W. T. (2020). The origin of the heartbeat and theories of muscle contraction. Physiological concepts and conflicts in the 19th century. *Progress in Biophysics and Molecular Biology*, *159*, 3–9. doi:10.1016/j.pbiomolbio.2020.05.009 PMID:32622835

Kumfor, F., Irish, M., Hodges, J. R., & Piguet, O. (2013). Discrete Neural Correlates for the Recognition of Negative Emotions: Insights from Frontotemporal Dementia. *PLoS One*, *8*(6), 457. doi:10.1371/journal.pone.0067457 PMID:23805313

Kunz, D., Stotz, S., de Zeeuw, J., Papakonstantinou, A., Dümchen, S., Haberecht, M., Plotkin, M., & Bes, F. (2023). Prognostic biomarkers in prodromal α-synucleinopathies: DAT binding and REM sleep without atonia. *Journal of Neurology, Neurosurgery, and Psychiatry*, *94*(7), 532–540. doi:10.1136/jnnp-2022-330048 PMID:36725328

Laisney, M., Bon, L., Guiziou, C., Daluzeau, N., Eustache, F., & Desgranges, B. (2013). Cognitive and affective theory of mind in mild to moderate Alzheimer's disease. *Journal of Neuropsychology*, *7*(1), 107–120. doi:10.1111/j.1748-6653.2012.02038.x PMID:23088554

Lally, J., Ajnakina, O., Stubbs, B., Cullinane, M., Murphy, K. C., & Gaughran, F. (2017). Remission and recovery from first-episode psychosis in adults: Systematic review and meta-analysis of long-term outcome studies. *Br J Psychiatry J Ment Sci. Dez*, *211*(6), 350–358. PMID:28982659

Lang, A. E., Siderowf, A. D., Macklin, E. A., Poewe, W., Brooks, D. J., Fernandez, H. H., Rascol, O., Giladi, N., Stocchi, F., Tanner, C. M., Postuma, R. B., Simon, D. K., Tolosa, E., Mollenhauer, B., Cedarbaum, J. M., Fraser, K., Xiao, J., Evans, K. C., Graham, D. L., ... Dam, T. (2022). Trial of Cinpanemab in Early Parkinson's Disease. *The New England Journal of Medicine*, *387*(5), 408–420. doi:10.1056/NEJMoa2203395 PMID:35921450

Lange, M., Joly, F., Vardy, J., Ahles, T., Dubois, M., Tron, L., Winocur, G., De Ruiter, M. B., & Castel, H. (2019). Cancer-related cognitive impairment: An update on state of the art, detection, and management strategies in cancer survivors. *Annals of Oncology : Official Journal of the European Society for Medical Oncology*, *30*(12), 1925–1940. doi:10.1093/annonc/mdz410 PMID:31617564

Langston, J. W. (2017). The MPTP Story. *Journal of Parkinson's Disease*, *7*(s1), S11–S19. doi:10.3233/JPD-179006 PMID:28282815

Lansford, J. E., Sharma, C., Malone, P. S., Woodlief, D., Dodge, K. A., Oburu, P., Pastorelli, C., Skinner, A. T., Sorbring, E., Tapanya, S., Tirado, L. M., Zelli, A., Al-Hassan, S. M., Alampay, L. P., Bacchini, D., Bombi, A. S., Bornstein, M. H., Chang, L., Deater-Deckard, K., & Di Giunta, L. (2014). Corporal punishment, maternal warmth, and child adjustment: a longitudinal study in eight countries. *Journal of Clinical Child and Adolescent Psychology*, *43*(4), 670–685. doi:10.1080/15374416.2014.893518

Lappin, J.M., Heslin, M., Lomas, B., Jones, P.B., Doody, G.A., & Reininghaus, U.A. (2019). Early sustained recovery following first episode psychosis: Evidence from the AESOP10 follow-up study. *Schizophr Res.*, *199*, 341–5.

Lawrence, J. A., Griffin, L., Balcueva, E. P., Groteluschen, D. L., Samuel, T. A., Lesser, G. J., Naughton, M. J., Case, L. D., Shaw, E. G., & Rapp, S. R. (2016). A study of donepezil in female breast cancer survivors with self-reported cognitive dysfunction 1 to 5 years following adjuvant chemotherapy. *Journal of Cancer Survivorship: Research and Practice*, *10*(1), 176–184. doi:10.1007/s11764-015-0463-x PMID:26130292

Lawton, M. P., & Brody, E. M. (1969). Assessment of older people: Self-maintaining and instrumental activities of daily living. *The Gerontologist*, *9*(3), 179–186. doi:10.1093/geront/9.3_Part_1.179 PMID:5349366

Leary, E. B., Moore, H. E. IV, Schneider, L. D., Finn, L. A., Peppard, P. E., & Mignot, E. (2018). Periodic limb movements in sleep: Prevalence and associated sleepiness in the Wisconsin Sleep Cohort. *Clinical Neurophysiology*, *129*(11), 2306–2314. doi:10.1016/j.clinph.2018.08.022 PMID:30243181

Lebrun, Y., & Buyssens, E. (1982). Metalanguage and speech pathology. *The British Journal of Disorders of Communication*, *17*(1), 21–25. doi:10.3109/13682828209011404 PMID:7044405

Lee, T.Y., Lee, J., Kim, M., Choe, E., & Kwon, J.S. (2018). Can We Predict Psychosis Outside the Clinical High-Risk State? A Systematic Review of Non-Psychotic Risk Syndromes for Mental Disorders. *Schizophr Bull.*, *44*(2), 276–85.

Lees, A. J., Tolosa, E., & Olanow, C. W. (2015). Four pioneers of L-dopa treatment: Arvid Carlsson, Oleh Hornykiewicz, George Cotzias, and Melvin Yahr. *Movement Disorders*, *30*(1), 19–36. doi:10.1002/mds.26120 PMID:25488030

Lee, V. M., Goedert, M., & Trojanowski, J. Q. (2001). Neurodegenerative tauopathies. *Annual Review of Neuroscience*, *24*(1), 1121–1159. doi:10.1146/annurev.neuro.24.1.1121 PMID:11520930

Lee, W. H., Doucet, G. E., Leibu, E., & Frangou, S. (2018). Resting-state network connectivity and metastability predict clinical symptoms in schizophrenia. *Schizophrenia Research, 201*, 208–216. doi:10.1016/j.schres.2018.04.029 PMID:29709491

Lensing, M. B., Zeiner, P., Sandvik, L., & Opjordsmoen, S. (2015). Quality of Life in Adults Aged 50+ With ADHD. *Journal of Attention Disorders, 19*(5), 405–413. doi:10.1177/1087054713480035 PMID:23514676

Lepre, R. M. & De Oliveira, J. (2022). Adolescência e construção da personalidade moral. *Dialogia,* e21333 . doi:10.5585/41.2022.21333

Lesage, S., Lunati, A., Houot, M., Romdhan, S. B., Clot, F., Tesson, C., Mangone, G., Toullec, B. L., Courtin, T., Larcher, K., Benmahdjoub, M., Arezki, M., Bouhouche, A., Anheim, M., Roze, E., Viallet, F., Tison, F., Broussolle, E., Emre, M., ... Brice, A. (2020). Characterization of Recessive Parkinson Disease in a Large Multicenter Study. *Annals of Neurology, 88*(4), 843–850. doi:10.1002/ana.25787 PMID:33045815

Levine, S. Z., Rotstein, A., Kodesh, A., Sandin, S., Lee, B. K., Weinstein, G., Schnaider Beeri, M., & Reichenberg, A. (2023). Adult Attention-Deficit/Hyperactivity Disorder and the Risk of Dementia. *JAMA Network Open, 6*(10), e2338088. doi:10.1001/jamanetworkopen.2023.38088 PMID:37847497

Leyton, C. E., Hsieh, S., Mioshi, E., & Hodges, J. R. (2013). Cognitive decline in logopenic aphasia: More than losing words. *Neurology, 80*(10), 897–93. doi:10.1212/WNL.0b013e318285c15b PMID:23390170

Libon, D. J., Xie, S. X., Wang, X., Massimo, L., Moore, P., Vesely, L., Khan, A., Chatterjee, A., Coslett, H. B., Hurtig, H. I., Liang, T. W., & Grossman, M. (2009). Neuropsychological decline in frontotemporal lobar degeneration: A longitudinal analysis. *Neuropsychology, 23*(3), 337–346. doi:10.1037/a0014995 PMID:19413447

Lieberman, J. (1993). Time Course and Biologic Correlates of Treatment Response in First-Episode Schizophrenia. *Arch Gen Psychiatry, 50*(5), 369.

Lieberman, J. A., Perkins, D., Belger, A., Chakos, M., Jarskog, F., Boteva, K., & Gilmore, J. (2001). The early stages of schizophrenia: Speculations on pathogenesis, pathophysiology, and therapeutic approaches. *Biological Psychiatry, 50*(11), 884–897. doi:10.1016/S0006-3223(01)01303-8 PMID:11743943

Lima, N.C., Kirov, R., & de Almondes, K.M. (2022). Impairment of executive functions due to sleep alterations: An integrative review on the use of P300. *Frontiers in Neuroscience, 22*, 16:906492. . doi:10.3389/fnins.2022.906492

Lindquist, K. A., & Barrett, L. F. (2012). A functional architecture of the human brain: Emerging insights from the science of emotion. *Trends in Cognitive Sciences, 16*(11), 533–540. doi:10.1016/j.tics.2012.09.005 PMID:23036719

Lipner, E., O'Brien, K. J., Pike, M. R., Ered, A., & Ellman, L. M. (2022). Environmental Risk Factors and Cognitive Outcomes in Psychosis: Pre, Perinatal, and Early Life Adversity. In D. M. Barch & J. W. Young (Eds.), *Cognitive Functioning in Schizophrenia: Leveraging the RDoC Framework* (Vol. 63, pp. 205–240). Springer International Publishing. doi:10.1007/7854_2022_378

Li, Q., Rajagopalan, C., & Clifford, G. D. (2014). A machine learning approach to multi-level ecg signal quality classification. *Computer Methods and Programs in Biomedicine, 117*(3), 435–447. doi:10.1016/j.cmpb.2014.09.002 PMID:25306242

Li, S., & Le, W. (2017). Milestones of Parkinson's Disease Research: 200 Years of History and Beyond. *Neuroscience Bulletin, 33*(5), 598–602. doi:10.1007/s12264-017-0178-2 PMID:28895075

Liu, L., Yang, J., Wang, J., Nie, M., Wang, Z., Guan, H., Hu, J., & Hong, F. (2022). Relationship between Sleep Duration and Stroke History in Middle-Aged and Elderly in Guiyang: A Cross-Sectional Survey. *Neuropsychiatric Disease and Treatment, 18*, 243–252. doi:10.2147/NDT.S340834 PMID:35185332

Livingston, G., Huntley, J., Sommerlad, A., Ames, D., Ballard, C., Banerjee, S., Brayne, C., Burns, A., Cohen-Mansfield, J., Cooper, C., Costafreda, S. G., Dias, A., Fox, N., Gitlin, L. N., Howard, R., Kales, H. C., Kivimäki, M., Larson, E. B., Ogunniyi, A., ... Mukadam, N. (2020). Dementia prevention, intervention, and care: 2020 report of the Lancet Commission. *Lancet, 396*(10248), 413–446. doi:10.1016/S0140-6736(20)30367-6 PMID:32738937

Lo Buglio, G., Pontillo, M., Cerasti, E., Polari, A., Schiano Lomoriello, A., Vicari, S., Lingiardi, V., Boldrini, T., & Solmi, M. (2022). A network analysis of anxiety, depressive, and psychotic symptoms and functioning in children and adolescents at clinical high risk for psychosis. *Front Psychiatry, 28*. . doi:10.3389/fpsyt.2022.1016154

Lobo, L. C. (2018). Inteligência artificial, o Futuro da Medicina e a Educação Médica. *Revista Brasileira de Educação Médica, 42*(3), 3–8. doi:10.1590/1981-52712015v42n3rb20180115editorial1

Loewy, R. L., Corey, S., Amirfathi, F., Dabit, S., Fulford, D., Pearson, R., Hua, J. P. Y., Schlosser, D., Stuart, B. K., Mathalon, D. H., & Vinogradov, S. (2019). Childhood trauma and clinical high risk for psychosis. *Schizophrenia Research, 205*, 10–14. doi:10.1016/j.schres.2018.05.003 PMID:29779964

Loh, K. P., Pandya, C., Zittel, J., Kadambi, S., Flannery, M., Reizine, N., Magnuson, A., Braganza, G., Mustian, K., Dale, W., Duberstein, P., & Mohile, S. G. (2017). Associations of sleep disturbance with physical function and cognition in older adults with cancer. *Supportive Care in Cancer, 25*(10), 3161–3169. doi:10.1007/s00520-017-3724-6 PMID:28455547

Lunardini, F., Luperto, M., Romeo, M., Basilico, N., Daniele, K., Azzolino, D., Damanti, S., Abbate, C., Mari, D., Cesari, M., Borghese, N. A., & Ferrante, S. (2020, September 21). Supervised Digital Neuropsychological Tests for Cognitive Decline in Older Adults: Usability and Clinical Validity Study. *JMIR mHealth and uHealth, 8*(9), e17963. doi:10.2196/17963 PMID:32955442

Lutsey, P. L., Misialek, J. R., Mosley, T. H., Gottesman, R. F., Punjabi, N. M., Shahar, E., MacLehose, R., Ogilvie, R. P., Knopman, D., & Alonso, A. (2018). Sleep characteristics and risk of dementia and Alzheimer's disease: The Atherosclerosis Risk in Communities Study. *Alzheimer's & Dementia*, *14*(2), 157–166. doi:10.1016/j.jalz.2017.06.2269 PMID:28738188

Lysen, T. S., Darweesh, S. K. L., Ikram, M. K., Luik, A. I., & Ikram, M. A. (2019). Sleep and risk of parkinsonism and Parkinson's disease: A population-based study. *Brain*, *142*(7), 2013–2022. doi:10.1093/brain/awz113 PMID:31038176

Maan, A., Mansour, M., Ruskin, J. N., & Heist, E. K. (2014). Impact of catheter ablation on p-wave parameters on 12-lead electrocardiogram in patients with atrial fibrillation. *Journal of Electrocardiology*, *47*(5), 725–733. doi:10.1016/j.jelectrocard.2014.04.010 PMID:24850319

Machulda, M.M., Whitwell, J.L., & Duffy, J.R. (2013). *Identification of an atypical variant of logopenic progressive aphasia.* Academic Press.

Macoir, J., Martel-Sauvageau, V., Bouvier, L., Laforce, R., & Monetta, L. (2021). Heterogeneity of repetition abilities in logopenic variant primary progressive aphasia. *Dementia & Neuropsychologia*, *15*(3), 405–412. doi:10.1590/1980-57642021dn15-030014 PMID:34630930

Magee, C., Oberle, E., Guhn, M., Gadermann, A., & Puyat, J. H. (2023). Risk of Diagnosed Adolescent-Onset Non-Affective Psychotic Disorder by Migration Background in British Columbia: A Retrospective Cohort Study. *Canadian Journal of Psychiatry*, *68*(1), 33–42. doi:10.1177/07067437221100351 PMID:35698751

Malaspina, D., Harlap, S., Fennig, S., Heiman, D., Nahon, D., Feldman, D., & Susser, E. S. (2001). Advancing Paternal Age and the Risk of Schizophrenia. *Archives of General Psychiatry*, *58*(4), 361. doi:10.1001/archpsyc.58.4.361 PMID:11296097

Malinowski, F.R., Tasso, B. de C., Ortiz, B.B., Higuchi, C.H., Noto, C., & Belangero, S.I. (1999). Schneider's first-rank symptoms as predictors of remission in antipsychotic-naive first episode psychosis. *Rev Bras Psiquiatria Sao Paulo Braz*, *42*(1), 22–6.

Malla, A. K., Norman, R. M. G., & Joober, R. (2005). First-episode psychosis, early intervention, and outcome: What have we learned? *Canadian Journal of Psychiatry*, *50*(14), 881–891. doi:10.1177/070674370505001402 PMID:16494257

Mancopes, R., & Schultz, F. (2008). Processing of metaphors in transcortical motor aphasia. *Dementia & Neuropsychologia*, *2*(4), 339–348. doi:10.1590/S1980-57642009DN20400019 PMID:29213596

Mansur, A. d. P., & Favarato, D. A. (2016). Trends in Mortality Rate from Cardiovascular Disease in Brazil, 1980-2012. *Arquivos Brasileiros de Cardiologia, 107*, 20 – 25.

Marasca, A., Yates, D., Schneider, A., Feijó, L., & Bandeira, D. (2020). Avaliação psicológica online: Considerações a partir da pandemia do novo coronavírus (COVID-19) para a prática e o ensino no contexto a distância. *Estudos de Psicologia*, *37*, e200085. doi:10.1590/1982-0275202037e200085

Margarido, I. T. M. M. (2020). *A Medicina do Futuro nos Dias de Hoje: Aplicações da Inteligência Artificial* [Doctoral dissertation]. Universidade de Lisboa.

Marras, C., Lang, A., van de Warrenburg, B. P., Sue, C. M., Tabrizi, S. J., Bertram, L., Mercimek-Mahmutoglu, S., Ebrahimi-Fakhari, D., Warner, T. T., Durr, A., Assmann, B., Lohmann, K., Kostic, V., & Klein, C. (2016). Nomenclature of genetic movement disorders: Recommendations of the international Parkinson and movement disorder society task force. *Movement Disorders, 31*(4), 436–457. doi:10.1002/mds.26527 PMID:27079681

Marras, C., Lohmann, K., Lang, A., & Klein, C. (2012). Fixing the broken system of genetic locus symbols. *Neurology, 78*(13), 1016–1024. doi:10.1212/WNL.0b013e31824d58ab PMID:22454269

Marschark, M., & Nall, L. (1985). Metaphoric competence in cognitive and language development. *Advances in Child Development and Behavior, 19*, 49–81. doi:10.1016/S0065-2407(08)60388-2 PMID:2418642

Marshall, C. R., Hardy, C. J. D., Volkmer, A., Russell, L. L., Bond, R. L., Fletcher, P. D., Clark, C. N., Mummery, C. J., Schott, J. M., Rossor, M. N., Fox, N. C., Crutch, S. J., Rohrer, J. D., & Warren, J. D. (2018). Primary progressive aphasia: A clinical approach. *Journal of Neurology, 265*(6), 1474–1490. doi:10.1007/s00415-018-8762-6 PMID:29392464

Marshall, M., Lewis, S., Lockwood, A., Drake, R., Jones, P., & Croudace, T. (2005). Association Between Duration of Untreated Psychosis and Outcome in Cohorts of First-Episode Patients: A Systematic Review. *Archives of General Psychiatry, 62*(9), 975–983. doi:10.1001/archpsyc.62.9.975 PMID:16143729

Martis, R. J., Acharya, U. R., Mandana, K., Ray, A., & Chakraborty, C. (2013). Cardiac decision making using higher order spectra. *Biomedical Signal Processing and Control, 8*(2), 193–203. doi:10.1016/j.bspc.2012.08.004

Masliah, E., Rockenstein, E., Adame, A., Alford, M., Crews, L., Hashimoto, M., Seubert, P., Lee, M., Goldstein, J., Chilcote, T., Games, D., & Schenk, D. (2005). Effects of alpha-synuclein immunization in a mouse model of Parkinson's disease. *Neuron, 46*(6), 857–868. doi:10.1016/j.neuron.2005.05.010 PMID:15953415

Massetti, G., Albini, F., Casati, C., Toneatto, C., Terruzzi, S., Etzi, R., Tesio, L., Gallace, A., & Vallar, G. (2023). Validation of "Neurit.Space": Three Digital Tests for the Neuropsychological Evaluation of Unilateral Spatial Neglect. *Journal of Clinical Medicine, 12*(8), 3042. doi:10.3390/jcm12083042 PMID:37109378

Mata, I. F., Lockhart, P. J., & Farrer, M. J. (2004). Parkin genetics: One model for Parkinson's disease. *Human Molecular Genetics, 13*(Spec No 1), R127–R133. doi:10.1093/hmg/ddh089 PMID:14976155

Mathers, C. D., & Loncar, D. (2006). Projections of global mortality and burden of disease from 2002 to 2030. *PLOS Medicine, 3*(11), 1–20. doi:10.1371/journal.pmed.0030442

Mazaheri, V., & Khodadadi, H. (2020). Heart arrhythmia diagnosis based on the combination of morphological, frequency and nonlinear features of ecg signals and metaheuristic feature selection algorithm. *Expert Systems with Applications*, *161*, 113697. doi:10.1016/j.eswa.2020.113697

McCleery, A., & Nuechterlein, K. H. (2019). Cognitive impairment in psychotic illness: Prevalence, profile of impairment, developmental course, and treatment considerations. *Dialogues in Clinical Neuroscience*, *21*(3), 239–248. doi:10.31887/DCNS.2019.21.3/amccleery PMID:31749648

McCracken, K., & Phillips, D. R. (2017). Demographic and Epidemiological Transition. *International Encyclopedia of Geography: People, the Earth. Environmental Technology*, •••, 1–8. doi:10.1002/9781118786352.wbieg0063

McCutcheon, R. A., Reis Marques, T., & Howes, O. D. (2020). Schizophrenia—An Overview. *JAMA Psychiatry*, *77*(2), 201. doi:10.1001/jamapsychiatry.2019.3360 PMID:31664453

McGorry, P. D., Nelson, B., Phillips, L. J., Yuen, H. P., Francey, S. M., Thampi, A., Berger, G. E., Amminger, G. P., Simmons, M. B., Kelly, D., Dip, G., Thompson, A. D., & Yung, A. R. (2013). Randomized controlled trial of interventions for young people at ultra-high risk of psychosis: Twelve-month outcome. *The Journal of Clinical Psychiatry*, *74*(4), 349–356. doi:10.4088/JCP.12m07785 PMID:23218022

Mendonca, F., Sudo, F. K., Santiago-Bravo, G., Oliveira, N., Assuncao, N., Rodrigues, F., Soares, R., Calil, V., Bernardes, G., Erthal, P., Drummond, C., Tovar-Moll, F., & Mattos, P. (2021). Mild Cognitive Impairment or Attention-Deficit/Hyperactivity Disorder in Older Adults? A Cross Sectional Study. *Frontiers in Psychiatry*, *12*, 737357. doi:10.3389/fpsyt.2021.737357 PMID:34616321

Menezes, P. R., Scazufca, M., Busatto, G. F., Coutinho, L. M., Mcguire, P. K., & Murray, R. M. (2007). Incidence of first-contact psychosis in São Paulo, Brazil. *The British Journal of Psychiatry*, *191*(S51), s102–s106. Advance online publication. doi:10.1192/bjp.191.51.s102 PMID:18055924

Mesholam-Gately, R. I., Giuliano, A. J., Goff, K. P., Faraone, S. V., & Seidman, L. J. (2009). Neurocognition in first-episode schizophrenia: A meta-analytic review. *Neuropsychology*, *23*(3), 315–336. doi:10.1037/a0014708 PMID:19413446

Mesulam, M. M. (1982). Slowly progressive aphasia without generalized dementia. *Annals of Neurology*, *11*(6), 592–598. doi:10.1002/ana.410110607 PMID:7114808

Mesulam, M. M. (2001). Article. *Primary Progressive Aphasia*, *49*, 425–432.

Mesulam, M. M. (2013). Primary progressive aphasia: A dementia of the language network. *Dementia & Neuropsychologia*, *7*(1), 2–9. doi:10.1590/S1980-57642013DN70100002 PMID:24707349

Mesulam, M. M., Weintraub, S., Rogalski, E. J., Wieneke, C., Geula, C., & Bigio, E. H. (2014). Asymmetry and heterogeneity of Alzheimer's and frontotemporal pathology in primary progressive aphasia. *Brain*, *137*(4), 1176–1192. doi:10.1093/brain/awu024 PMID:24574501

Mesulam, M. M., Wieneke, C., Cynthia Thompson, C., Rogalski, E., & Weintraub, S. (2012). Quantitative classification of primary progressive aphasia at early and mild impairment stages. *Brain*, *135*(5), 1537–1553. doi:10.1093/brain/aws080 PMID:22525158

Mesulam, M., Coventry, C., Bigio, E. H., Geula, C., Thompson, C., Bonakdarpour, B., Gefen, T., Rogalski, E. J., & Weintraub, S. (2021). Nosology of Primary Progressive Aphasia and the Neuropathology of Language. *Advances in Experimental Medicine and Biology*, *1281*, 33–49. doi:10.1007/978-3-030-51140-1_3 PMID:33433867

Michielsen, M., Comijs, H. C., Aartsen, M. J., Semeijn, E. J., Beekman, A. T. F., Deeg, D. J. H., & Kooij, J. J. S. (2015). The Relationships Between ADHD and Social Functioning and Participation in Older Adults in a Population-Based Study. *Journal of Attention Disorders*, *19*(5), 368–379. doi:10.1177/1087054713515748 PMID:24378286

Michielsen, M., De Kruif, J. Th. C. M., Comijs, H. C., Van Mierlo, S., Semeijn, E. J., Beekman, A. T. F., Deeg, D. J. H., & Kooij, J. J. S. (2018). The Burden of ADHD in Older Adults: A Qualitative Study. *Journal of Attention Disorders*, *22*(6), 591–600. doi:10.1177/1087054715610001 PMID:26515893

Michielsen, M., Semeijn, E., Comijs, H. C., Van De Ven, P., Beekman, A. T. F., Deeg, D. J. H., & Kooij, J. J. S. (2012). Prevalence of attention-deficit hyperactivity disorder in older adults in the Netherlands. *The British Journal of Psychiatry*, *201*(4), 298–305. doi:10.1192/bjp.bp.111.101196 PMID:22878132

Miglis, M. G., Adler, C. H., Antelmi, E., Arnaldi, D., Baldelli, L., Boeve, B. F., Cesari, M., Dall'Antonia, I., Diederich, N. J., Doppler, K., Dušek, P., Ferri, R., Gagnon, J.-F., Gan-Or, Z., Hermann, W., Högl, B., Hu, M. T., Iranzo, A., Janzen, A., ... Oertel, W. H. (2021). Biomarkers of conversion to α-synucleinopathy in isolated rapid-eye-movement sleep behaviour disorder. *Lancet Neurology*, *20*(8), 671–684. doi:10.1016/S1474-4422(21)00176-9 PMID:34302789

Miller, I. N., & Cronin-Golomb, A. (2010). Gender differences in Parkinson's disease: Clinical characteristics and cognition. *Movement Disorders*, *25*(16), 2695–2703. doi:10.1002/mds.23388 PMID:20925068

Miner, B., & Kryger, M. H. (2017). Sleep in the Aging Population. *Sleep Medicine Clinics*, *12*(1), 31–38. doi:10.1016/j.jsmc.2016.10.008 PMID:28159095

Miotto, E. C., Lúcia, M. C. S., & Scaff, M. (2012). *Neuropsicologia Clínica*. Roca.

Molenberghs, P., Cunnington, R., & Mattingley, J. B. (2012). Brain regions with mirror properties: A meta-analysis of 125 human fMRI studies. *Neuroscience and Biobehavioral Reviews*, *36*(1), 341–349. doi:10.1016/j.neubiorev.2011.07.004 PMID:21782846

Moody, G. B., & Mark, R. G. (1983). *A new method for detecting atrial fibrillation using R-R intervals*. Academic Press.

Moody, G. B., Peng, C.-K., & Stanley, H. E. (2000b). Physiobank, physiotoolkit, and physionet. *Circulation*, *101*(23), e215–e220. PMID:10851218

Moreno-Küstner, B., Martín, C., & Pastor, L. (2018) Prevalence of psychotic disorders and its association with methodological issues. A systematic review and meta-analyses. *PLoS One, 13*(4), e0195687. . doi:10.1371/journal.pone.0195687

Moretti, R., Torre, P., Antonello, R. M., Cattaruzza, T., Cazzato, G., & Bava, A. (2004). Rivastigmine in Frontotemporal Dementia. *Drugs & Aging, 21*(14), 931–937. doi:10.2165/00002512-200421140-00003 PMID:15554751

Morgan, C., Lappin, J., Heslin, M., Donoghue, K., Lomas, B., & Reininghaus, U. (2014). Reappraising the long-term course and outcome of psychotic disorders: The AESOP-10 study. *Psychol Med., 44*(13), 2713–26.

Morgan, E., Schumm, L. P., McClintock, M., Waite, L., & Lauderdale, D. S. (2017). Sleep characteristics and daytime cortisol levels in older adults. *Sleep (Basel), 40*(5), zsx043. Advance online publication. doi:10.1093/sleep/zsx043 PMID:28329370

Morrison, P. D., Zois, V., McKeown, D. A., Lee, T. D., Holt, D. W., Powell, J. F., Kapur, S., & Murray, R. M. (2009). The acute effects of synthetic intravenous Δ9-tetrahydrocannabinol on psychosis, mood and cognitive functioning. *Psychological Medicine, 39*(10), 1607. doi:10.1017/S0033291709005522 PMID:19335936

Morton, J., & Frith, U. (1995). Causal modelling: a structural approach to developmental psychopathology. In D. Cicchetti & D. Cohen (Eds.), *Developmental Psychopathology* (pp. 357–390). Wiley.

Müller, M., Turner, D., Barra, S., Rösler, M., & Retz, W. (2023). ADHD and associated psychopathology in older adults in a German community sample. *Journal of Neural Transmission (Vienna, Austria), 130*(3), 313–323. doi:10.1007/s00702-022-02584-4 PMID:36617378

Murray, R. M., Paparelli, A., Morrison, P. D., Marconi, A., & Di Forti, M. (2013). What can we learn about schizophrenia from studying the human model, drug-induced psychosis? *American Journal of Medical Genetics. Part B, Neuropsychiatric Genetics, 162*(7), 661–670. doi:10.1002/ajmg.b.32177 PMID:24132898

Nakamura, K. (2022). Information seeking criteria: Artificial intelligence, economics, psychology, and neuroscience. *Reviews in the Neurosciences, 33*(1), 31–41. doi:10.1515/revneuro-2020-0137 PMID:33855841

Nasreddine, Z. S., Phillips, N. A., Bédirian, V., Charbonneau, S., Whitehead, V., Collin, I., Cummings, J. L., & Chertkow, H. (2005). The Montreal Cognitive Assessment, MoCA: A Brief Screening Tool For Mild Cognitive Impairment. *Journal of the American Geriatrics Society, 53*(4), 695–699. doi:10.1111/j.1532-5415.2005.53221.x PMID:15817019

National Centre of Excellence in Youth Mental Health. (2016). *Early Psychosis Guidelines Writing Group and EPPIC National Support Program, Australian Clinical Guidelines for Early Psychosis* (2nd ed.). Orygen.

Neary, D., Snowden, J. S., Gustafson, L., Passant, U., Stuss, D., Black, S., Freedman, M., Kertesz, A., Robert, P. H., Albert, M., Boone, K., Miller, B. L., Cummings, J., & Benson, D. F. (1998). Frontotemporal lobar degeneration: A consensus on clinical diagnostic criteria. *Neurology*, *51*(6), 1546–1554. doi:10.1212/WNL.51.6.1546 PMID:9855500

Nishioka, K., Hayashi, S., Farrer, M. J., Singleton, A. B., Yoshino, H., Imai, H., Kitami, T., Sato, K., Kuroda, R., Tomiyama, H., Mizoguchi, K., Murata, M., Toda, T., Imoto, I., Inazawa, J., Mizuno, Y., & Hattori, N. (2006). Clinical heterogeneity of alpha-synuclein gene duplication in Parkinson's disease. *Annals of Neurology*, *59*(2), 298–309. doi:10.1002/ana.20753 PMID:16358335

Nitrini, R., Brucki, S. M. D., Yassuda, M. S., Fichman, H. C., & Caramelli, P. (2021). The Figure Memory Test: Diagnosis of memory impairment in populations with heterogeneous educational background. *Dementia & Neuropsychologia*, *15*(2), 173–185. doi:10.1590/1980-57642021dn15-020004 PMID:34345358

Nyström, A., Petersson, K., & Janlöv, A.-C. (2020). Being Different but Striving to Seem Normal: The Lived Experiences of People Aged 50+ with ADHD. *Issues in Mental Health Nursing*, *41*(6), 476–485. doi:10.1080/01612840.2019.1695029 PMID:32267788

O'Neill, A., Mechelli, A., & Bhattacharyya, S. (2019). Dysconnectivity of Large-Scale Functional Networks in Early Psychosis: A Meta-analysis. *Schizophrenia Bulletin*, *45*(3), 579–590. doi:10.1093/schbul/sby094 PMID:29982729

Obeso, J. A., Stamelou, M., Goetz, C. G., Poewe, W., Lang, A. E., Weintraub, D., Burn, D., Halliday, G. M., Bezard, E., Przedborski, S., Lehericy, S., Brooks, D. J., Rothwell, J. C., Hallett, M., DeLong, M. R., Marras, C., Tanner, C. M., Ross, G. W., Langston, J. W., ... Stoessl, A. J. (2017). Past, present, and future of Parkinson's disease: A special essay on the 200th Anniversary of the Shaking Palsy. *Movement Disorders*, *32*(9), 1264–1310. doi:10.1002/mds.27115 PMID:28887905

Oh, S. Y., Jeong, E., Pahk, K., Cheon, G. J., Park, K. H., & Choe, J. G. (2012). FDG PET-CT of chemobrain: An SPM analysis in patients with breast cancer before and after neoadjuvant chemotherapy. *Journal of Nuclear Medicine*, *53*(supplement 1), 1922–1922. https://jnm.snmjournals.org/cgi/content/short/53/supplement_1/1922

Oliveira, L., & Junior, L. (2020). Aplicabilidade da inteligência artificial na psiquiatria: Uma revisão de ensaios clínicos. *Revista Debates em Psiquiatria*, *10*(1), 14–25. doi:10.25118/2236-918X-10-1-2

Oliver, D., Davies, C., Zelaya, F., Selvaggi, P., De Micheli, A., Catalan, A., Baldwin, H., Arribas, M., Modinos, G., Crossley, N. A., Allen, P., Egerton, A., Jauhar, S., Howes, O. D., McGuire, P., & Fusar-Poli, P. (2023). Parsing neurobiological heterogeneity of the clinical high-risk state for psychosis: A pseudo-continuous arterial spin labelling study. *Frontiers in Psychiatry*, *14*, 1092213. doi:10.3389/fpsyt.2023.1092213 PMID:36970257

Onwumere, J., Bebbington, P., & Kuipers, E. (2011). Family interventions in early psychosis: Specificity and effectiveness. *Epidemiology and Psychiatric Sciences*, *20*(2), 113–119. doi:10.1017/S2045796011000187 PMID:21714356

Osher, J. E., Wicklund, A. H., Rademaker, A., Johnson, N., & Weintraub, S. (2007). The mini-mental state examination in behavioral variant frontotemporal dementia and primary progressive aphasia. *American Journal of Alzheimer's Disease and Other Dementias, 22*(6), 468–473. doi:10.1177/1533317507307173 PMID:18166606

Ossenkoppele, R., Jansen, W. J., Rabinovici, G. D., Knol, D. L., van der Flier, W. M., van Berckel, B. N. M., Scheltens, P., Visser, P. J., Verfaillie, S. C. J., Zwan, M. D., Adriaanse, S. M., Lammertsma, A. A., Barkhof, F., Jagust, W. J., Miller, B. L., Rosen, H. J., Landau, S. M., Villemagne, V. L., Rowe, C. C., ... Brooks, D. J. (2015). Prevalence of amyloid PET positivity in dementia syndromes: A meta-analysis. *Journal of the American Medical Association, 313*(19), 1939–1949. doi:10.1001/jama.2015.4669 PMID:25988463

Ota, S., Kanno, S., Morita, A., Narita, W., Kawakami, N., Kakinuma, K., Saito, Y., Kobayashi, E., Baba, T., Iizuka, O., Nishio, Y., Matsuda, M., Odagiri, H., Endo, K., Takanami, K., Mori, E., & Suzuki, K. (2021). Echolalia in patients with primary progressive aphasia. *European Journal of Neurology, 28*(4), 1113–1122. doi:10.1111/ene.14673 PMID:33305428

Ourry, V., Rehel, S., André, C., Mary, A., Paly, L., Delarue, M., Requier, F., Hendy, A., Collette, F., Marchant, N. L., Felisatti, F., Palix, C., Vivien, D., de la Sayette, V., Chételat, G., Gonneaud, J., & Rauchs, G. (2023). Medit-Ageing Research Group. Effect of cognitive reserve on the association between slow wave sleep and cognition in community-dwelling older adults. *Aging (Albany NY), 15*(18), 9275–9292. doi:10.18632/aging.204943 PMID:37770186

Outeiro, T. F., Alcalay, R. N., Antonini, A., Attems, J., Bonifati, V., Cardoso, F., Chesselet, M.-F., Hardy, J., Madeo, G., McKeith, I., Mollenhauer, B., Moore, D. J., Rascol, O., Schlossmacher, M. G., Soreq, H., Stefanis, L., & Ferreira, J. J. (2023). Defining the Riddle in Order to Solve It: There Is More Than One "Parkinson's Disease.". *Movement Disorders, 38*(7), 1127–1142. doi:10.1002/mds.29419 PMID:37156737

Owens, T.E., Machulda, M.M., & Duffy, J. (2018). Patterns of neuropsychological dysfunction and cortical volume changes in logopenic aphasia. *J Alzheimer's Dis, 66*(3), 1015-1025. doi:10.3233/JAD-171175

Pace-Schott, E. F., & Spencer, R. M. (2011). Age-related changes in the cognitive function of sleep. *Progress in Brain Research, 191*, 75–89. doi:10.1016/B978-0-444-53752-2.00012-6 PMID:21741545

Pagano, G., Taylor, K. I., Anzures-Cabrera, J., Marchesi, M., Simuni, T., Marek, K., Postuma, R. B., Pavese, N., Stocchi, F., Azulay, J.-P., Mollenhauer, B., López-Manzanares, L., Russell, D. S., Boyd, J. T., Nicholas, A. P., Luquin, M. R., Hauser, R. A., Gasser, T., Poewe, W., ... Bonni, A. (2022). Trial of Prasinezumab in Early-Stage Parkinson's Disease. *The New England Journal of Medicine, 387*(5), 421–432. doi:10.1056/NEJMoa2202867 PMID:35921451

Paisán-Ruiz, C., Lewis, P. A., & Singleton, A. B. (2013). LRRK2: Cause, Risk, and Mechanism. *Journal of Parkinson's Disease, 3*(2), 85–103. doi:10.3233/JPD-130192 PMID:23938341

Palagini, L., Geoffroy, P. A., Miniati, M., Perugi, G., Biggio, G., Marazziti, D., & Riemann, D. (2022). Insomnia, sleep loss, and circadian sleep disturbances in mood disorders: A pathway toward neurodegeneration and neuroprogression? A theoretical review. *CNS Spectrums*, *27*(3), 298–308. doi:10.1017/S1092852921000018 PMID:33427150

Pan, J., & Tompkins, W. J. (1985). A real-time QRS detection algorithm (Vol. 32). *IEEE Transactions on Biomedical Engineering*, *32*(3), 230–236. doi:10.1109/TBME.1985.325532 PMID:3997178

Parkinson, J. (2002). An essay on the shaking palsy (Reprinted). *The Journal of Neuropsychiatry and Clinical Neurosciences*, *14*(2), 223–236. doi:10.1176/jnp.14.2.223 PMID:11983801

Parvaneh, S., Rubin, J., Babaeizadeh, S., & Xu-Wilson, M. (2019). Cardiac arrhythmia detection using deep learning: A review. *Journal of Electrocardiology*, *57*, S70–S74. doi:10.1016/j.jelectrocard.2019.08.004 PMID:31416598

Patel, B., Irwin, D. J., Kaufer, D., Boeve, B. F., Taylor, A., & Armstrong, M. J. (2022). Outcome Measures for Dementia With Lewy Body Clinical Trials: A Review. *Alzheimer Disease and Associated Disorders*, *36*(1), 64–72. doi:10.1097/WAD.0000000000000473 PMID:34393189

Patel, D., Steinberg, J., & Patel, P. (2018). Insomnia in the Elderly: A Review. *Journal of Clinical Sleep Medicine*, *14*(6), 1017–1024. doi:10.5664/jcsm.7172 PMID:29852897

Patterson, C. (2018). The state of the art of dementia research: New frontiers. *World Alzheimer Report*. https://www.alz.co.uk/research/world-report-2018

Pendergrass, J. C., Targum, S. D., & Harrison, J. E. (2018). Cognitive Impairment Associated with Cancer: A Brief Review. *Innovations in Clinical Neuroscience*, *15*(1–2), 36–44. https://www.ncbi.nlm.nih.gov/pubmed/29497579 PMID:29497579

Petersen, L., Mortensen, P. B., & Pedersen, C. B. (2011). Paternal Age at Birth of First Child and Risk of Schizophrenia. *The American Journal of Psychiatry*, *168*(1), 82–88. doi:10.1176/appi.ajp.2010.10020252 PMID:20952457

Petrėnas, L. S. A., & Vaidotas, M. (2015). Low-complexity detection of atrial fibrillation in continuous long-term monitoring. *Computers in Biology and Medicine,* 1 – 8. . doi:10.1016/j.compbiomed.2015.01.019

Petroi, D., Duffya, J. R., Borgertb, A., Stranda, E. A., Machuldac, M. M., Senjemd, M. L., Jack, C. R. Jr, Josephsa, K. A., & Whitwelld, J. L. (2020). Neuroanatomical correlates of phonologic errors in logopenic progressive aphasia. *Brain and Language*, *204*, 104773. doi:10.1016/j.bandl.2020.104773 PMID:32114145

Pfeffer, R. I., Kurosaki, T. T., Harrah, C. H., Chance, J. M., & Filos, S. (1982). Measurement of Functional Activities in Older Adults in the Community. *Journal of Gerontology*, *37*(3), 323–329. doi:10.1093/geronj/37.3.323 PMID:7069156

Philipp-Wiegmann, F., Retz-Junginger, P., Retz, W., & Rösler, M. (2016). The intraindividual impact of ADHD on the transition of adulthood to old age. *European Archives of Psychiatry and Clinical Neuroscience*, *266*(4), 367–371. doi:10.1007/s00406-015-0644-7 PMID:26438010

Pick, E., Kleinbub, J. R., Mannarini, S., & Palmieri, A. (2019). Empathy In Neurodegenerative Diseases: A Systematic Review. *Neuropsychiatric Disease and Treatment*, *15*, 3287–3304. doi:10.2147/NDT.S225920 PMID:31819455

Pieri, L., Tosi, G., & Romano, D. (2023). Virtual reality technology in neuropsychological testing: A systematic review. *Journal of Neuropsychology*, *17*(2), 382–399. doi:10.1111/jnp.12304 PMID:36624041

Player, L., Mackenzie, L., Willis, K., & Loh, S. Y. (2014). Women's experiences of cognitive changes or 'chemobrain' following treatment for breast cancer: A role for occupational therapy? *Australian Occupational Therapy Journal*, *61*(4), 230–240. doi:10.1111/1440-1630.12113 PMID:24499127

Poddar, M., Birajdar, A. C., & Virmani, J. (2019). Chapter 5 - automated classification of hypertension and coronary artery disease patients by pnn, knn, and svm classifiers using hrv analysis. In *Machine Learning in Bio-Signal Analysis and Diagnostic Imaging*. Academic Press. https://www.sciencedirect.com/science/article/pii/B9780128160862000059

Polymeropoulos, M. H., Lavedan, C., Leroy, E., Ide, S. E., Dehejia, A., Dutra, A., Pike, B., Root, H., Rubenstein, J., Boyer, R., Stenroos, E. S., Chandrasekharappa, S., Athanassiadou, A., Papapetropoulos, T., Johnson, W. G., Lazzarini, A. M., Duvoisin, R. C., Di Iorio, G., Golbe, L. I., & Nussbaum, R. L. (1997). Mutation in the alpha-synuclein gene identified in families with Parkinson's disease. *Science*, *276*(5321), 2045–2047. doi:10.1126/science.276.5321.2045 PMID:9197268

Ponjoan, A., Garre-Olmo, J., Blanch, J., Fages, E., Alves-Cabratosa, L., Martí-Lluch, R., Comas-Cufí, M., Parramon, D., Garcia-Gil, M., & Ramos, R. (2019). Epidemiology of dementia: Prevalence and incidence estimates using validated electronic health records from primary care. *Clinical Epidemiology*, *11*, 217–228. doi:10.2147/CLEP.S186590 PMID:30881138

Posselt, C.M., Albert, N., Nordentoft, M., & Hjorthøj, C. (2021). The Danish OPUS Early Intervention Services for First-Episode Psychosis: A Phase 4 Prospective Cohort Study With Comparison of Randomized Trial and Real-World Data. *Am J Psychiatry, 178*(10), 941-951. doi:10.1176/appi.ajp.2021.20111596

Postuma, R. B., Berg, D., Stern, M., Poewe, W., Olanow, C. W., Oertel, W., Obeso, J., Marek, K., Litvan, I., Lang, A. E., Halliday, G., Goetz, C. G., Gasser, T., Dubois, B., Chan, P., Bloem, B. R., Adler, C. H., & Deuschl, G. (2015). MDS clinical diagnostic criteria for Parkinson's disease. *Movement Disorders*, *30*(12), 1591–1601. doi:10.1002/mds.26424 PMID:26474316

Premack, D., & Woodruff, G. (1978). Does the chimpanzee have a "theory of mind"? *Behavioral and Brain Sciences*, *4*(4), 515–526. doi:10.1017/S0140525X00076512

Prete, M., Cellini, N., Ronconi, L., & Di Rosa, E. (2023). Cognitive reserve moderates the relationship between sleep difficulties and executive functions in patients with Parkinson's disease. *Sleep Medicine*, *111*, 82–85. doi:10.1016/j.sleep.2023.09.009 PMID:37738789

Primi, R. (2018). Avaliação psicológica no século XXI: De onde viemos e para onde vamos. *Psicologia (Conselho Federal de Psicologia), 38*(spe), 87–97. doi:10.1590/1982-3703000209814

Prince, M., Bryce, R., Albanese, E., Wimo, A., Ribeiro, W., & Ferri, C. P. (2013). The global prevalence of dementia: A systematic review and metaanalysis. *Alzheimer's & Dementia, 9*(1), 63–75. doi:10.1016/j.jalz.2012.11.007 PMID:23305823

Prodan, C. I., Monnot, M., & Ross, E. D. (2009). Behavioural abnormalities associated with rapid deterioration of language functions in semantic dementia respond to sertraline. *Journal of Neurology, Neurosurgery, and Psychiatry, 80*(12), 1416–1417. doi:10.1136/jnnp.2009.173260 PMID:19917830

Proverbio, A. M., Crotti, N., Zani, A., & Adorni, R. (2009). The role of left and right hemispheres in the comprehension of idiomatic language: An electrical neuroimaging study. *BMC Neuroscience, 10*(1), 116. doi:10.1186/1471-2202-10-116 PMID:19754932

Pullens, M. J. J., De Vries, J., Van Warmerdam, L. J. C., Van De Wal, M. A., & Roukema, J. A. (2013). Chemotherapy and cognitive complaints in women with breast cancer. *Psycho-Oncology, 22*(8), 1783–1789. doi:10.1002/pon.3214 PMID:23109296

Queiroz, J. A., Junior, A., Lucena, F., & Barros, A. K. (2017). Diagnostic decision support systems for atrial fibrillation based on a novel electrocardiogram approach. *Journal of Electrocardiology*. https://www.sciencedirect.com/science/article/pii/S0022073617304259

Queiroz, J. A., Azoubel, L., & Barros, A. K. (2019). Support system for classification of beat-to-beat arrhythmia based on variability and morphology of electrocardiogram. *EURASIP Journal on Advances in Signal Processing, 2019*(1), 16. Advance online publication. doi:10.1186/s13634-019-0613-9

Querfurth, H. W., & LaFerla, F. M. (2010). Mechanisms of disease. *The New England Journal of Medicine, 362*(4), 329–344. doi:10.1056/NEJMra0909142 PMID:20107219

Qurraie, S. S., & Afkhami, R. G. (2017). Ecg arrhythmia classification using time frequency distribution techniques. *Biomedical Engineering Letters, 7*(4), 325–332. Advance online publication. doi:10.1007/s13534-017-0043-2 PMID:30603183

Rahman, A. A., & Morrison, B. E. (2019). Contributions of VPS35 Mutations to Parkinson's Disease. *Neuroscience, 401*, 1–10. doi:10.1016/j.neuroscience.2019.01.006 PMID:30660673

Rahman, S., Sahakian, B. J., Hodges, J. R., Rogers, R. D., & Robbins, T. W. (1999). Specific cognitive deficits in mild frontal variant frontotemporal dementia. *Brain, 122*(Pt 8), 1469–1493. doi:10.1093/brain/122.8.1469 PMID:10430832

Rahul, D.R., & Ponniah, R.J. (2019). Language impairment in primary progressive aphasia and other neurodegenerative diseases. *J Genetics, 98*(95). https://doi. 12041-019-1139-5 doi:org/10.1007/s

Raj, S., Ray, K. C., & Shankar, O. (2016). Cardiac arrhythmia beat classification using dost and psotuned svm. *Computer Methods and Programs in Biomedicine, 136*, 163–177. doi:10.1016/j.cmpb.2016.08.016 PMID:27686713

Rambeau, A., Beauplet, B., Laviec, H., Licaj, I., Leconte, A., Chatel, C., Le Bon, P., Denhaerynck, J., Clarisse, B., Frenkiel, N., Lange, M., & Joly, F. (2019). Prospective comparison of the Montreal Cognitive Assessment (MoCA) and the Mini Mental State Examination (MMSE) in geriatric oncology. *Journal of Geriatric Oncology, 10*(2), 235–240. doi:10.1016/j.jgo.2018.08.003 PMID:30150019

Rankin, K. P., Kramer, J. H., & Miller, B. L. (2005). Patterns of cognitive and emotional empathy in frontotemporal lobar degeneration. *Cognitive and Behavioral Neurology: Official Journal of the Society for Behavioral and Cognitive Neurology, 18*(1), 28–36. https://doi.org/ doi:10.1097/01.wnn.0000152225.05377.ab

Rapp, A. M., & Wild, B. (2011). Nonliteral language in Alzheimer dementia: A review. *Journal of the International Neuropsychological Society, 17*(2), 207–218. doi:10.1017/S1355617710001682 PMID:21241530

Reeve, S., Sheaves, B., & Freeman, D. (2019). Sleep Disorders in Early Psychosis: Incidence, Severity, and Association With Clinical Symptoms. *Schizophr Bull., 45*(2), 287-295. doi:10.1093/schbul/sby129

Reilly, J., Rodriguez, A. D., Lamy, M., & Neils-Strunjas, J. (2010). Cognition, language, and clinical pathological features of non-Alzheimer's dementias: An overview. *Journal of Communication Disorders, 43*(5), 438–452. doi:10.1016/j.jcomdis.2010.04.011 PMID:20493496

Reis, C., Dias, S., Rodrigues, A. M., Sousa, R. D., Gregório, M. J., Branco, J., Canhão, H., & Paiva, T. (2018). Sleep duration, lifestyles and chronic diseases: A cross-sectional population-based study. *Sleep Science, 11*(4), 217–230. doi:10.5935/1984-0063.20180036 PMID:30746039

Repici, M., & Giorgini, F. (2019). DJ-1 in Parkinson's Disease: Clinical Insights and Therapeutic Perspectives. *Journal of Clinical Medicine, 8*(9), 1377. doi:10.3390/jcm8091377 PMID:31484320

Ribeiro da Costa, C., Soares, J. M., Oliveira-Silva, P., Sampaio, A., & Coutinho, J. F. (2021). Interplay Between the Salience and the Default Mode Network in a Social-Cognitive Task Toward a Close Other. *Frontiers in Psychiatry, 12*, 718400. doi:10.3389/fpsyt.2021.718400 PMID:35197871

Riboldi, G. M., & Di Fonzo, A. B. (2019). GBA, Gaucher Disease, and Parkinson's Disease: From Genetic to Clinic to New Therapeutic Approaches. *Cells, 8*(4), 364. doi:10.3390/cells8040364 PMID:31010158

Rimvall, M. K., van Os, J., Verhulst, F., Wolf, R. T., Larsen, J. T., Clemmensen, L., Skovgaard, A. M., Rask, C. U., & Jeppesen, P. (2020). Mental Health Service Use and Psychopharmacological Treatment Following Psychotic Experiences in Preadolescence. *The American Journal of Psychiatry, 177*(4), 318–326. doi:10.1176/appi.ajp.2019.19070724 PMID:32098486

Rimvall, M. K., Wolf, R. T., Olsen, E. M., Skovgaard, A. M., Clemmensen, L., Oxholm, A. S., Verhulst, F., Rask, C. U., van Os, J., & Jeppesen, P. (2021). Healthcare Costs, School Performance, and Health-related Quality of Life in Adolescence Following Psychotic Experiences in Preadolescence: A Longitudinal Cohort Study. *Schizophrenia Bulletin, 47*(3), 682–691. doi:10.1093/schbul/sbaa175 PMID:33345286

Rizzo, A. S., & Koenig, S. T. (2017). Is clinical virtual reality ready for primetime? *Neuropsychology*, *31*(8), 877–899. doi:10.1037/neu0000405 PMID:29376669

Robinson, D.G., Schooler, N.R., Marcy, P., Gibbons, R.D., Hendricks Brown, C., John, M., Mueser, K.T., Penn, D.L., Rosenheck, R.A., Addington, J., Brunette, M.F., Correll, C.U., Estroff, S.E., Mayer-Kalos, P.S., Gottlieb, J.D., Glynn, S.M., Lynde, D.W., Gingerich, S., Pipes, R., Miller, A.L., Severe, J.B., & Kane, J.M. (2022). Outcomes During and After Early Intervention Services for First-Episode Psychosis: Results Over 5 Years From the RAISE-ETP Site-Randomized Trial. *Schizophr Bull., 48*(5), 1021-1031. . doi:10.1093/schbul/sbac053

Rodrigues, F. (2021). Inteligência Artificial E Doenças Mentais. *Recisatec - Revista Científica Saúde e Tecnologia, 1*(5).

Rodriguez, J., Goni, A., & Illarramendi, A. (2005). Real-time classification of ecgs on a pda. *IEEE Transactions on Information Technology in Biomedicine*, *9*(1), 23–34. doi:10.1109/TITB.2004.838369 PMID:15787004

Rogalski, E. J., & Mesulam, M. M. (2009). Clinical Trajectories and Biological Features of Primary Progressive Aphasia (PPA). *Current Alzheimer Research*, *6*(4), 331–336. doi:10.2174/156720509788929264 PMID:19689231

Rogalski, E., Cobia, D., Harrison, T. M., Wieneke, C., Thompson, C. K., Weintraub, S., & Mesulam, M.-M. (2011, March). Anatomy of Language Impairments in Primary Progressive Aphasia. *The Journal of Neuroscience : The Official Journal of the Society for Neuroscience*, *31*(9), 3344–3350. doi:10.1523/JNEUROSCI.5544-10.2011 PMID:21368046

Rogalski, E., Sridhar, J., Rader, B., Martersteck, A., Chen, K., Cobia, D., Thompson, C. K., Weintraub, S., Bigio, E. H., & Mesulam, M.-M. (2016). Aphasic variant of Alzheimer disease. Clinical, anatomic, and genetic features. *Neurology*, *87*(13), 1–7. doi:10.1212/WNL.0000000000003165 PMID:27566743

Roher, J.D., Caso, F., & Mahoney, C. (2013). *Patterns of longitudinal brain atrophy in the logopenic variant of primary progressive aphasia.* Academic Press.

Ropper, A. H., Samuels, M. A., Klein, J. P., & Prasad, S. (2023). *Adams and Victor's Principles of Neurology* (12th ed.). McGrawHill.

Rousseaux, M., Sève, A., Vallet, M., Pasquier, F., & Mackowiak-Cordoliani, M. A. (2010). An analysis of communication in conversation in patients with dementia. *Neuropsychologia*, *48*(13), 3884–3890. doi:10.1016/j.neuropsychologia.2010.09.026 PMID:20888846

Ruff, R. M., Light, R. H., Parker, S. B., & Levin, H. S. (1996). Benton controlled Oral Word Association Test: Reliability and updated norms. *Archives of Clinical Neuropsychology*, *11*(4), 329–338. doi:10.1093/arclin/11.4.329 PMID:14588937

Rund, B.R., Barder, H.E., Evensen, J., Haahr, U., ten Velden Hegelstad, W., Joa, I., Johannessen, J.O., Langeveld, J., Larsen, T.K., Melle, I., Opjordsmoen, S., Røssberg, J.I., Simonsen, E., Sundet, K., Vaglum, P., McGlashan, T., & Friis, S. (2016). Neurocognition and Duration of Psychosis: A 10-year Follow-up of First-Episode Patients. *Schizophr Bull., 42*(1), 87-95. doi:10.1093/schbul/sbv083

Rutigliano, G., Valmaggia, L., Landi, P., Frascarelli, M., Cappucciati, M., Sear, V., Rocchetti, M., De Micheli, A., Jones, C., Palombini, E., McGuire, P., & Fusar-Poli, P. (2016). Persistence or recurrence of non-psychotic comorbid mental disorders associated with 6-year poor functional outcomes in patients at ultra high risk for psychosis. *Journal of Affective Disorders, 203*, 101–110. doi:10.1016/j.jad.2016.05.053 PMID:27285723

Sabatinelli, D., Fortune, E. E., Li, Q., Siddiqui, A., Krafft, C., Oliver, W. T., Beck, S., & Jeffries, J. (2011). Emotional perception: Meta-analyses of face and natural scene processing. *NeuroImage, 54*(3), 2524–2533. doi:10.1016/j.neuroimage.2010.10.011 PMID:20951215

Sahoo, S., Dash, M., Behera, S., & Sabut, S. (2020). Machine learning approach to detect cardiac arrhythmias in ecg signals: A survey (Vol. 41). *Ingénierie et Recherche Biomédicale : IRBM = Biomedical Engineering and Research, 41*(4), 185–194. doi:10.1016/j.irbm.2019.12.001

Salazar de Pablo, G., Davies, C., de Diego, H., Solmi, M., Shin, J. I., Carvalho, A. F., Radua, J., & Fusar-Poli, P. (2021). Transition to psychosis in randomized clinical trials of individuals at clinical high risk of psychosis compared to observational cohorts: A systematic review and meta-analysis. *European Psychiatry, 64*(1), e51. doi:10.1192/j.eurpsy.2021.2222 PMID:34315555

Salazar de Pablo, G., Woods, S. W., Drymonitou, G., de Diego, H., & Fusar-Poli, P. (2021). Prevalence of Individuals at Clinical High-Risk of Psychosis in the General Population and Clinical Samples: Systematic Review and Meta-Analysis. *Brain Sciences, 11*(11), 1544. doi:10.3390/brainsci11111544 PMID:34827543

Sameni, R., Shamsollahi, M. B., Jutten, C., & Babaie-Zade, M. (2005). Filtering noisy ecg signals using the extended kalman filter based on a modified dynamic ecg model. Computers in Cardiology, 1017–1020. doi:10.1109/CIC.2005.1588283

Sánchez-Gutiérrez, T., Fernandez-Castilla, B., Barbeito, S., González-Pinto, A., Becerra-García, J. A., & Calvo, A. (2020). Cannabis use and nonuse in patients with first-episode psychosis: A systematic review and meta-analysis of studies comparing neurocognitive functioning. *European Psychiatry, 63*(1), e6. doi:10.1192/j.eurpsy.2019.9 PMID:32093788

Sánchez-Torres, A.M., Moreno-Izco, L., Lorente-Omeñaca, R., Cabrera, B., Lobo, A., González-Pinto, A.M., Merchán-Naranjo, J., Corripio, I., Vieta, E., de la Serna, E., Butjosa, A., Contreras, F., Sarró, S., Mezquida, G., Ribeiro, M., Bernardo, M., & Cuesta, M.J. (2018). Individual trajectories of cognitive performance in first episode psychosis: a 2-year follow-up study. *Eur Arch Psychiatry Clin Neurosci., 268*(7), 699-711.

Santos-Lobato, B. L., Schumacher-Schuh, A., Mata, I. F., Letro, G. H., Braga-Neto, P., Brandão, P. R. P., Godeiro-Junior, C. O., Coletta, M. V. D., Camargos, S. T., Borges, V., Rieder, C. R. M., & Tumas, V. (2021). Genetics of Parkinson's disease in Brazil: A systematic review of monogenic forms. *Arquivos de Neuro-Psiquiatria*, *79*(7), 612–623. doi:10.1590/0004-282x-anp-2020-0409 PMID:34468500

Santos, M. T. F., Sougey, E. B., & Alchieri, J. C. (2009). Validity and reliability of the screening test for Alzheimer's disease with proverbs (STADP) for the elderly. *Arquivos de Neuro-Psiquiatria*, *67*(3b, n.3B), 836–842. doi:10.1590/S0004-282X2009000500011 PMID:19838514

Sasai-Sakuma, T., Takeuchi, N., Asai, Y., Inoue, Y., & Inoue, Y. (2020). Prevalence and clinical characteristics of REM sleep behavior disorder in Japanese elderly people. *Sleep (Basel)*, *43*(8), zsaa024. Advance online publication. doi:10.1093/sleep/zsaa024 PMID:32064524

Sassone, J., Reale, C., Dati, G., Regoni, M., Pellecchia, M. T., & Garavaglia, B. (2021). The Role of VPS35 in the Pathobiology of Parkinson's Disease. *Cellular and Molecular Neurobiology*, *41*(2), 199–227. doi:10.1007/s10571-020-00849-8 PMID:32323152

Satpute, A. B., Kang, J., Bickart, K. C., Yardley, H., Wager, T. D., & Barrett, L. F. (2015). Involvement of Sensory Regions in Affective Experience: A Meta-Analysis. *Frontiers in Psychology*, *6*, 1860. doi:10.3389/fpsyg.2015.01860 PMID:26696928

Satterthwaite, T. D., & Baker, J. T. (2015). How can studies of resting-state functional connectivity help us understand psychosis as a disorder of brain development? *Current Opinion in Neurobiology*, *30*, 85–91. doi:10.1016/j.conb.2014.10.005 PMID:25464373

Saviola, F., Bellani, M., Perlini, C., Squarcina, L., Maggioni, E., Zacà, D., Lasalvia, A., Dusi, N., Bonetto, C., Cristofalo, D., Alessandrini, F., Zoccatelli, G., Ciceri, E., Mesiano, L., Semrov, E., Lo Parrino, R., Furlato, K., Pratelli, M., Ruggeri, M., ... Jovicich, J. (2021). First-episode psychosis: Structural covariance deficits in salience network correlate with symptoms severity. *Journal of Psychiatric Research*, *136*, 409–420. doi:10.1016/j.jpsychires.2021.01.044 PMID:33647856

Schimmelmann, B. G., Conus, P., Cotton, S., McGorry, P. D., & Lambert, M. (2007). Pretreatment, baseline, and outcome differences between early-onset and adult-onset psychosis in an epidemiological cohort of 636 first-episode patients. *Schizophrenia Research*, *95*(1-3), 1–8. doi:10.1016/j.schres.2007.06.004 PMID:17628441

Schimmelpfennig, J., Topczewski, J., Zajkowski, W., & Jankowiak-Siuda, K. (2023). The role of the salience network in cognitive and affective deficits. *Frontiers in Human Neuroscience*, *17*, 1133367. doi:10.3389/fnhum.2023.1133367 PMID:37020493

Schliebs, R., & Arendt, T. (2006). The significance of the cholinergic system in the brain during aging and in Alzheimer's disease. *Journal of Neural Transmission (Vienna, Austria)*, *113*(11), 1625–1644. doi:10.1007/s00702-006-0579-2 PMID:17039298

Schneider, M., Debbané, M., Bassett, A. S., Chow, E. W. C., Fung, W. L. A., van den Bree, M. B. M., Owen, M., Murphy, K. C., Niarchou, M., Kates, W. R., Antshel, K. M., Fremont, W., McDonald-McGinn, D. M., Gur, R. E., Zackai, E. H., Vorstman, J., Duijff, S. N., Klaassen, P. W. J., Swillen, A., ... Eliez, S. (2014). Psychiatric Disorders From Childhood to Adulthood in 22q11.2 Deletion Syndrome: Results From the International Consortium on Brain and Behavior in 22q11.2 Deletion Syndrome. *The American Journal of Psychiatry*, *171*(6), 627–639. doi:10.1176/appi.ajp.2013.13070864 PMID:24577245

Schoeler, T., Monk, A., Sami, M. B., Klamerus, E., Foglia, E., Brown, R., Camuri, G., Altamura, A. C., Murray, R., & Bhattacharyya, S. (2016). Continued versus discontinued cannabis use in patients with psychosis: A systematic review and meta-analysis. *The Lancet. Psychiatry*, *3*(3), 215–225. doi:10.1016/S2215-0366(15)00363-6 PMID:26777297

Schroyen, G., Blommaert, J., van Weehaeghe, D., Sleurs, C., Vandenbulcke, M., Dedoncker, N., Hatse, S., Goris, A., Koole, M., Smeets, A., van Laere, K., Sunaert, S., & Deprez, S. (2021). Neuroinflammation and Its Association with Cognition, Neuronal Markers and Peripheral Inflammation after Chemotherapy for Breast Cancer. *Cancers (Basel)*, *13*(16), 4198. doi:10.3390/cancers13164198 PMID:34439351

Schroyen, G., Sleurs, C., Bartsoen, E., Smeets, D., van Weehaeghe, D., Van Laere, K., Smeets, A., Deprez, S., & Sunaert, S. (2022). Neuroinflammation as potential precursor of leukoencephalopathy in early-stage breast cancer patients: A cross-sectional PET-MRI study. *The Breast*, *62*, 61–68. doi:10.1016/j.breast.2022.02.001 PMID:35131644

Schumacher, J., Peraza, L. R., Firbank, M., Thomas, A. J., Kaiser, M., Gallagher, P., O'Brien, J. T., Blamire, A. M., & Taylor, J. P. (2019). Dysfunctional brain dynamics and their origin in Lewy body dementia. *Brain*, *142*(6), 1767–1782. doi:10.1093/brain/awz069 PMID:30938426

Schurz, M., Radua, J., Aichhorn, M., Richlan, F., & Perner, J. (2014). Fractionating theory of mind: A meta-analysis of functional brain imaging studies. *Neuroscience and Biobehavioral Reviews*, *42*, 9–34. doi:10.1016/j.neubiorev.2014.01.009 PMID:24486722

Secher, R.G., Hjorthøj, C.R., Austin, S.F., Thorup, A., Jeppesen, P., Mors, O., & Nordentoft, M. (2015). Ten-year follow-up of the OPUS specialized early intervention trial for patients with a first episode of psychosis. *Schizophr Bull.*, *41*(3), 617-26.

Seda, G., & Han, T. S. (2020). Effect of Obstructive Sleep Apnea on Neurocognitive Performance. *Sleep Medicine Clinics*, *15*(1), 77–85. doi:10.1016/j.jsmc.2019.10.001 PMID:32005352

Seeley, W. W., Menon, V., Schatzberg, A. F., Keller, J., Glover, G. H., Kenna, H., Reiss, A. L., & Greicius, M. D. (2007). Dissociable intrinsic connectivity networks for salience processing and executive control. *The Journal of Neuroscience : The Official Journal of the Society for Neuroscience*, *27*(9), 2349–2356. doi:10.1523/JNEUROSCI.5587-06.2007 PMID:17329432

Seigers, R., Loos, M., Van Tellingen, O., Boogerd, W., Smit, A. B., & Schagen, S. B. (2016). Neurobiological changes by cytotoxic agents in mice. *Behavioural Brain Research*, *299*, 19–26. doi:10.1016/j.bbr.2015.10.057 PMID:26602283

Senaha, M. L. H., Caramelli, P., Brucki, S. M. D., Smid, J., Takada, L. T., Porto, C. S., César, K. G., Matioli, M. N. P., Soares, R. T., Mansur, L. L., & Nitrini, R. (2013). Primary progressive aphasia Classification of variants in 100 consecutive Brazilian cases. *Dementia & Neuropsychologia*, *7*(1), 110–1121. doi:10.1590/S1980-57642013DN70100017 PMID:29213827

Senaratna, C. V., Perret, J. L., Lodge, C. J., Lowe, A. J., Campbell, B. E., Matheson, M. C., Hamilton, G. S., & Dharmage, S. C. (2017). Prevalence of obstructive sleep apnea in the general population: A systematic review. *Sleep Medicine Reviews*, *34*, 70–81. doi:10.1016/j.smrv.2016.07.002 PMID:27568340

Setién-Suero, E., Murillo-García, N., Sevilla-Ramos, M., Abreu-Fernández, G., Pozueta, A., & Ayesa-Arriola, R. (2022). Exploring the Relationship Between Deficits in Social Cognition and Neurodegenerative Dementia: A Systematic Review. *Frontiers in Aging Neuroscience*, *14*, 778093. doi:10.3389/fnagi.2022.778093 PMID:35572150

Shah, R. R. (2005). Mechanistic basis of adverse drug reactions: The perils of inappropriate dose schedules. Expert Opinion on Drug Safety. doi:10.1517/14740338.4.1.103

Shany-Ur, T., Poorzand, P., Grossman, S. N., Growdon, M. E., Jang, J. Y., Ketelle, R. S., Miller, B. L., & Rankin, K. P. (2012). Comprehension of insincere communication in neurodegenerative disease: Lies, sarcasm, and theory of mind. *Cortex*, *48*(10), 1329–1341. doi:10.1016/j.cortex.2011.08.003 PMID:21978867

Sharifi, V., Eaton, W. W., Wu, L. T., Roth, K. B., Burchett, B. M., & Mojtabai, R. (2015). Psychotic experiences and risk of death in the general population: 24–27 year follow-up of the Epidemiologic Catchment Area study. *The British Journal of Psychiatry*, *207*(1), 30–36. doi:10.1192/bjp.bp.113.143198 PMID:25953893

Sharma, M. J., Lavoie, S., & Callahan, B. L. (2021). A Call for Research on the Validity of the Age-of-Onset Criterion Application in Older Adults Being Evaluated for ADHD: A Review of the Literature in Clinical and Cognitive Psychology. *The American Journal of Geriatric Psychiatry*, *29*(7), 669–678. doi:10.1016/j.jagp.2020.10.016 PMID:33191098

Sheffield, J.M., Karcher, N.R., & Barch, D.M. (2018). Cognitive Deficits in Psychotic Disorders: A Lifespan Perspective. *Neuropsychol Rev., 28*(4), 509-533. doi:10.1007/s11065-018-9388-2

Shi, L., Chen, S.-J., Ma, M.-Y., Bao, Y.-P., Han, Y., Wang, Y.-M., Shi, J., Vitiello, M. V., & Lu, L. (2018). Sleep disturbances increase the risk of dementia: A systematic review and meta-analysis. *Sleep Medicine Reviews*, *40*, 4–16. doi:10.1016/j.smrv.2017.06.010 PMID:28890168

Shimokawa, A., Yatomi, N., Anamizu, S., Ashikari, I., Kohno, M., Maki, Y., Torii, S., Isono, H., Sugai, Y., Koyama, N., & Matsuno, Y. (2000). Comprehension of emotions: Comparison between Alzheimer type and vascular type dementias. *Dementia and Geriatric Cognitive Disorders*, *11*(5), 268–274. doi:10.1159/000017249 PMID:10940678

Sidransky, E., Nalls, M. A., Aasly, J. O., Aharon-Peretz, J., Annesi, G., Barbosa, E. R., Bar-Shira, A., Berg, D., Bras, J., Brice, A., Chen, C.-M., Clark, L. N., Condroyer, C., De Marco, E. V., Dürr, A., Eblan, M. J., Fahn, S., Farrer, M., Fung, H.-C., ... Ziegler, S. G. (2009). Multi-center analysis of glucocerebrosidase mutations in Parkinson disease. *The New England Journal of Medicine*, *361*(17), 1651–1661. doi:10.1056/NEJMoa0901281 PMID:19846850

Sieradzka, D., Power, R.A., Freeman, D., Cardno, A.G., McGuire, P., & Plomin, R. (2014). Are Genetic Risk Factors for Psychosis Also Associated with Dimension-Specific Psychotic Experiences in Adolescence? *PLOS ONE*, *9*(4), e94398.

Singh, S., Pandey, S. K., Pawar, U., & Janghel, R. R. (2018). Classification of ecg arrhythmia using recurrent neural networks (Vol. 132). *Procedia Computer Science, 132,* 1290 – 1297. https://www.sciencedirect.com/science/article/pii/S1877050918307774

Singh, S., Strong, R. W., Jung, L., Li, F. H., Grinspoon, L., Scheuer, L. S., Passell, E. J., Martini, P., Chaytor, N., Soble, J. R., & Germine, L. (2021). The TestMyBrain Digital Neuropsychology Toolkit: Development and Psychometric Characteristics. *Journal of Clinical and Experimental Neuropsychology*, *43*(8), 786–795. doi:10.1080/13803395.2021.2002269 PMID:34907842

Singleton, A. B., Farrer, M., Johnson, J., Singleton, A., Hague, S., Kachergus, J., Hulihan, M., Peuralinna, T., Dutra, A., Nussbaum, R., Lincoln, S., Crawley, A., Hanson, M., Maraganore, D., Adler, C., Cookson, M. R., Muenter, M., Baptista, M., Miller, D., ... Gwinn-Hardy, K. (2003). Alpha-Synuclein locus triplication causes Parkinson's disease. *Science*, *302*(5646), 841. doi:10.1126/science.1090278 PMID:14593171

Small, S.L., Hart, J., Nguyen, T., & Gordon, B. (1995). *Distributed representations of semantic knowledge in the brain.* Academic Press.

Smith, L., & Schapira, A. H. V. (2022). GBA Variants and Parkinson Disease: Mechanisms and Treatments. *Cells*, *11*(8), 1261. doi:10.3390/cells11081261 PMID:35455941

Snitselaar, M. A., Smits, M. G., Van Der Heijden, K. B., & Spijker, J. (2017). Sleep and Circadian Rhythmicity in Adult ADHD and the Effect of Stimulants: A Review of the Current Literature. *Journal of Attention Disorders*, *21*(1), 14–26. doi:10.1177/1087054713479663 PMID:23509113

Snowed, J. S., Goulding, P. J., & Neary, D. (1989). Semantic Dementia: A Form of Circumscribed Cerebral Atrophy. *Behavioural Neurology*, *2*, 167–182. doi:10.1155/1989/124043

Solanto, M. V. (2011). *Cognitive-behavioral therapy for adult ADHD: Targeting executive dysfunction.* The Guilford Press.

Sönmez, N., Romm, K. L., Østefjells, T., Grande, M., Jensen, L. H., Hummelen, B., Tesli, M., Melle, I., & Røssberg, J. I. (2020). Cognitive behavior therapy in early psychosis with a focus on depression and low self-esteem: A randomized controlled trial. *Comprehensive Psychiatry*, *97*, 152157. doi:10.1016/j.comppsych.2019.152157 PMID:31935529

Sousa, H., Almeida, S., Bessa, J., & Pereira, M. G. (2020). The Developmental Trajectory of Cancer-Related Cognitive Impairment in Breast Cancer Patients: A Systematic Review of Longitudinal Neuroimaging Studies. *Neuropsychology Review, 30*(3), 287–309. doi:10.1007/s11065-020-09441-9 PMID:32607817

Sousa, R. M., Ferri, C. P., Acosta, D., Albanese, E., Guerra, M., Huang, Y., Jacob, K. S., Jotheeswaran, A. T., Rodriguez, J. J., Pichardo, G. R., Rodriguez, M. C., Salas, A., Sosa, A. L., Williams, J., Zuniga, T., & Prince, M. (2009). Contribution of chronic diseases to disability in elderly people in countries with low and middle incomes: A 10/66 Dementia Research Group population-based survey. *Lancet, 374*(9704), 1821–1830. doi:10.1016/S0140-6736(09)61829-8 PMID:19944863

Spillantini, M. G., Schmidt, M. L., Lee, V. M., Trojanowski, J. Q., Jakes, R., & Goedert, M. (1997). Alpha-synuclein in Lewy bodies. *Nature, 388*(6645), 839–840. doi:10.1038/42166 PMID:9278044

Spinelli, E. G., Caso, F., Agosta, F., Gambina, G., Magnani, G., Canu, E., Blasi, V., Perani, D., Comi, G., Falini, A., Gorno-Tempini, M. L., & Filippi, M. (2015). A multimodal neuroimaging study of a case of crossed nonfluent/agrammatic primary progressive aphasia. *Journal of Neurology, 262*(10), 2336–2345. doi:10.1007/s00415-015-7845-x PMID:26194195

Spitsyna, G., Warren, J. E., Scoot, S. K., Turkheimer, F. E., & Wise, R. J. S. (2006). Covering Language Streams in the Human Temporal Lobe. *The Journal of Neuroscience : The Official Journal of the Society for Neuroscience, 26*(28), 7328–7336. doi:10.1523/JNEUROSCI.0559-06.2006 PMID:16837579

Staines, L., Healy, C., Coughlan, H., Clarke, M., Kelleher, I., Cotter, D., & Cannon, M. (2022). Psychotic experiences in the general population, a review; definition, risk factors, outcomes and interventions. *Psychological Medicine, 52*(15), 3297–3308. doi:10.1017/S0033291722002550 PMID:36004805

Stankeviciute, L., Falcon, C., Operto, G., Garcia, M., Shekari, M., Iranzo, Á., Niñerola-Baizán, A., Perissinotti, A., Minguillón, C., Fauria, K., Molinuevo, J. L., Zetterberg, H., Blennow, K., Suárez-Calvet, M., Cacciaglia, R., Gispert, J. D., & Grau-Rivera, O. (2023). Differential effects of sleep on brain structure and metabolism at the preclinical stages of AD. *Alzheimer's & Dementia, 19*(12), 5371–5386. doi:10.1002/alz.13102 PMID:37194734

Stanton, K. J., Denietolis, B., Goodwin, B. J., & Dvir, Y. (2020). Childhood Trauma and Psychosis: An Updated Review. *Child and Adolescent Psychiatric Clinics of North America, 29*(1), 115–129. doi:10.1016/j.chc.2019.08.004 PMID:31708041

Starzer, M., Hansen, H.G., Hjorthøj, C., Albert, N., Nordentoft, M., & Madsen, T. (2023). 20-year trajectories of positive and negative symptoms after the first psychotic episode in patients with schizophrenia spectrum disorder: results from the OPUS study. *World Psychiatry, 22*(3), 424-432. doi:10.1002/wps.21121

Steele, P., Cheng, N., Phillips, L. J., Bryce, S., Alvarez-Jimenez, M., & Allott, K. (2021). Cognitive strengths in first episode psychosis: A thematic analysis of clinicians' perspectives. *BMC Psychiatry*, *21*(1), 612. doi:10.1186/s12888-021-03627-y PMID:34876068

Stefanis, N.C., Hanssen, M., Smirnis, N.K., Avramopoulos, D.A., Evdokimidis, I.K., Stefanis, C.N., Verdoux, H., & Van Os, J. (2002). Evidence that three dimensions of psychosis have a distribution in the general population. *Psychol Med.*, *32*(2), 347-58.

Stern, Y. (2012). Cognitive reserve in ageing and Alzheimer's disease. *Lancet Neurology*, *11*(11), 1006–1012. doi:10.1016/S1474-4422(12)70191-6 PMID:23079557

Stevens, K. N., Creanor, S., Jeffery, A., Whone, A., Zajicek, J., Foggo, A., Jones, B., Chapman, R., Cocking, L., Wilks, J., Webb, D., Carroll, C., Inches, J., Underwood, D., Frost, J., James, A., Schofield, C., James, R., O'Reilly, C., ... Rose, J. (2022). Evaluation of Simvastatin as a Disease-Modifying Treatment for Patients With Parkinson Disease. *JAMA Neurology*, *79*(12), 1232–1241. doi:10.1001/jamaneurol.2022.3718 PMID:36315128

Stickley, A., Koyanagi, A., Takahashi, H., Ruchkin, V., & Kamio, Y. (2017). Attention-deficit/hyperactivity disorder symptoms and loneliness among adults in the general population. *Research in Developmental Disabilities*, *62*, 115–123. doi:10.1016/j.ridd.2017.01.007 PMID:28131008

Stilo, S. A., & Murray, R. M. (2010). The epidemology of schizophrenia: Replacing dogma with knowledge. *Dialogues in Clinical Neuroscience*, *12*(3), 305–315. doi:10.31887/DCNS.2010.12.3/sstilo PMID:20954427

Sullivan, P. F., Kendler, K. S., & Neale, M. C. (2003). Schizophrenia as a Complex Trait: Evidence From a Meta-analysis of Twin Studies. *Archives of General Psychiatry*, *60*(12), 1187. doi:10.1001/archpsyc.60.12.1187 PMID:14662550

Sumerall, S. W., Timmons, P. L., James, A. L., Ewing, M. J. M., & Oehlert, M. E. (1997). Expanded norms for the controlled oral word association test. *Journal of Clinical Psychology*, *53*(5), 517–521. Advance online publication. doi:10.1002/(SICI)1097-4679(199708)53:5<517::AID-JCLP14>3.0.CO;2-H PMID:9257231

Sun, X., He, C., & Qu, H. (2023). Bibliometric review on sleep and Alzheimer disease between 1986 and 2023. *Medicine*, *102*(44), e35764. doi:10.1097/MD.0000000000035764 PMID:37932981

Suvisaari, J., Mantere, O., Keinänen, J., Mäntylä, T., Rikandi, E., & Lindgren, M. (2018). Is It Possible to Predict the Future in First-Episode Psychosis? *Front Psychiatry, 9*(580).

Tanaka, K., Suzuki, T., Chiba, T., Shimura, H., Hattori, N., & Mizuno, Y. (2001). Parkin is linked to the ubiquitin pathway. *Journal of Molecular Medicine (Berlin, Germany)*, *79*(9), 482–494. doi:10.1007/s001090100242 PMID:11692161

Tansey, M. G., Wallings, R. L., Houser, M. C., Herrick, M. K., Keating, C. E., & Joers, V. (2022). Inflammation and immune dysfunction in Parkinson disease. *Nature Reviews. Immunology*, *22*(11), 11. Advance online publication. doi:10.1038/s41577-022-00684-6 PMID:35246670

Tchen, N., Juffs, H. G., Downie, F. P., Yi, Q. L., Hu, H., Chemerynsky, I., Clemons, M., Crump, M., Goss, P. E., Warr, D., Tweedale, M. E., & Tannock, I. F. (2003). Cognitive function, fatigue, and menopausal symptoms in women receiving adjuvant chemotherapy for breast cancer. *Journal of Clinical Oncology*, *21*(22), 4175–4183. Advance online publication. doi:10.1200/JCO.2003.01.119 PMID:14615445

Teichmann, M. (2021). The current international consensus criteria can lead to under and over-diagnosis of primary progressive aphasia variants. *Revue Neurologique, 177*, 370–375.

Teichmann, M., Kas, A., Boutet, C., Ferrieux, S., Nogues, M., Samri, D., Rogan, C., Dormont, D., Dubois, B., & Migliaccio, R. (2013). Deciphering logopenic primary progressive aphasia: A clinical, imaging and biomarker investigation. *Brain, 136*(11), 3474–3488. doi:10.1093/brain/awt266 PMID:24108322

Thacker, E. L., & Ascherio, A. (2008). Familial aggregation of Parkinson's disease: A meta-analysis. *Movement Disorders, 23*(8), 1174–1183. doi:10.1002/mds.22067 PMID:18442112

Thompson, A. D., Nelson, B., Yuen, H. P., Lin, A., Amminger, G. P., McGorry, P. D., Wood, S. J., & Yung, A. R. (2014). Sexual Trauma Increases the Risk of Developing Psychosis in an Ultra High-Risk "Prodromal" Population. *Schizophrenia Bulletin, 40*(3), 697–706. doi:10.1093/schbul/sbt032 PMID:23455040

Thompson, C. K. (2019). Neurocognitive Recovery of Sentence Processing in Aphasia. *Journal of Speech, Language, and Hearing Research: JSLHR, 62*(11), 3947–3972. doi:10.1044/2019_JSLHR-L-RSNP-19-0219 PMID:31756151

Thorell, L. B., Holst, Y., & Sjöwall, D. (2019). Quality of life in older adults with ADHD: Links to ADHD symptom levels and executive functioning deficits. *Nordic Journal of Psychiatry, 73*(7), 409–416. doi:10.1080/08039488.2019.1646804 PMID:31380715

Tombaugh, T. N. (2004). Trail Making Test A and B: Normative data stratified by age and education. *Archives of Clinical Neuropsychology, 19*(2), 203–214. Advance online publication. doi:10.1016/S0887-6177(03)00039-8 PMID:15010086

Tong, T., Lu, H., Zong, J., Lv, Q., & Chu, X. (2020). Chemotherapy-related cognitive impairment in patients with breast cancer based on MRS and DTI analysis. *Breast Cancer (Tokyo, Japan), 27*(5), 893–902. doi:10.1007/s12282-020-01094-z PMID:32301096

Torgalsbøen, A.K., Mohn, C., Larøi, F., Fu, S., & Czajkowski, N. (2023). A ten-year longitudinal repeated assessment study of cognitive improvement in patients with first-episode schizophrenia and healthy controls: The Oslo Schizophrenia Recovery (OSR) study. *Schizophr Res., 260*, 92-98.

Torralva, T., Roca, M., Gleichgerrcht, E., Bekinschtein, T., & Manes, F. (2009). A neuropsychological battery to detect specific executive and social cognitive impairments in early frontotemporal dementia. *Brain, 132*(Pt 5), 1299–1309. doi:10.1093/brain/awp041 PMID:19336463

Torres, B., Santos, R. L., Barroso de Sousa, M. F., Simoes Neto, J. P., Lima Nogueira, M. M., & Belfort, T. T. (2015). Facial expression recognition in Alzheimer's disease: A longitudinal study. *Arquivos de Neuro-Psiquiatria*, *73*(5), 383–389. doi:10.1590/0004-282X20150009 PMID:26017202

Trovão, L. D. O., Alves, G. S., Carrilho, C. G., Ricci, T. G., Bittencourt, L., Alves, C., Brito, N. C., Nardi, A. E., Malaspina, D., & Veras, A. B. (2022). Early trauma and schizophrenia onset: Preliminary results of an outpatient cohort in Brazil. *Trends in Psychiatry and Psychotherapy*. Advance online publication. doi:10.47626/2237-6089-2020-0024 PMID:33872478

Tsipouras, M. G., & Fotiadis, D. I. (2004). Automatic arrhythmia detection based on time and time–frequency analysis of heart rate variability. *Computer Methods and Programs in Biomedicine*, *74*(2), 95–108. doi:10.1016/S0169-2607(03)00079-8 PMID:15013592

Turkington, D., Kingdon, D., & Weiden, P. J. (2006). Cognitive Behavior Therapy for Schizophrenia. *The American Journal of Psychiatry*, *163*(3), 365–373. doi:10.1176/appi.ajp.163.3.365 PMID:16513854

Tysnes, O.-B., & Storstein, A. (2017). Epidemiology of Parkinson's disease. *Journal of Neural Transmission, 124*(8), 901–905. doi:10.1007/s00702-017-1686-y

Tzeng, N.-S., Chung, C.-H., Lin, F.-H., Yeh, C.-B., Huang, S.-Y., Lu, R.-B., Chang, H.-A., Kao, Y.-C., Yeh, H.-W., Chiang, W.-S., Chou, Y.-C., Tsao, C.-H., Wu, Y.-F., & Chien, W.-C. (2017). Risk of Dementia in Adults With ADHD: A Nationwide, Population-Based Cohort Study in Taiwan. *Journal of Attention Disorders*. doi:10.1177/1087054717714057 PMID:28629260

Uddin, L. Q. (2015). Salience processing and insular cortical function and dysfunction. *Nature Reviews. Neuroscience*, *16*(1), 55–61. doi:10.1038/nrn3857 PMID:25406711

Ulugut, H., Stek, S., & Wageman, L. E. E. (2022). The natural history of primary progressive aphasia: Beyond aphasia. *Journal of Neurology*, *269*(3), 1375–1385. doi:10.1007/s00415-021-10689-1 PMID:34216263

United Nations & Human Rights Office of the High Commisioner. (2021). *Update to the 2012 Analytical Outcome Study on the normative standards in international human rights law in relation to older persons. Working paper prepared by the Office of the High Commissioner for Human Rights*. United Nations. https://www.ohchr.org/en/documents/outcome-documents/ohchr-working-paper-update-2012-analytical-outcome-study-normative

Valente, E. M., Abou-Sleiman, P. M., Caputo, V., Muqit, M. M. K., Harvey, K., Gispert, S., Ali, Z., Del Turco, D., Bentivoglio, A. R., Healy, D. G., Albanese, A., Nussbaum, R., González-Maldonado, R., Deller, T., Salvi, S., Cortelli, P., Gilks, W. P., Latchman, D. S., Harvey, R. J., ... Wood, N. W. (2004). Hereditary early-onset Parkinson's disease caused by mutations in PINK1. *Science*, *304*(5674), 1158–1160. doi:10.1126/science.1096284 PMID:15087508

Van Os, J., Linscott, R.J., Myin-Germeys, I., Delespaul, P., & Krabbendam, L. (2009). A systematic review and meta-analysis of the psychosis continuum: evidence for a psychosis proneness-persistence-impairment model of psychotic disorder. *Psychol Med., 39*(2), 179-95. doi:10.1017/S0033291708003814

van Os, J., Rutten, B. P., & Poulton, R. (2008). Gene–Environment Interactions in Schizophrenia: Review of Epidemiological Findings and Future Directions. *Schizophrenia Bulletin, 34*(6), 1066–1082. doi:10.1093/schbul/sbn117 PMID:18791076

Vandenberghe, R. (2016). Classification of the primary progressive aphasias: Principles and review of progress since 2011. *Alzheimer's Research & Therapy, 8*(1), 16. doi:10.1186/s13195-016-0185-y PMID:27097664

Verdolini, N., Amoretti, S., Mezquida, G., Cuesta, M.J., Pina-Camacho, L., García-Rizo, C., Lobo, A., González-Pinto, A., Merchán-Naranjo, J., Corripio, I., Salagre, E., Baeza, I., Bergé, D., Garriga, M., Bioque, M., Vallespir, C., Serra, M., Vieta, E., & Bernardo, M. (2021). The effect of family environment and psychiatric family history on psychosocial functioning in first-episode psychosis at baseline and after 2 years. *Eur Neuropsychopharmacol., 49,* 54-68.

Victor, S., & Nayak, V. M. (2003). Every heart beat is under neural command: An hypothesis relating to the cardiac rhythm. *Heart Lung and Circulation, 12*(1), 11–17. doi:10.1046/j.1444-2892.2003.00178.x PMID:16352101

Videtič Paska, A., & Kouter, K. (2021). Machine learning as the new approach in understanding biomarkers of suicidal behavior. *Bosnian Journal of Basic Medical Sciences, 21*(4), 398–408. doi:10.17305/bjbms.2020.5146 PMID:33485296

Vita, A., Gaebel, W., Mucci, A., Sachs, G., Barlati, S., Giordano, G.M., Nibbio, G., Nordentoft, M., Wykes, T., & Galderisi, S. (2022). European Psychiatric Association guidance on treatment of cognitive impairment in schizophrenia. *European Psychiatry, 65*(1), e57.

Vizziello, M., Borellini, L., Franco, G., & Ardolino, G. (2021). Disruption of Mitochondrial Homeostasis: The Role of PINK1 in Parkinson's Disease. *Cells, 10*(11), 3022. doi:10.3390/cells10113022 PMID:34831247

Vogel, M., Meier, J., Grönke, S., Waage, M., Schneider, W., Freyberger, H. J., & Klauer, T. (2011). Differential effects of childhood abuse and neglect: Mediation by posttraumatic distress in neurotic disorder and negative symptoms in schizophrenia? *Psychiatry Research, 189*(1), 121–127. doi:10.1016/j.psychres.2011.01.008 PMID:21295351

von Coelln, R., Dawson, V. L., & Dawson, T. M. (2004). Parkin-associated Parkinson's disease. *Cell and Tissue Research, 318*(1), 175–184. doi:10.1007/s00441-004-0924-4 PMID:15503153

Vuilleumier, P., & Pourtois, G. (2007). Distributed and interactive brain mechanisms during emotion face perception: Evidence from functional neuroimaging. *Neuropsychologia, 45*(1), 174–194. doi:10.1016/j.neuropsychologia.2006.06.003 PMID:16854439

Wagner, L. I., Sweet, J., Butt, Z., Lai, J., & Cella, D. (2009). Measuring patient self-reported cognitive function: Development of the Functional Assessment of Cancer Therapy – Cognitive Function instrument. *The Journal of Supportive Oncology*.

Wang, C., & Holtzman, D. M. (2020). Bidirectional relationship between sleep and Alzheimer's disease: Role of amyloid, tau, and other factors. *Neuropsychopharmacology*, *45*(1), 104–120. doi:10.1038/s41386-019-0478-5 PMID:31408876

Wang, N., Zhu, P., Huang, R., Wang, C., Sun, L., Lan, B., He, Y., Zhao, H., & Gao, Y. (2020). PINK1: The guard of mitochondria. *Life Sciences*, *259*, 118247. doi:10.1016/j.lfs.2020.118247 PMID:32805222

Warrington, E. K., & Shallice, T. (1984). Category Specific Semantic Impairments. *Brain*, *107*(3), 829–854. doi:10.1093/brain/107.3.829 PMID:6206910

Wefel, J. S., Vardy, J., Ahles, T., & Schagen, S. B. (2011). International Cognition and Cancer Task Force recommendations to harmonise studies of cognitive function in patients with cancer. *The Lancet. Oncology*, *12*(7), 703–708. doi:10.1016/S1470-2045(10)70294-1 PMID:21354373

Westwood, A. J., Beiser, A., Jain, N., Himali, J. J., DeCarli, C., Auerbach, S. H., Pase, M. P., & Seshadri, S. (2017). Prolonged sleep duration as a marker of early neurodegeneration predicting incident dementia. *Neurology*, *88*(12), 1172–1179. doi:10.1212/WNL.0000000000003732 PMID:28228567

Whalley, K. (2019). A glial contribution to 'chemobrain.'. *Nature Reviews. Neuroscience*, *20*(2), 67–67. doi:10.1038/s41583-018-0115-z PMID:30602722

Whittaker, A. L., George, R. P., & O'Malley, L. (2022). Prevalence of cognitive impairment following chemotherapy treatment for breast cancer: A systematic review and meta-analysis. *Scientific Reports*, *12*(1), 2135. doi:10.1038/s41598-022-05682-1 PMID:35136066

Williams, E. T., Chen, X., & Moore, D. J. (2017). VPS35, the Retromer Complex and Parkinson's Disease. *Journal of Parkinson's Disease*, *7*(2), 219–233. doi:10.3233/JPD-161020 PMID:28222538

Wilson, S. M., Henry, M. L., Besbris, M., Ogar, J. M., Dronkers, N. F., Jarrold, W., Miller, B. L., & Gorno-Tempini, M. L. (2010). Connected speech production in three variants of primary progressive aphasia. *Brain*, *133*(7), 2069–2088. doi:10.1093/brain/awq129 PMID:20542982

Winocur, G., Henkelman, M., Wojtowicz, J. M., Zhang, H., Binns, M. A., & Tannock, I. F. (2012). The effects of chemotherapy on cognitive function in a mouse model: A prospective study. *Clinical Cancer Research*, *18*(11), 3112–3121. doi:10.1158/1078-0432.CCR-12-0060 PMID:22467680

Wong, Y. C., & Krainc, D. (2017). α-synuclein toxicity in neurodegeneration: Mechanism and therapeutic strategies. *Nature Medicine*, *23*(2), 1–13. doi:10.1038/nm.4269 PMID:28170377

World Economic Situation and Prospects as of mid-2023. (2023). https://www.un.org/development/desa/dpad/publication/world-economic-situation-and-prospects-as-of-mid-2023/

World Health Organization. (2004). *The Atlas of Heart Disease and Stroke*. Author.

World Health Organization. (2018). *Aging and health.* https://www.who.int/news-room/fact-sheets/detail/ageing-and-health

World Health Organization. (2018). *International statistical classification of diseases and related health problems (11th Revision).* https://icd.who.int/browse11/l-m/en

Yaffe, K., Laffan, A. M., Harrison, S. L., Redline, S., Spira, A. P., Ensrud, K. E., Ancoli-Israel, S., & Stone, K. L. (2011). Sleep-disordered breathing, hypoxia, and risk of mild cognitive impairment and dementia in older women. *Journal of the American Medical Association, 306*(6), 613–619. doi:10.1001/jama.2011.1115 PMID:21828324

Yao, Q., Wang, R., Fan, X., Liu, J., & Li, Y. (2020). Multiclass arrhythmia detection from 12-lead varied-length ecg using attention-based time-incremental convolutional neural network. *Information Fusion, 53,* 174–182. doi:10.1016/j.inffus.2019.06.024

Yates, K., Lång, U., Cederlöf, M., Boland, F., Taylor, P., Cannon, M., McNicholas, F., DeVylder, J., & Kelleher, I. (2019). Association of Psychotic Experiences With Subsequent Risk of Suicidal Ideation, Suicide Attempts, and Suicide Deaths: A Systematic Review and Meta-analysis of Longitudinal Population Studies. *JAMA Psychiatry, 76*(2), 180–189. doi:10.1001/jamapsychiatry.2018.3514 PMID:30484818

Ye, H., Robak, L. A., Yu, M., Cykowski, M., & Shulman, J. M. (2023). Genetics and Pathogenesis of Parkinson's Syndrome. *Annual Review of Pathology, 18*(1), 95–121. doi:10.1146/annurev-pathmechdis-031521-034145 PMID:36100231

Youmans, G., & Bourgeois, M. (2010). Theory of mind in individuals with Alzheimer-type dementia. *Aphasiology, 24*(4), 515–534. doi:10.1080/02687030902801573

Yuan, J., Libon, D. J., Karjadi, C., Ang, A. F. A., Devine, S., Auerbach, S. H., Au, R., & Lin, H. (2021). Association Between the Digital Clock Drawing Test and Neuropsychological Test Performance: Large Community-Based Prospective Cohort (Framingham Heart Study). *Journal of Medical Internet Research, 23*(6), e27407. doi:10.2196/27407 PMID:34100766

Yung, A. R. (2017). Treatment of people at ultra-high risk for psychosis. *World Psychiatry; Official Journal of the World Psychiatric Association (WPA), 16*(2), 207–208. doi:10.1002/wps.20424 PMID:28498602

Yung, A. R., McGorry, P. D., McFarlane, C. A., Jackson, H. J., Patton, G. C., & Rakkar, A. (1996). Monitoring and care of young people at incipient risk of psychosis. *Schizophrenia Bulletin, 22*(2), 283–303. doi:10.1093/schbul/22.2.283 PMID:8782287

Zaki, J., & Ochsner, K. N. (2012). The neuroscience of empathy: Progress, pitfalls and promise. *Nature Neuroscience, 15*(5), 675–680. doi:10.1038/nn.3085 PMID:22504346

Zanelli, J., Mollon, J., Sandin, S., Morgan, C., Dazzan, P., Pilecka, I., Reis Marques, T., David, A. S., Morgan, K., Fearon, P., Doody, G. A., Jones, P. B., Murray, R. M., & Reichenberg, A. (2019). Cognitive Change in Schizophrenia and Other Psychoses in the Decade Following the First Episode. *The American Journal of Psychiatry*, *176*(10), 811–819. doi:10.1176/appi.ajp.2019.18091088 PMID:31256609

Zarei, A., & Asl, B. M. (2020). Automatic classification of apnea and normal subjects using new features extracted from hrv and ecg-derived respiration signals. *Biomedical Signal Processing and Control*, *59*, 101927. doi:10.1016/j.bspc.2020.101927

Zarranz, J. J., Alegre, J., Gómez-Esteban, J. C., Lezcano, E., Ros, R., Ampuero, I., Vidal, L., Hoenicka, J., Rodriguez, O., Atarés, B., Llorens, V., Gomez Tortosa, E., del Ser, T., Muñoz, D. G., & de Yebenes, J. G. (2004). The new mutation, E46K, of alpha-synuclein causes Parkinson and Lewy body dementia. *Annals of Neurology*, *55*(2), 164–173. doi:10.1002/ana.10795 PMID:14755719

Zavecz, Z., Shah, V. D., Murillo, O. G., Vallat, R., Mander, B. A., Winer, J. R., Jagust, W. J., & Walker, M. P. (2023). NREM sleep as a novel protective cognitive reserve factor in the face of Alzheimer's disease pathology. *BMC Medicine*, *21*(1), 156. doi:10.1186/s12916-023-02811-z PMID:37138290

Zhang, L., Reif, A., Du Rietz, E., Lagerberg, T., Butwicka, A., D'Onofrio, B. M., Johnell, K., Pedersen, N. L., Larsson, H., & Chang, Z. (2021). Comedication and Polypharmacy With ADHD Medications in Adults: A Swedish Nationwide Study. *Journal of Attention Disorders*, *25*(11), 1519–1528. doi:10.1177/1087054720923725 PMID:32478603

Zhao, Y., Ficek, B., Webster, K., Frangakis, C., Caffo, B., Hillis, A. E., Faria, A., & Tsapkini, K. (2021). White-matter integrity predicts electrical stimulation (tDCS) and language therapy effects in primary progressive aphasia. *Neurorehabilitation and Neural Repair*, *35*(1), 44–57. doi:10.1177/1545968320971741 PMID:33317422

Zhou, S., Zhao, J., & Zhang, L. (2022). Application of Artificial Intelligence on Psychological Interventions and Diagnosis: An Overview. *Frontiers in Psychiatry*, *13*, 811665. doi:10.3389/fpsyt.2022.811665 PMID:35370846

Zimmerman, M. E., Bigal, M. E., Katz, M. J., Derby, C. A., & Lipton, R. B. (2013). Are sleep onset/maintenance difficulties associated with medical or psychiatric comorbidities in nondemented community-dwelling older adults? *Journal of Clinical Sleep Medicine*, *9*(4), 363–369. doi:10.5664/jcsm.2590 PMID:23585752

About the Contributors

Katie Moraes de Almondes is an Associate Professor of the Department of Psychology and vice-coordinator of the Post-Graduate Program in Psychobiology (Capes 6 concept) of the Federal University of Rio Grande do Norte. Master's and Doctorate in Psychobiology from the Federal University of Rio Grande do Norte. Post-Doctorate by the Molecular Medicine Program at the Federal University of Minas Gerais and by the Department of Personality, Evaluation and Psychological Treatment of the University of Salamanca - Spain. CNPq Productivity Fellow. Vice-president of the Brazilian Psychology Society. Coordinator of the Sleep Clinic - AMBSONO and of the Aging Neuropsychology Service - SENE. Consultant and reviewer of the MHPSS.net of the IASC (Inter-Agency Standing Committee), established by the United Nations. Sleep Psychologist certified by the Brazilian Sleep Association. Technical training in Cultural Neuropsychology from North Carolina University Wilmington and the American Psychological Association.

* * *

Paulo Roberto de Brito-Marques is an adjunct professor, Head of the Neurology discipline at the Faculty of Medical Sciences at the University of Pernambuco (UPE). Coordinator of the Lato Senso postgraduate course at the Specialization level in Cognitive and Behavioral Neurology at UPE. Preceptor of the Neurology Residency at the Oswaldo Cruz University Hospital of UPE. Member of the Alzheimer's and Parkinson's Disease Committee for Technical-Scientific Advice and Honorary Provision – of the Pharmaceutical Assistance Management, of the State of Pernambuco Department of Health.

Sonia Brucki is Coordinator of Cognitive and Behavioral Neurology Group at Hospital das Clínicas da Faculdade de Medicina da Universidade de São Paulo (HC-FMUSP).

Leonardo Caixeta has a degree in medicine from UFG, a specialization in Neurology and a medical residency in Psychiatry from the Institute of Psychiatry at USP, a master's degree in Neurology from the University of São Paulo (FM-USP), and a doctorate in Neurology from the University of São Paulo (2001), a fellow in Behavioral Neurology from the University of Manchester in England. Professor of Neurology at the Department of Clinical Medicine at the Faculty of Medicine at the Federal University of Goiás (UFG). Professor of the Postgraduate Course in Health Sciences at UFG.

Marcelo de Brito is a neurologist at Cognitive and Behavioral Neurology Group at Hospital das Clínicas da Faculdade de Medicina da Universidade de São Paulo (HC-FMUSP). Neurologist at Instituto do Câncer do Estado de São Paulo (ICESP).

Tiago Figueiredo is a psychiatrist at the Institute of Psychiatry of the Federal University of Rio de Janeiro (IPUB/UFRJ), Rio de Janeiro, Brazil. Master's in Clinical Psychiatry from IPUB/UFRJ, Rio de Janeiro, Brazil. PhD in Medical Sciences from Instituto D'Or de Pesquisa e Ensino, Rio de Janeiro, Brazil. Post-Doctoral Residency in the Child Health Program at the Faculty of Medicine of UFMG, Minas Gerais, Brazil.

Ricardo Nitrini is Senior Professor of Neurology at Faculdade de Medicina da Universidade de São Paulo (FMUSP)

Givago Souza is a physiotherapist and biologist from Federal University of Para. His research is focused in the development of technological solutions for health.

Index

T

V

Recommended Reference Books

IGI Global's reference books are available in three unique pricing formats:
Print Only, E-Book Only, or Print + E-Book.

Order direct through IGI Global's Online Bookstore at **www.igi-global.com** or through your preferred provider.

ISBN: 9781799886341
EISBN: 9781799886365
© 2022; 414 pp.
List Price: US$ **380**

ISBN: 9781799886891
EISBN: 9781799886914
© 2022; 261 pp.
List Price: US$ **325**

ISBN: 9781799878889
EISBN: 9781799878896
© 2022; 242 pp.
List Price: US$ **325**

ISBN: 9781799887836
EISBN: 9781799887843
© 2022; 377 pp.
List Price: US$ **325**

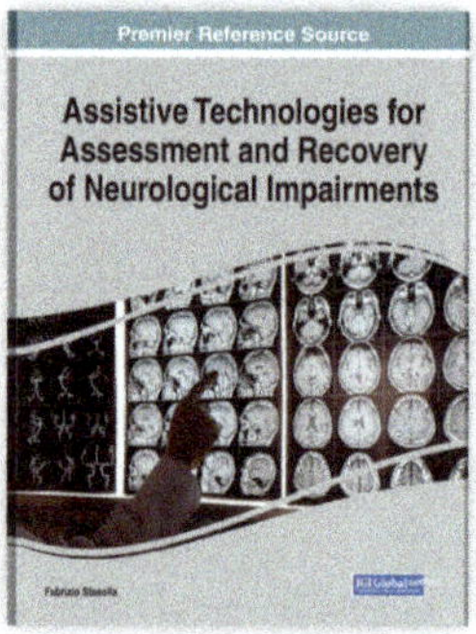

ISBN: 9781799874300
EISBN: 9781799874317
© 2022; 396 pp.
List Price: US$ **325**

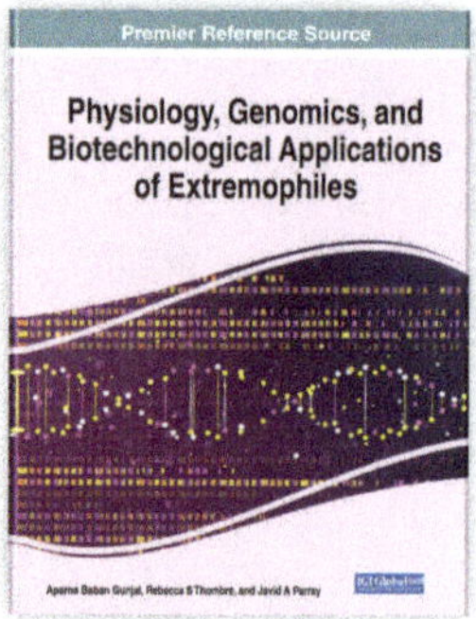

ISBN: 9781799891444
EISBN: 9781799891468
© 2022; 473 pp.
List Price: US$ **380**

Do you want to stay current on the latest research trends, product announcements, news, and special offers?
Join IGI Global's mailing list to receive customized recommendations, exclusive discounts, and more.
Sign up at: **www.igi-global.com/newsletters.**

Publisher of Timely, Peer-Reviewed Inclusive Research Since 1988

IGI Global
PUBLISHER of TIMELY KNOWLEDGE

www.igi-global.com Sign up at www.igi-global.com/newsletters facebook.com/igiglobal twitter.com/igiglobal

www.ingramcontent.com/pod-product-compliance
Lightning Source LLC
LaVergne TN
LVHW081527300926
849161LV00006B/1329